G. I. Marchuk A. S. Sarkisyan

Mathematical Modelling of Ocean Circulation

With 62 Figures

Springer-Verlag
Berlin Heidelberg New York
London Paris Tokyo

Academician G. I. Marchuk
The Presidium of the USSR
Academy of Sciences
Leninsky Prospekt 14
117901, B-71, Moscow, USSR

A. S. Sarkisyan
Correspondent-Member of the USSR
Department of Numerical Mathematics of the USSR
Academy of Sciences
Gorky Street 11
103009 Moscow, USSR

Title of the original Russian edition:
Matematicheskoye modelirovanie tsirkulytsii okeana
© by Hydro Metrological Publishing House, Moscow 1986

ISBN 3-540-18925-4 Springer-Verlag Berlin Heidelberg New York
ISBN 0-387-18925-4 Springer-Verlag New York Berlin Heidelberg

Library of Congress Cataloging-in-Publication Data. Matematicheskie modeli tsirkuliatsii v okeane. English. The mathematical modeling of the ocean circulation : with 83 figures / [edited by] G. I. Marchuk, A. S. Sarkisyan. p. cm. Translation of: Matematicheskie modeli tsirkuliatsii v okeane. Bibliography: p. Includes index. ISBN 0-387-18925-4 (U.S.) 1. Ocean circulation–Mathematical models. I. Marchuk, G. I. (Guriĭ Ivanovich), 1925-. II. Sarkisian, A. S. (Artem Sarkisovich) III. Title. GC228.5.M3813 1988 551.47'072'4–dc 19 88-15948

Typesetting: ASCO Trade Typesetting Ltd.; North Point, Hong Kong
Printing: Druckhaus Beltz, Hemsbach; binding: J. Schäffer, Grünstadt
2132/3130-543210 – Printed on acid-free paper

Preface

The problems of ocean dynamics present more and more complex tasks for investigators, based on the continuously sophistication of theoretical models, which are applied with the help of universal and efficient algorithms of numerical mathematics.

The present level of our knowledge in the field of mathematical physics and numerical mathematics allows one to give rather complete theoretical analysis of basic statements of problems as well as numerical algorithms. Our task is to perform such analysis and also to analyze the results of calculations in order to improve our knowledge of the mechanism of large-scale hydrological processes occurring in the World Ocean. The new level of numerical mathematics has essentially influenced , the formation of new solution methods of ocean dynamics problems, among which an important one is the splitting method, which has been already widely practised in various fields of science and engineering.

A number of monographs by N.N. Yanenko, A.A. Samarsky, G.I. Marchuk (Rozhdestvensky and Yanenko 1968; Samarsky and Andreyev 1976; Marchuk 1970, 1980b) and others are devoted to the description of this methods. But the methods of the splitting theory require extensive creative work for their application to concrete problems, which are peculiar, as a rule, in problem formulation. The success of the application of these methods is related to the deep understanding of the essence of the described processes.

In the last decades fundamental works of Arakawa, K. Bryan, R. Haney and others have contributed greatly to the development of theoretical oceanology and mathematical methods of modelling. A number of investigations have been carried out, using observed data over temperature and salinity for restoration, initialization and four-dimensional analysis (Sarkisyan et al. 1985 a, b). The authors of this monograph consisting of five chapters, present numerical algorithms for solving ocean dynamics problems and the application of these algorithms to concrete calculations of the SSI, horizontal and vertical components of climatic currents and other characteristics, taking into account the baroclinicity of sea water, the forms of the coast line, the bottom relief and the tangential wind stress.

Chapter 1 is devoted to the general formulation of the problem, the estimation of characteristic values, simplification of initial equations and construction of simple, quasi geostrophic models of ocean circulation. These estimations have two peculiarities.

1. There are not only external parameters, but also a characteristic density anomaly in the formulas for characteristic values. It stresses the necessity to consider the baroclinicity of sea water in even the simplest problem of real ocean circulation.

2. On the basis of these estimations, it is shown that there is a minimal number of factors, the consideration of which is absolutely necessary; otherwise the problem under consideration is not of geophysical interest.

Chapter 2 is devoted to the simplest methods of difference approximation, the construction of difference gridnets and the solution of ocean thermodynamics equations. They include the method of directed differences, the method of enhancing the approximation order, the Fiadeiro-Veronis method and others.

Furthermore Chapter 2 is an introduction to well-developed and established numerical methods, which are presented in Chapters 3 and 4. Here, together with the splitting method, other methods of constructing difference approximations of quasi-linear problems of ocean hydrodynamics are applied, in the combination with which it is possible to construct algorithms of the second-order approximation by all independent variables, absolutely stable with respect to calculation and satisfying the basic conservation laws.

Let us dwell briefly on the basic peculiarities of these two chapters. In Chapter 3 the series of possible correct formulations of non-stationary problems of ocean circulation is formulated and the methods of their solution are given. In particular, a non-linear problem of the calculation of all ocean thermohydrodynamic characteristics, with the exception of turbulence, is considered. After the transition from physical values to their deviations and corresponding simplifications, the theorem of the solution uniqueness of the formulated problem is proved. Further, the condition is determined, which is necessary for the construction of stable difference splitting schemes of the second-order accuracy by time. Such schemes are constructed on the basis of the two-cyclic method of the component by component splitting. At the same time the absolute stability of splitting schemes of the second-order accuracy by all coordinates (in the case of a regular spacial gridnet) is proved. Besides the establishment of the splitting method, the practical criterion for the choice of the time interval, while solving the ocean dynamics problems, is obtained.

Chapter 4 is devoted to stationary problems of ocean dynamics. From the physical point of view the expediency of such a problem is explained by the necessity to solve the problem of initial data in the ocean. Thus, the modelling of the ocean circulation is a problem-solving stage of the ocean-atmosphere interaction and the forecast of hydrometeorological characteristics. Therefore the problem involves the generation of the currents and density fields in the ocean upper layers under the influence of weather conditions in the atmosphere. The latter represents two problems. First, it is necessary to solve the problem of the stationary (climatic) state of the ocean, and then, choosing every time the climatic condition as "initial" and, using actual data about weather conditions near the ocean surface, to solve the problem of ocean circulation perturbations under the influence of meteorological situations. The constructed fields of the hydrological elements, taking into account the real meteorological processes in the atmosphere, can be used as initial data while solving the problems of weather forecast in the atmosphere and the circulation in the ocean.

First, a linearized formulation of the ocean climatic state, taking into account the turbulence, is considered. Thus the solution uniqueness of this problem is proved, the approximation methods are considered and iterative methods in solving difference equations are worked out (in particular, the method of minimal residuals); more over the methods of convergency acceleration and the criterion of the iterative process convergency are obtained. In Chapter 4 some non-linear formulations of problems are considered.

The last chapter (Chap. 5) is devoted to the performance of the determined algorithms for the calculation of stationary and seasonal climatic characteristics of concrete geophysical objects (the World Ocean, various oceanic basins, etc.).

Diagnostic calculations of World Ocean dynamics are carried out by the temperature and salinity fields, specified in 1° squares. The maps of horizontal and vertical currents for various levels and the four seasons, as well as maps for the World Ocean level, are also presented in the monograph. The obtained results are compared with the available maps and atlases of currents and the realistic character of the obtained results is shown. However, diagnostic calculations have certain drawbacks, therefore, a considerable part of Chapter 5 is devoted to the discussion of the problem, to the methods of solution and results of concrete calculations on adjustment and initialization. Maps and curves are also presented, which reflect the evolution of kinetic energy and other characteristics in the adjustment process. The results of these calculations show the reliability of current fields, obtained by using observed data. Next to adjustment, the most interesting procedure of using observed data is the four-dimensional analysis. As it is known, since the beginning of 1981, research and expedition work on the multinational program "Sections" is being carried out. The scientific basis of this program, the first results and examples of the four-dimensional analysis of oceanographic observed data have been already published in English (Dymnikov et al. 1983; Marchuk et al. 1983, 1984). The international program WOCE is devoted to the problem of obtaining and analyzing oceanographic data. Thus, the diagnostic and adjustment calculations presented in Chapter 5, as well as the analogous calculations by other models [for example, the β-spiral method (Stommel and Schott 1977)] are the basis of future methods of theoretical analysis of oceanographic observed data. The authors express their gratitude to L.I. Egorova for translating the manuscript from Russian into English.

G.I. Marchuk, A.S. Sarkisyan

Contents

Denotation List

$\boldsymbol{u} = u\boldsymbol{i} + v\boldsymbol{j} + w\boldsymbol{k}$	The vector of the current velocity;
u, v, w	The components of the vector of the current velocity by the x-, y-, z-axes of the Cartesian coordinate system; the x-axis is directed to the east; the y-axis to the north; the z-axis vertically downwards;
t	Time;
ϱ_1, P_1	Density and pressure in the sea water
$\varrho_1 = \varrho_0 + \varrho$	where ϱ_0 is a constant value and ϱ is a density anomaly;
$P_1 = \varrho_0 g z + P$	where P is a pressure anomaly;
P_a	Atmospheric pressure at sea level;
T_1, S_1	Temperature and salinity of the sea water;
$S_1 = S_0 + S,\ T_1 = T_0 + T$	where T_0, S_0 are constant values, and T, S are the anomalies of temperature and salinity;
$T = \bar{T}(z) + T'$, $S = \bar{S}(z) + S'$, $\varrho = \bar{\varrho}(z) + \varrho'$, $P = \bar{P}(z) + P'$	where $\bar{T}(z)$, $\bar{S}(z)$, $\bar{\varrho}(z)$, $\bar{P}(z)$ are some averages for the ocean water area under consideration of distributions of anomalies of temperature, salinity, density and pressure, and T', S', ϱ', and P' their corresponding deviations;
μ, ν	The coefficients of horizontal and vertical turbulent impulse exchange;
μ_1, ν_1	The coefficients of horizontal and vertical turbulent diffusion of heat and salts;
τ_x, τ_y	The components of tangential wind stress;
Q, Q_1	Heat and salt fluxes through the ocean surface;
Q', Q_1'	Heat and salt fluxes through the lateral boundaries of the basin;
$\bar{u}, \bar{v}$	Average by height of the horizontal components of the current velocity;
T_b, S_b	Temperature and salinity at the lateral boundary of the basin;
v_λ, v_θ, v_z	The components of the current velocity by the λ, θ, z axes of the spherical coordinate system;
$\theta = 90° - \varphi$	The addition to the latitude φ (the polar angle);
$\alpha = \sqrt{\dfrac{\omega \sin\varphi}{\nu}}$	The Ekman parameter for the ocean;

δ — δ equals 1 or 0, depending on non-slip or slip conditions at the ocean bottom;

$S_x = \int_0^H u\,dz,\ S_y = \int_0^H v\,dz$ — The components of the total mass transport;

Δ — Laplacian;

J — Jacobian;

L_H — The characteristic horizontal scale in the bottom layer of the ocean

$$A = \frac{\partial u}{\partial t} + \frac{\partial u^2}{\partial x} + \frac{\partial uv}{\partial y} + \frac{\partial uw}{\partial z} - A_l \Delta u$$

$$B = \frac{\partial v}{\partial t} + \frac{\partial uv}{\partial x} + \frac{\partial v^2}{\partial y} + \frac{\partial vw}{\partial z} - A_l \Delta v\,;$$

$\operatorname{rot}\tau = \frac{\partial \tau_y}{\partial x} - \frac{\partial \tau_x}{\partial y}$ — The vertical component of the tangential wind stress vorticity;

P_s — The water pressure anomaly on the ocean surface (at $z = 0$);

JEBAR — The joint effect of the baroclinicity and bottom relief 1 Sverdrup $= 10^{12}\,\mathrm{cm^3\,s^{-1}}$;

ν' — The coefficient of the vertical turbulent impulse exchange in the atmosphere;

n — The vector, normal to the ocean bottom;

N — Normal to the lateral boundary;

H — The ocean depth;

a — Average radius of the Earth;

ω — The angular velocity of the Earth's rotation;

g — The gravity acceleration;

ζ_1 — The elevation of the ocean's free surface;

$\zeta = \zeta_1 + \frac{P_a}{\varrho_0 g}$ — The transformed sea-surface topography (SST); positive values ζ and ζ_1 correspond to the sea surface elevation;

l — The Coriolis parameter;

$\beta = \frac{dl}{dy}$ — The variation of the Coriolis parameter with latitude;

L_0 — The characteristic horizontal scale;

h_0 — The characteristic vertical scale (the characteristic thickness of the baroclinic layer);

$\mathscr{P}_0, (\delta\varrho)_0$ — The characteristic values of the pressure and density anomaly, the subscript "0" denotes the characteristic value of the corresponding parameter;

u_g, v_g — The components of the velocity of the geostrophic current;

u^d, v^d, w^d — The components of the velocity of the drift current;

$u_H, v_H, v_\theta^{(H)}, v_\lambda^{(H)}$ — The sub- or superscript "H" corresponds to the value of the given characteristics on the ocean bottom;

$K_0 = \dfrac{v_0}{l_0 L_0}$	The Kibel (Rossby) number;
$E_v = \dfrac{v}{l_0 h_0^2}$, $E_m = \dfrac{\mu}{l_0 L_0^2}$	The Ekman numbers, corresponding to the vertical and horizontal turbulent impuls exchange;
$Pe = \dfrac{w_0 h_0}{v_1}$, $\tilde{P}e = \dfrac{v_0 L_0}{\mu_1}$	The Peclet numbers for the vertical and horizontal turbulent diffusion;
$\alpha' = \sqrt{\dfrac{\omega \sin\varphi}{v'}}$	The Ekman parameter for atmosphere;
H_1	The reference level in the dynamic method and the depth of the baroclinicity in the theory of the ocean baroclinic layer;
S_y^d	The integral meridional transport in the ocean baroclinic layer H_1 by the dynamic method;
Ψ_d	The total mass transport stream function calculated by the dynamic method, when the ocean bottom is the reference level;
h_z	The thickness of the upper, quasi-homogeneous layer;
W	The wind velocity;
G	The pressure vector gradient;
SST	The sea-surface topography;
ACG	The anti-cyclonic gyre;
CG	The cyclonic gyre;
ACC	The Antartic circumpolar current;
MBF	The main Black Sea flow.

Chapter 1

Formulation of the Problem, Transformation of Equations and Elaboration of Ocean Circulation Models

1.1 Initial Equations, Their Simplification and Transformations

1.1.1 Initial Equations and Boundary Conditions

A system of governing equations and appropriate boundary conditions are necessary for calculating the ocean currents. In studying large-scale circulation we have to take into account the earth's sphericity, however, rectangular coordinates are more demonstrative and simple. Therefore, we use the Cartesian system to describe the equations and their transformations for the Northern Hemisphere. A set of main relations in spherical coordinates and the equation for the Southern Hemisphere are presented at the end of the next section. To investigate the large-scale currents in a baroclinic ocean (more exactly, in an ocean with variable water density) we use the Boussinesq and quasi-static approximations. Thus, we use the following system of equations for describing the thermohydrodynamics equations of motion for the ocean:

$$\frac{\partial u}{\partial t} + u\frac{\partial u}{\partial x} + v\frac{\partial u}{\partial y} + w\frac{\partial u}{\partial z} - lv = -\frac{1}{\rho_0}\frac{\partial P_1}{\partial x} + \frac{\partial}{\partial z}\nu\frac{\partial u}{\partial z} + \frac{\partial}{\partial x}\mu\frac{\partial u}{\partial x} + \frac{\partial}{\partial y}\mu\frac{\partial u}{\partial y}\,; \tag{1.1}$$

$$\frac{\partial v}{\partial t} + u\frac{\partial v}{\partial x} + v\frac{\partial v}{\partial y} + w\frac{\partial v}{\partial z} + lu = -\frac{1}{\rho_0}\frac{\partial P_1}{\partial y} + \frac{\partial}{\partial z}\nu\frac{\partial v}{\partial z} + \frac{\partial}{\partial x}\mu\frac{\partial v}{\partial x} + \frac{\partial}{\partial y}\mu\frac{\partial v}{\partial y}\,; \tag{1.2}$$

equation of statics:

$$\frac{\partial P_1}{\partial z} = \rho_1 g\,; \tag{1.3}$$

equation of continuity of incompressible liquid:

$$\frac{\partial u}{\partial x} + \frac{\partial v}{\partial y} + \frac{\partial w}{\partial z} = 0\,; \tag{1.4}$$

equation of transfer of heat and salts:

$$\frac{\partial T}{\partial t}+u\frac{\partial T}{\partial x}+v\frac{\partial T}{\partial y}+w\frac{\partial T}{\partial z}=\frac{\partial}{\partial z}\nu_1\frac{\partial T}{\partial z}+\frac{\partial}{\partial x}\mu_1\frac{\partial T}{\partial x}+\frac{\partial}{\partial y}\mu_1\frac{\partial T}{\partial y}\,; \tag{1.5}$$

$$\frac{\partial s}{\partial t}+u\frac{\partial s}{\partial x}+v\frac{\partial s}{\partial y}+w\frac{\partial s}{\partial z}=\frac{\partial}{\partial z}\nu_1\frac{\partial s}{\partial z}+\frac{\partial}{\partial x}\mu_1\frac{\partial s}{\partial x}+\frac{\partial}{\partial y}\mu_1\frac{\partial s}{\partial y}\,; \tag{1.6}$$

and equation of state (Bryan and Cox 1972):

$$\rho = a_{1k}T + a_{2k}S + a_{3k}T^2 + a_{4k}S^2 + a_{5k}TS + a_{6k}T^3 + a_{7k}S^2T + \cdots \tag{1.7}$$

The system (1.1)–(1.7) contains seven unknown variables: the components of flow velocity u, v, w along the axes 0_x, 0_y, 0_z, pressure P_1, density $\rho_1 = \rho + \rho_0$, temperature T and salinity S. To solve any problem concerning non-stationary ocean circulation it is sufficient to have the initial values of only four functions u, v, T and S, from which the other three initial fields are determined. We shall dwell on the parameterization of turbulent exchange and the choice of the coefficients ν, μ, ν_1, μ_1, in Chapter 5.

The boundary conditions along the vertical axis are: on the ocean surface, for $z = -\zeta_1(x, y, t)$:

$$P_1 = P_a \tag{1.8}$$

$$\rho_0\nu\frac{\partial u}{\partial z} = -\tau_x\,, \qquad \rho_0\nu\frac{\partial u}{\partial z} = -\tau y \tag{1.9}$$

$$w = -\left(\frac{\partial\zeta_1}{\partial t}+u\frac{\partial\zeta_1}{\partial x}+v\frac{\partial\zeta_1}{\partial y}\right) \tag{1.10}$$

$$\frac{\partial T}{\partial z} = Q\,; \tag{1.11}$$

or

$$T = T(x, y, t) \tag{1.12}$$

$$\frac{\partial S}{\partial z} = Q_1\,; \tag{1.13}$$

or

$$S = S(x, y, t)\,. \tag{1.14}$$

At the ocean bottom, for $z = H(x, y)$, the non-slip condition:

$$u = v = w = 0\,; \tag{1.15}$$

or the free-slip condition:

$$\frac{\partial u}{\partial n} = \frac{\partial v}{\partial n} = 0\,; \tag{1.16}$$

$$w = u_H\frac{\partial H}{\partial x} + v_H\frac{\partial H}{\partial y}\,; \tag{1.17}$$

is applied for the flow velocity.

For temperature and salinity we use the boundary conditions:

$$\frac{\partial T}{\partial n} = \frac{\partial S}{\partial n} = 0 \, ; \tag{1.18}$$

or

$$T = T_H \, ; \qquad S = S_H \, , \tag{1.19}$$

where n is the normal to the ocean bottom. For generality, some of the boundary conditions are given in two versions. The appropriate condition to be chosen depends on the problem under consideration.

We now formulate the side boundary conditions. The lateral boundaries of a domain are assumed to be vertical. The flow velocity components u and v on the liquid part of the lateral boundaries are given as functions of coordinates and time, while on the solid part we apply either the non-slip or the free-slip condition. We shall, however, solve some problems without any regard to the lateral exchange. In this case it is sufficient to specify the velocity component normal to the boundary. In studying large-and medium-scale currents often the mean vertical velocity is specified at the side boundaries. This "smoothed" boundary condition gives rise to certain errors in the fields of coastal currents, but its effect far from the coasts is probably small. The boundary in the horizontal plane is usually approximated by a broken line, each segment being parallel to one of the coordinate axes. Thus, for u and v the following boundary conditions are applied:

$$\frac{1}{H}\int_0^H u dz = \bar{u} \, ;$$

$$\frac{1}{H}\int_0^H v dz = \bar{v} \, . \tag{1.20}$$

On the solid part of the boundary and on the island contours we have $\bar{u} = \bar{v} = 0$. On the lateral boundaries for temperature and salinity we assume:

$$\frac{\partial T}{\partial N} = Q' \, ; \qquad \frac{\partial S}{\partial N} = Q'_1 \tag{1.21}$$

or

$$T = T_b \, ; \qquad S = S_b \, , \tag{1.22}$$

where N is the normal to the lateral boundary. On the solid part of the lateral boundary $Q' = Q'_1 = 0$.

1.1.2 Simplification of Equations and Boundary Conditions for Large-Scale Flows

In general, the system of equations and boundary conditions given above can represent the main equations for any problem on ocean current dynamics. However, the number of such realizations is small since we have to deal with a complicated system of non-linear equations which can be solved only with the help of a sophisticated computer. It is therefore natural to simplify these equations so that they can be solved without any perceptible loss of accuracy, despite the limited capabilities of modern computers.

We first transform the equation of statics (1.3) by integrating over the vertical from $-\zeta_1$ to z under the boundary conditions (1.8):

$$\begin{aligned} P_1 &= P_a + g\int_{-\zeta_1}^{z} \rho_1 dz = P_a + g\int_{-\zeta_1}^{0} \rho_1 dz + g\int_0^z \rho_1 dz \\ &\approx P_a + g\int_{-\zeta_1}^{0} \rho_0 dz + g\int_0^z (\rho_0 + \rho) dz \\ &= P_a + \rho_0 g\zeta_1 + \rho_0 g z + g\int_0^z \rho dz\,. \end{aligned}$$

Disregarding the pressure of a uniform liquid column, we obtain the following relation for the pressure anomaly:

$$P = P_a + \rho_0 g\zeta_1 + g\int_0^z \rho dz\,. \tag{1.23}$$

In place of the physical topography it is more convenient to use the conventional sea-surface topography:

$$\zeta = \zeta_1 + \frac{P_a}{\rho_0 g}\,. \tag{1.24}$$

Thus, the pressure anomaly Eq. (1.3) and the boundary condition (1.8) are simplified as follows:

$$P = \rho_0 g\zeta + g\int_0^z \rho dz\,. \tag{1.25}$$

In order to evaluate the order of magnitude of the terms in the initial relations we use dimensionless variables. A sufficient number of cases with regard to the application of dimensionless quantities have been given in the literature. Therefore there is no apparent need here to do this once again. Later, however, after the analysis of dimensionless equations, we shall demonstrate the significance and the peculiarity of our dimensionless relations.

We have:

$$\begin{aligned} &x = L_0\bar{x}\,; \quad y = L_0\bar{y}\,; \quad z = h_0\bar{z}\,; \quad u = v_0\bar{u}\,; \quad v = v_0\bar{v}\,; \\ &w = w_0\bar{w}\,; \quad t = t_0\bar{t}\,; \quad P = \mathscr{P}_0\bar{P}\,; \quad \rho = (\delta\rho)_0\bar{\rho}\,; \\ &\zeta = \zeta_0\bar{\zeta}\,; \quad l = l_0\bar{l}\,; \quad \beta = \beta_0\bar{\beta}\,. \end{aligned} \tag{1.26}$$

We first consider large-scale motion in a baroclinic ocean layer outside the equator. The characteristic horizontal scale L_0, the characteristic thickness of the baroclinic layer h_0 and the characteristic density anomaly $(\delta\rho)_0$ are assumed to be given. All other quantities, even the characteristic horizontal flow velocity v_0, will be determined through these three parameters, l_0, β_0 and known constants. The ratio of the characteristic horizontal scale L_0 to the characteristic disturbance propagation velocity will be taken as the characteristic time. This means that we deal only with those hydrodynamic processes for which the disturbance spreading velocity is of the same order as the particle velocity. Thus,

$$t_0 = \frac{L_0}{v_0} \,. \tag{1.27}$$

Furthermore, in order to make rough estimates we first evaluate the vertical velocity component from the continuity equation:

$$w_0 = \frac{h_0}{L_0} v_0 \,. \tag{1.28}$$

Now, in Eq. (1.1) by changing over to dimensionless variables, dividing all terms by $l_0 v_0$ and then using Eqs. (1.27) and (1.28), we obtain:

$$K_0\left(\frac{\partial \bar{u}}{\partial \bar{t}} + \bar{u}\frac{\partial \bar{u}}{\partial \bar{x}} + \bar{v}\frac{\partial \bar{u}}{\partial \bar{y}} + \bar{w}\frac{\partial \bar{u}}{\partial \bar{z}}\right) - \bar{l}\,\bar{v} = -\frac{\mathscr{P}_0}{\rho_0 L_0 l_0 v_0}\frac{\partial \bar{P}}{\partial \bar{x}} + E_v\frac{\partial^2 \bar{u}}{\partial \bar{z}^2} + E_m \bar{\Delta}\,\bar{u}\,, \tag{1.29}$$

where

$$K_0 = \frac{v_0}{l_0 L_0} \tag{1.30}$$

is the Kibel (Rossby) number, and

$$E_v = \frac{\nu}{l_0 h_0^2}\,; \qquad E_m = \frac{\mu}{l_0 L_0^2} \tag{1.31}$$

are the Ekman numbers for the vertical and horizontal turbulent viscosities respectively. In dimensionless variables the heat transfer equation (1.5) takes the form:

$$\frac{\partial \bar{T}}{\partial \bar{t}} + \bar{u}\frac{\partial \bar{T}}{\partial \bar{x}} + \bar{v}\frac{\partial \bar{T}}{\partial \bar{y}} + \bar{w}\frac{\partial \bar{T}}{\partial \bar{z}} = \frac{1}{Pe}\frac{\partial^2 \bar{T}}{\partial \bar{z}^2} + \frac{1}{\tilde{P}e}\bar{\Delta}\,\bar{T}\,, \tag{1.32}$$

where

$$Pe = \frac{w_0 h_0}{\nu_1}\,; \qquad \tilde{P}e = \frac{v_0 L_0}{\mu_1} \tag{1.33}$$

are the Péclet numbers for the vertical and horizontal diffusions respectively.

In rough estimation we take the following limits for certain parameters: $L_0 = 100 - 1000\,\text{km}$, $v_0 = 1 - 10\,\text{cm s}^{-1}$, $\nu = 1 - 100\,\text{cm}^2\,\text{s}^{-1}$ $\mu = 10^6 - 10^9\,\text{cm}^2\,\text{s}^{-1}$ and $l_0 = 10^{-4}\,\text{s}^{-1}$. It is readily seen that even when the parameters vary over such a wide range, the numbers K_0, E_v and E_m are less than one by several orders of magnitude. Then from Eq. (1.29) it necessarily follows that the dimensionless factor of the pressure gradient should be of the order of one:

$$\frac{\mathscr{P}_0}{\rho_0 L_0 l_0 v_0} = 1 . \tag{1.34}$$

On the Other hand, from Eq. (1.25) we have

$$\mathscr{P}_0 = \rho_0 g \zeta_0 = g(\delta\rho)_0 h_0 . \tag{1.35}$$

The relations (1.27), (1.28), (1.34) and (1.35) yield the following expressions for the characteristic scales:

$$v_0 = \frac{g h_0 (\delta\rho)_0}{\rho_0 L_0 l_0} ; \qquad w_0 = \frac{g(\delta\rho)_0}{\rho_0 l_0}\left(\frac{h_0}{L_0}\right)^2 ;$$

$$t_0 = \frac{\rho_0 l_0 L_0^2}{g h_0 (\delta\rho)_0} ; \qquad \mathscr{P}_0 = g h_0 (\delta\rho)_0 ; \qquad \zeta_0 = \frac{(\delta\rho)_0 h_0}{\rho_0} . \tag{1.36}$$

Let us take $h_0 = 500\,\mathrm{m} = 5 \times 10^4\,\mathrm{cm}$, $L_0 = 10^3\,\mathrm{km} = 10^8\,\mathrm{cm}$, $(\delta\rho)_0 = 10^{-3}\,\mathrm{g\,cm^{-3}}$, $\rho_0 = 1\,\mathrm{g\,cm^{-3}}$, $g = 10^3\,\mathrm{cm\,s^{-2}}$. According to the formulas (1.36) we obtain (all quantities in CGS units):

$$\mathscr{P}_0 = 5 \times 10^4 ; \quad \zeta_0 = 50 ; \quad v_0 = 5 ; \quad w_0 = 2.5 \times 10^{-3} ; \quad t_0 = 2 \times 10^7 . \tag{1.37}$$

The dimensionless forms of Eqs. (1.2) and (1.6) are the same as those of Eqs. (1.29) and (1.32) and are therefore not given here. The dimensionless factors in these equations are small. For example, for the characteristic values given in Eq. (1.37) and for the turbulance coefficients $v = 10^2$, $\mu = 10^7$, $v_1 = 1$, $\mu_1 = 10^6$, we have $E_v = 4 \times 10^{-4}$, $E_m = 10^{-5}$, $\tilde{P}e^{-1} = 0.8 \times 10^{-2}$, $\tilde{P}e^{-1} = 2 \times 10^{-3}$ and $K_0 = 5 \times 10^{-4}$. This means that in the equations of motion, geostrophic balance is the dominating factor, while in Eqs. (1.5) and (1.6) all terms on the left-hand sides are important. Turbulence plays a prominent role only in thin boundary layers. Even for such intensive flows as the Gulf Stream, the Kibel (Rossby) number is small. For instance, for $v_0 = 50\,\mathrm{cm^2\,s^{-2}}$, $L_0 = 50\,\mathrm{km} = 5 \times 10^6\,\mathrm{cm}$, we have $K_0 = 0.1$, i.e. for strong jetlike currents, it is possible to take into account the non-linear terms in Eqs. (1.1) and (1.2) by successive approximations.

We evaluate the characteristic value of w once again. Since the geostrophic balance predominates in the equations of motion we cannot generally assume that all the terms in the continuity equation are of the same order of magnitude. Hence, Eq. (1.28), which is derived from (1.4), may not be exact. If in Eq. (1.4) u and v are replaced by:

$$u_g = -\frac{1}{\rho_0 l}\frac{\partial P}{\partial y} ; \qquad v_g = \frac{1}{\rho_0 l}\frac{\partial P}{\partial x} , \tag{1.38}$$

we then obtain:

$$\frac{\partial w}{\partial z} = -\frac{\beta}{l} v_g .$$

Hence, the characteristic value w_0 in a baroclinic ocean layer, for $\beta_0 = 2 \times 10^{-3}$, is equal to:

$$w_0 = \frac{\beta_0 v_0 L_0}{l_0} = 5 \times 10^{-4}\,, \tag{1.39}$$

i.e. one-fifth of the previous value. This means that all the estimates obtained earlier with the help of Eq. (1.28) are still valid. For deep layers, however, the characteristic vertical velocity has to be found from the boundary condition (Eq. 1.17). Sea water in deep layers is almost homogeneous, therefore, there is no need to determine the characteristic flow velocity in terms of the baroclinic layer thickness. The horizontal flow velocity at deep ocean layers is far less than in the baroclinic layer, but vertical flows are strongly influenced by the bottom topography. Assuming that the characteristic variation in the ocean bottom depth is $H_0 = 1$ km, the characteristic velocity $v_H = 1\,\mathrm{cm\,s^{-1}}$ and the characteristic horizontal scale $L_H = 500$ km, we obtain $w_H = \dfrac{v_H H_0}{L_H} = 2 \times 10^{-3}\,\mathrm{cm\,s^{-1}}$, i.e. it is a quantity of the same order of magnitude as in Eq. (1.37). Since Eq. (1.28) yields much simpler equations of the type (1.29) and (1.32), and a variation of w_0 by severalfold does not affect the qualitative conclusions, we retain formula (1.36) for calculating w_0. Now we estimate the value of w at the ocean surface, using the boundary conditions (1.10) and substituting wind drift velocity $v_d = 25\,\mathrm{cm\,s^{-1}}$ for the characteristic velocity v_0. Despite the large value of v_d, we find that $w_0 = \dfrac{v_d \zeta_0}{L_0} = 10^{-5}$. This is two orders of magnitude less than the characteristic vertical velocity found above. Therefore Eq. (1.10) can be replaced by a simpler "rigid lid" condition: $w = 0$ at $z = 0$. Hence, we can use the formulas (1.36) to estimate the characteristic large-scale, deep-layer circulation, substituting h_0 for H_0, L_0 for L_H and reducing $(\delta\rho)_0$ by one order of magnitude. The characteristic values thus found would be close to the values for the baroclinic layer: v_0 would slightly decrease, w_0 would increase, $\mathscr{P}_0$ would decrease and so on.

Finally, we mention the basic property of formulas (1.36). In order to find v_0, w_0, t_0, $\mathscr{P}_0$ and ζ_0 by these formulas, we have to specify, in addition to the usual parameters, also the "internal" characteristics $(\delta\rho)_0$ and h_0 of the baroclinic ocean. The unknown quantities could as well be expressed only through external characteristics with the help of the boundary conditions, but we do not consider this worthwhile. Boundary conditions of the type (1.9), (1.11) are suitable only in estimating the characteristics in the upper boundary layer of approximately 30-m thickness and not for the baroclinic layer. On the other hand, the characteristic values $(\delta\rho)_0$ and h_0 are known to a similar accuracy as that of tangential wind stress or heat and salt fluxes. In particular, using any atlas of hydrological elements, we can find h_0 with the help of the following criterion: the density gradient at the lower boundary of the baroclinic layer is less by one order of magnitude than the gradient on the ocean surface. By the formulas (1.36) we emphasize the following. Inhomogeneity of sea water is a decisive factor in ocean dynamics. Therefore, all characteristic values of large-scale currents are to be estimated not from the tangential wind stress, but in terms of the thermal factors, i.e. $(\delta\rho)_0$. In the sequel we shall return to this point repeatedly with concrete examples.

The estimates made above lead to one more simplification. After replacing Eq. (1.3) by (1.23) or (1.25) and the condition (1.10) by the "rigid lid" condition, we can apply the boundary conditions (1.9), (1.11)–(1.14) to the undisturbed ocean surface $z = 0$.

We have already noted that the formulas (1.36) are suitable for estimating not only the characteristic values for large-scale currents but also those of meso-scale intensive currents with a characteristic horizontal scale of 50 km; the only exception being the equatorial currents. At the equator $l = 0$, and in a narrow equatorial belt a few hundred kilometres wide, we have $|\sin\varphi| \ll 1$, therefore, there is no geostrophic balance in this belt. In the equatorial zone the pressure gradient is balanced mainly by non-linear inertial terms. On equating the characteristic values of the terms in Eq. (1.2) we obtain:

$$\frac{v_0}{t_0} = \frac{v_0^2}{L_0} = \frac{w_0 v_0}{h_0} = \frac{1}{\rho_0}\frac{\mathscr{P}_0}{L_0}\,. \tag{1.40}$$

As before, we calculate the characteristic values $\mathscr{P}_0$ and ζ_0 from Eq. (1.36). Thus, for the equatorial belt we obtain:

$$\mathscr{P}_0 = gh_0(\delta\rho)_0\,; \qquad \zeta_0 = \frac{(\delta\rho)_0}{\rho_0}h_0\,; \qquad v_0 = \left(\frac{gh_0(\delta\rho)_0}{\rho_0}\right)^{1/2};$$

$$w_0 = \frac{h_0}{L_0}\left(\frac{gh_0(\delta\rho)_0}{\rho_0}\right)^{1/2}; \qquad t_0 = \frac{L_0}{v_0} = L_0\left(\frac{\rho_0}{gh_0(\delta\rho)_0}\right)^{1/2}. \tag{1.41}$$

We consider the characteristic values L_0 and h_0 as one order less in magnitude than those for the large-scale off-equatorial processes because we are dealing with a narrow jet of intensive equatorial currents. This jet spreads over the whole ocean and therefore it may seem that we have to choose different characteristic scales for x and y. But the jet meanders about the equator northward and southward under the influence of inertia forces (Kolesmikov et al. 1968; Philander and Pacanowski 1980; O'Brien 1979). The same characteristic scale L_0 can therefore be used for both directions. Observations show that the equatorial currents have a complicated vertical structure and may quite often reverse their directions (Kolesmikov et al. 1968). Therefore, for h_0, we take a value of one order of magnitude less than the depth of the baroclinic layer at middle latitudes. The value of $(\delta\rho)_0$ is also less than that at the middle latitudes but not so much as by one order of magnitude. Thus, for the equatorial current we take $L_0 = 10^7$, $h_0 = 5 \times 10^3$, $(\delta\rho)_0 = 10^{-3}$ and according to Eq. (1.41) we obtain:

$$\mathscr{P}_0 = 5 \times 10^3\,; \quad \zeta_0 = 5\,; \quad v_0 \approx 70\,; \quad w_0 \approx 3.5 \times 10^{-2}\,; \quad t_0 \approx 1.4 \times 10^5\,. \tag{1.42}$$

Comparing Eqs. (1.42) and (1.37), we find that the anomaly of pressure and sea-surface topography is one order of magnitude less than the values at the middle latitudes. And the characteristic horizontal scale is also one order less in magnitude, therefore, the pressure gradients and sea-surface topography are of the same order of magnitude as at the middle latitudes. Consequently, the same pressure gradient gives rise to a velocity at the equator one order of magnitude greater than at the middle latitudes. For these values of L_0 and v_0, the value of t_0 is almost two orders less than at the middle latitudes. Possibly, the time derivatives at the equator represent the small differences of large quantities and therefore v_0/t_0 should not be equated to the characteristic values of other quantities.

Furthermore, as in Eqs. (1.36), the inhomogeneity of sea water plays a decisive role also in Eq. (1.41), but here the density anomaly results in a greater flow velocity. At the equator the inertial terms cannot be disregarded as they play here a more important role than in the jetlike currents at middle latitudes.

Let us now mention the role played by turbulent viscosity at the equator. In the equation, analogous to Eq. (1.29), for the equator the term due to the vertical turbulent viscosity will contain a factor Pr/Pe, where $Pr = \nu/\nu_1$ is the Prandtl number. Since the characteristic vertical scale is small, we take $\nu = 10\,\mathrm{cm}^2\,\mathrm{s}^{-1}$. This is the usual value used for ν at the equatorial zone (O'Brien 1979). For $\nu = 10$ and for the characteristic values given in Eq. (1.42), we obtain $Pr/Pe < 0.1$. Thus, the vertical turbulent viscosity at the equatorial zone is one order of magnitude less than the interial terms or the pressure gradient. Its role, however, is more perceptible at the equator than at the middle latitudes. Similarly, we readily find that for $\mu = 10^6\,\mathrm{cm}^2\,\mathrm{s}^{-1}$ the effect of the lateral friction is one order less than that of the vertical viscosity.

1.1.3 Velocity in the Quasi-Geostrophic Model: Calculation Formulas

Neglecting the inertial terms and the lateral mixing effects, and after certain simple transformations, let us represent Eqs. (1.1) and (1.2) as follows:

$$\nu\frac{\partial^2 M}{\partial z^2} - ilM = \frac{1}{\rho_0}\left(\frac{\partial P}{\partial x} + i\frac{\partial P}{\partial y}\right), \tag{1.43}$$

where

$$M = u + iv\,. \tag{1.44}$$

Assume that the pressure gradients on the right-hand side of Eq. (1.43) do not depend on z. Then the equation can be easily solved for the complex velocity:

$$M = \frac{1}{l\rho_0}\left(i\frac{\partial P}{\partial x} - \frac{\partial P}{\partial y}\right) + c_1 e^{-\alpha(1+i)z} + c_2 e^{\alpha(1+i)z}\,;$$

$$\alpha = \sqrt{\frac{\omega \sin\varphi}{\nu}} = \sqrt{\frac{l}{2\nu}}\,. \tag{1.45}$$

Expression (1.45) can be used in solving certain auxiliary problems. Now consider the vertical distribution of wind in the lower atmospheric boundary layer. Let us take the origin of the coordinates at the sea surface and let 0_z be directed upward. Replace P by the atmospheric pressure P_a and take the following boundary conditions for determining the integration constants: at the sea surface the wind velocity is zero and at the top of the boundary layer it is equal to the geostrophic wind velocity. Hence, from Eq. (1.45) we easily obtain:

$$u' + iv' = \frac{1}{l\rho_0}\left(i\frac{\partial P_a}{\partial x} - \frac{\partial P_a}{\partial y}\right)(1 - e^{-(1+i)\alpha' z})\,, \tag{1.46}$$

where $\alpha' = \sqrt{\frac{l}{2\nu'}}$, ν' is the vertical turbulent viscosity coefficient for air and ρ_0 the mean air density at sea level. The vertical distribution thus obtained for the wind is the Akerblom model. In this model we are only interested in the tangential wind stress at sea level. Differentiating Eq. (1.46) with respect to z, we obtain:

$$\tau_x + i\tau_y = \rho_0 \nu' \frac{\partial M'}{\partial z}\bigg|_{z=0} = \frac{\nu'}{l}\alpha'(1+i)\left(i\frac{\partial P_a}{\partial x} - \frac{\partial P_a}{\partial y}\right) \tag{1.47}$$

or

$$\tau_x = \frac{1}{2\alpha'}\left(\frac{\partial P_a}{\partial x} + \frac{\partial P_a}{\partial y}\right); \quad \tau_y = \frac{1}{2\alpha'}\left(\frac{\partial P_a}{\partial x} - \frac{\partial P_a}{\partial y}\right). \tag{1.48}$$

We now apply Eq. (1.45) to describe the vertical distribution of currents at the near-surface Ekman boundary layer. As before, we take the coordinate origin on the sea surface, direct 0_z vertically downward, apply condition (1.9) at the sea surface and the geostrophic current condition at the lower boundary of the friction layer. Thus, we obtain:

$$M = \frac{1}{l\rho_0}\left(i\frac{\partial P}{\partial x} - \frac{\partial P}{\partial y}\right)_{z=0} + \frac{(1-i)}{2\rho_0 \nu\alpha}(\tau_x + i\tau_y)e^{-(1+i)\alpha z}. \tag{1.49}$$

The pressure gradients in this expression are the values at the sea surface. The formula itself is meaningful only in a thin boundary layer about 25–30 m thick; the second term describes the Ekman spiral. This formula is used particularly in determining the vertical turbulent viscosity in the Ekman near-surface layer:

$$\nu\frac{\partial^2 M}{\partial z^2}\bigg|_E = \frac{\alpha(1+i)}{\rho_0}(\tau_x + i\tau_y)e^{-(1+i)\alpha z}. \tag{1.50}$$

Finally, we apply Eq. (1.45) to the bottom boundary layer, using two types of boundary conditions for the ocean bottom: either the non-slip condition (1.15) or the free-slip condition (1.16). We apply the geostrophic flow condition at the upper boundary of the bottom stress layer. Under these boundary conditions, Eq. (1.45) yields:

$$M = \frac{1}{l\rho_0}\left(i\frac{\partial P}{\partial x} - \frac{\partial P}{\partial y}\right)_{z=H}\left[1 - \delta e^{(1+i)\alpha(z-H)}\right]. \tag{1.51}$$

Where $\delta = 1$ for non-slip and $\delta = 0$ for free-slip conditions at the sea bottom. From this relation we readily obtain an expression for the bottom stress:

$$\tau_x^H + i\tau_y^H = \rho_0 \nu \frac{\partial M}{\partial z}\bigg|_{z=H} = -\frac{(1+i)\alpha\delta}{l\rho_0}\left(i\frac{\partial P}{\partial x} - \frac{\partial P}{\partial y}\right)_{z=H} \tag{1.52}$$

or

$$\tau_x^H = \frac{\delta}{2\alpha}\left(\frac{\partial P}{\partial x} + \frac{\partial P}{\partial y}\right)_{z=H}; \quad \tau_y^H = \frac{\delta}{2\alpha}\left(\frac{\partial P}{\partial y} - \frac{\partial P}{\partial x}\right)_{z=H}. \tag{1.53}$$

From Eq. (1.51) we can also obtain an expression for the force of vertical turbulent viscosity at the bottom friction layer:

$$\left(\nu\frac{\partial^2 M}{\partial z^2}\right)_n = \frac{\delta}{\rho_0}\left(\frac{\partial P}{\partial x} + i\frac{\partial P}{\partial y}\right)_{z=H} e^{(1+i)\alpha(z-H)} . \tag{1.54}$$

Now let us rewrite Eq. (1.43) as follows:

$$M = \frac{1}{\rho_0 l}\left(i\frac{\partial P}{\partial x} - \frac{\partial P}{\partial y}\right) - \frac{i\nu}{l}\frac{\partial^2 M}{\partial z^2} , \tag{1.55}$$

and then simplify it. We take into account the vertical turbulent viscosity in a thin surface friction layer by the approximate formula (1.50), in the bottom friction layer by formula (1.54) and neglect it in the main ocean body. Thus, we obtain:

$$M = \frac{1}{\rho_0 l}\left(i\frac{\partial P}{\partial x} - \frac{\partial P}{\partial y}\right) + \frac{(1-i)}{2\rho_0 \nu\alpha}(\tau_x + i\tau_y)e^{-(1+i)\alpha z}$$

$$- \frac{\delta}{\rho_0 l}\left(i\frac{\partial P}{\partial x} - \frac{\partial P}{\partial y}\right)_{z=H} e^{(1+i)\alpha(z-H)} . \tag{1.56}$$

This formula gives the current velocity in terms of the pressure gradient and the specified tangential wind stress field. This formula has been derived by neglecting the inertial terms, lateral effects and by making approximate allowance for the vertical turbulent viscosity. It can also be derived rigorously by the asymptotic boundary layer correction method.

1.1.4 Equations for Integral Functions in Off-Equatorial Current Models

In Section 1.1.2 the pressure gradient was intentionally chosen as a measure in evaluating the characteristic quantities. At middle latitudes pressure is "counteracted" by the Coriolis force and at the equator by the inertial force. For some reason a pressure anomaly arises in the ocean. This anomaly initiates motion and as a result of the latter and other factors (rotation of the Earth, internal friction, etc.) forces emerge which balance the pressure field. In this way the pressure anomaly is the basic dynamic characteristic of currents. Were there no pressure anomaly, there would be only a pure wind drift current due to the entraining action of wind over a homogeneous, infinite ocean. But even in this case inhomogeneity in the wind field would give rise to sea-surface topography gradients. Therefore, the determination of the pressure field is a central problem in ocean dynamics. Various trends in ocean dynamics originate due the different methods used to determine the pressure field. In the theory of sea currents this problem is generally reduced to the determination of some integral function. This section deals with the construction of two differential equations for these integral functions.

Let us rewrite Eqs. (1.1) and (1.2) as follows:

$$\nu\frac{\partial^2 u}{\partial z^2} + lv = \frac{1}{\rho_0}\frac{\partial P}{\partial x} + A ; \tag{1.57}$$

$$\nu\frac{\partial^2 v}{\partial z^2} - lu = \frac{1}{\rho_0}\frac{\partial P}{\partial y} + B . \tag{1.58}$$

Here, A and B denote the sum of the inertial terms and the lateral exchange effect respectively:

$$A = \frac{\partial u}{\partial t} + u\frac{\partial u}{\partial x} + v\frac{\partial u}{\partial y} + w\frac{\partial u}{\partial z} - \mu \Delta u ; \tag{1.59}$$

$$B = \frac{\partial v}{\partial t} + u\frac{\partial v}{\partial x} + v\frac{\partial v}{\partial y} + w\frac{\partial v}{\partial z} - \mu \Delta v . \tag{1.60}$$

Integrating these equations vertically over the ocean thickness under the boundary conditions (1.9), and by virtue of Eq. (1.53), we obtain:

$$lS_y = \frac{1}{\rho_0}\int_0^H \frac{\partial P}{\partial x}dz - \frac{\tau_x}{\rho_0} - \frac{1}{2\alpha\rho_0}\left(\frac{\partial P}{\partial x} + \frac{\partial P}{\partial y}\right)_{z=H} + \int_0^H A dz ;$$

$$lS_x = -\frac{1}{\rho_0}\int_0^H \frac{\partial P}{\partial y}dz + \frac{\tau_y}{\rho_0} + \frac{1}{2\alpha\rho_0}\left(\frac{\partial P}{\partial y} - \frac{\partial P}{\partial x}\right)_{z=H} - \int_0^H B dz , \tag{1.61}$$

where

$$S_x = \int_0^H u dz ; \qquad S_y = \int_0^H v dz . \tag{1.62}$$

Equation (1.4), as is known, by integrating over z under the boundary conditions assumed for w [i.e. "rigid lid" on the surface and condition (1.15) at the ocean bottom], yields:

$$\frac{\partial S_x}{\partial x} + \frac{\partial S_y}{\partial y} = 0 . \tag{1.63}$$

By virtue of Eq. (1.63), after cross-differentiation, Eqs. (1.61) and (1.62) yield:

$$\beta S_y = -\frac{1}{\rho_0}J(H, P)_{z=H} + \frac{1}{\rho_0}\operatorname{rot}\tau - \frac{1}{2\alpha\rho_0}(\Delta P)_{z=H} + \frac{\partial}{\partial y}\int_0^H A dz - \frac{\partial}{\partial x}\int_0^H B dz , \tag{1.64}$$

where $\beta = \frac{dl}{dy}$; J is the Jacobian and $\operatorname{rot}\tau$ the vertical component of the vortex due to tangential wind stress:

$$J(H, P)_{z=H} = \left(\frac{\partial H}{\partial x}\frac{\partial P}{\partial y} - \frac{\partial H}{\partial y}\frac{\partial P}{\partial x}\right)_{z=H} ; \qquad \operatorname{rot}\tau = \frac{\partial \tau_y}{\partial x} - \frac{\partial \tau_x}{\partial y} . \tag{1.65}$$

We have taken $\alpha = \text{const}$ in converting Eqs. (1.61) and (1.63) into Eq. (1.64). This simplification is justified not only because the root of l occurs in the expression for α, but first of all because the term with the factor $\frac{1}{2}\alpha$ itself is small. In differentiating the bottom pressure gradient we have also disregarded the variability of H. In equations containing small terms we shall often assume $H = \text{const}$ and thereby neglect the secondary β-effect and the secondary bottom topography effect. In other words, in small terms we retain only the terms with higher derivatives. With regards to the main effects due to the Earth's sphericity and bottom relief variability, we not only retain them but also demonstrate their importance. The main β-effect forms the left-hand side and the bottom relief effect the first term on the right-hand side of Eq. (1.64). We simplify the last two terms on the right-hand side of Eq. (1.64) as follows. In the expressions for A and B we neglect the terms containing w and in the terms containing the time derivatives we replace u and v by their approximate expressions found from Eq. (1.56) for $\delta = 0$. Finally, in all other terms we replace u and v by the geostrophic approximations given by the first term on the right-hand side of Eq. (1.56). Thus, neglecting the secondary effects due to the variability of l and H, we obtain:

$$\frac{\partial}{\partial y}\int_0^H A dz - \frac{\partial}{\partial x}\int_0^H B dz \approx \frac{1}{\rho_0 l}\frac{\partial}{\partial t}\operatorname{div}\boldsymbol{\tau} - \int_0^H\left(\frac{\partial \Omega}{\partial t} + u_g\frac{\partial \Omega}{\partial x} + v_g\frac{\partial \Omega}{\partial y}\right)dz + \mu\int_0^H \Delta\Omega dz\,,$$

where

$$\operatorname{div}\boldsymbol{\tau} = \frac{\partial \tau_x}{\partial x} + \frac{\partial \tau_y}{\partial y}\,; \quad \Omega = \frac{\partial v_g}{\partial x} - \frac{\partial u_g}{\partial y} \approx \frac{1}{\rho_0 l}\Delta P\,. \tag{1.66}$$

Finally, we obtain:

$$\frac{\partial}{\partial y}\int_0^H A dz - \frac{\partial}{\partial x}\int_0^H B dz \approx \frac{1}{\rho_0 l}\frac{\partial}{\partial t} div\,\boldsymbol{\tau} - \frac{1}{\rho_0 l}\int_0^H\left[\frac{\partial}{\partial t}\Delta P + \frac{1}{\rho_0 l}J(P, \Delta P)\right]dz + \frac{\mu}{\rho_0 l}\int_0^H \Delta\Delta P dz\,. \tag{1.67}$$

Substituting Eq. (1.67) into Eq. (1.64), we obtain an approximate expression for the vorticity equation:

$$\beta S_y = -\frac{1}{\rho_0}J(H, P)_{z=H} + \frac{1}{\rho_0}\operatorname{rot}\boldsymbol{\tau} + \frac{1}{\rho_0 l}\frac{\partial}{\partial t}\operatorname{div}\boldsymbol{\tau} - \frac{1}{2\alpha\rho_0}(\Delta P)_{z=H} - \frac{1}{\rho_0 l}\int_0^H\left[\frac{\partial}{\partial t}\Delta P + \frac{1}{\rho_0 l}J(P, \Delta P)\right]dz + \frac{\mu}{\rho_0 l}\int_0^H \Delta\Delta P dz\,. \tag{1.68}$$

Equation (1.68) is the initial relation from which we shall derive two equations for the integral functions. We shall take the sea-surface topography ζ and the total mass transport stream function ψ as the integral functions. The sea-surface topography, up to a constant factor, is the pressure anomaly on the ocean surface $P_s = \rho_0 g\zeta$. From formula (1.25) we obtain a relation between the sea-surface slope, density gradient and the pressure gradient: for example,

$$\left(\frac{\partial P}{\partial x}\right)_{z=H} = \rho_0 g\frac{\partial \zeta}{\partial x} + g\int_0^H \frac{\partial \rho}{\partial x}dz = \frac{\partial P_s}{\partial x} + g\int_0^H \frac{\partial \rho}{\partial x}dz\,; \tag{1.69}$$

$$\left(\frac{\partial P}{\partial x}\right)_{z=H} = \left.\frac{\partial P}{\partial x}\right|_z - g\int_z^H \frac{\partial \rho}{\partial x}dz\,. \tag{1.70}$$

The relation between the y-derivatives is expressed in a similar manner.

It is a simple matter to derive an approximate formula connecting the pressure anomaly to the total mass transport stream function. The relation (1.63) permits the introduction of a simple dependence between the total mass transport stream function and the components of the integral mass transport:

$$S_x = -\frac{\partial \psi}{\partial y}\,; \qquad S_y = \frac{\partial \psi}{\partial y}\,, \tag{1.71}$$

and Eq. (1.61) can be rewritten as follows:

$$\frac{\partial \psi}{\partial x} = \frac{1}{\rho_0 l}\int_0^H \frac{\partial P}{\partial x} dz - \frac{\tau_x}{\rho_0 l} - \frac{1}{2\alpha\rho_0 l}\left(\frac{\partial P}{\partial x} + \frac{\partial P}{\partial y}\right)_{z=H} + \frac{1}{l}\int_0^H A dz\,; \tag{1.72}$$

$$\frac{\partial \psi}{\partial y} = \frac{1}{\rho_0 l}\int_0^H \frac{\partial P}{\partial y} dz - \frac{\tau_y}{\rho_0 l} - \frac{1}{2\alpha\rho_0 l}\left(\frac{\partial P}{\partial y} - \frac{\partial P}{\partial x}\right)_{z=H} + \frac{1}{l}\int_0^H B dz\,. \tag{1.73}$$

The first terms on the right-hand sides of these equations are important, whereas others can be neglected in the first approximation. In making rough estimates we may also neglect the variability of l and H in the first terms. Thus, we obtain an approximate expression:

$$\psi \approx \frac{1}{\rho_0 l}\int_0^H P dz\,, \tag{1.74}$$

which reveals the physical meaning of the connection between ψ and P: up to a constant factor the total mass transport stream function is the integral of the pressure anomaly over height.

An equation for the sea-surface topography is obtained from Eq. (1.68) with the help of relations of the type (1.69) and (1.70). Thus, we have:

$$\underset{\text{VI}}{-\mu\Delta\Delta\zeta} + \underset{\text{V}}{\frac{\partial}{\partial t}\Delta\zeta} + \underset{\text{IV}}{\frac{g}{l}J(\zeta,\Delta\zeta)} + \underset{\text{III}}{\frac{1}{2\alpha H}\Delta\zeta} + \underset{\text{II}}{\frac{l}{H}J(H,\zeta)}$$

$$+\underset{\text{I}}{\beta\frac{\partial \zeta}{\partial x}} = \underset{\text{VII}}{\frac{l}{\rho_0 g H}\operatorname{rot}\tau} + \frac{1}{\rho_0 g H}\left[\underset{\text{VIII}}{\beta\tau_x} + \underset{\text{IX}}{\frac{\partial}{\partial t}\operatorname{div}\tau}\right] + f_1\,, \tag{1.75}$$

where

$$f_1 = \underbrace{\frac{1}{H}\left(\frac{g}{\rho_0 l}\right)^2 J\left(\int_0^H (H-z)\rho dz, \int_0^H (H-z)\Delta\rho dz\right) - \frac{g}{\rho_0 l H}\int_0^H (H-z)J(\zeta,\Delta\rho)dz}_{\text{IV}}$$

$$\underbrace{-\frac{1}{\rho_0 H}\int_0^H (H-z)\frac{\partial}{\partial t}\Delta\rho dz}_{\text{V}} \underbrace{- \frac{g}{\rho_0^2 l H}\int_0^H J\left(\int_0^z \rho dz, \int_0^z \Delta\rho dz\right)dz}_{\text{IV}} \underbrace{- \frac{l}{2\alpha\rho_0 H}\int_0^H \Delta\rho dz}_{\text{III}}$$

$$\underbrace{-\frac{l}{\rho_0 H}\int_0^H J(H,\rho)dz}_{\text{II}} \underbrace{- \frac{\beta}{\rho_0 H}\int_0^H (H-z)\frac{\partial P}{\partial x}dz}_{\text{I}} + \underbrace{\frac{\mu}{H\rho_0}\int_0^H (H-z)\Delta\Delta\rho dz}_{\text{VI}}\,. \tag{1.76}$$

In order to derive an equation for ψ we turn our attention to the relations (1.72) and (1.73). Retaining only the first two terms on the right-hand sides of these equations, we change over from the pressure gradients to sea-surface slopes with the help of formulas of the type (1.69). Thus, we obtain:

$$\frac{\partial \zeta}{\partial x} = \frac{l}{gH}\frac{\partial \psi}{\partial x} - \frac{1}{\rho_0 H}\int_0^H (H-z)\frac{\partial P}{\partial x}dz + \frac{\tau_x}{\rho_0 g H}, \tag{1.77}$$

$$\frac{\partial \zeta}{\partial y} = \frac{l}{gH}\frac{\partial \psi}{\partial y} - \frac{1}{\rho_0 H}\int_0^H (H-z)\frac{\partial P}{\partial y}dz + \frac{\tau_y}{\rho_0 g H}. \tag{1.78}$$

The equation for ψ is derived from Eq. (1.75) as follows. Using Eqs. (1.77) and (1.78), we replace the first-order derivatives of ζ by the respective first-order derivatives of ψ. To replace the second-order derivatives we have to differentiate Eq. (1.77) with respect to x and Eq. (1.78) with respect to y, disregarding the variability of l and H (i.e. neglecting the secondary effects of the variability of these parameters). Thus, we obtain:

$$\underbrace{-\mu\Delta\Delta\psi}_{\text{VI}} + \underbrace{\frac{\partial}{\partial t}\Delta\psi}_{\text{V}} + \underbrace{\frac{1}{H}J(\psi,\Delta\psi)}_{\text{IV}} + \underbrace{\frac{l}{2\alpha H}\Delta\psi}_{\text{III}} + \underbrace{\frac{l}{H}J(H,\psi)}_{\text{II}} + \underbrace{\beta\frac{\partial\psi}{\partial x}}_{\text{I}}$$

$$= \underbrace{\frac{1}{\rho_0}\operatorname{rot}\tau}_{\text{VII}} + \underbrace{\frac{1}{\rho_0 H}\left(\frac{\partial H}{\partial y}\tau_x - \frac{\partial H}{\partial x}\tau_y\right)}_{\text{X}} + f_2, \tag{1.79}$$

where

$$f_2 = \underbrace{\frac{1}{H}\left(\frac{g}{\rho_0 l}\right)^2 J\left(\int_0^H (H-z)\rho dz, \int_0^H (H-z)\Delta\rho dz\right) - \left(\frac{g}{\rho_0 l}\right)^2 \int_0^H J\left(\int_0^z \rho dz, \int_0^z \Delta\rho dz\right)dz}_{\text{IV}}$$

$$\underbrace{-\frac{g}{2\alpha\rho_0 H}\int_0^H z\Delta\rho dz}_{\text{III}} \underbrace{- \frac{g}{\rho_0 H}\int_0^H zJ(H,\rho)dz}_{\text{II}}. \tag{1.80}$$

In deriving Eq. (1.79) we have also neglected certain small-order terms due to the tangential wind stress. Despite these simplifications, we have obtained a general, though cumbersome, Eq. (1.79) from which we can derive, for particular cases, the main relations of the many well-known models of the total mass transport theory.

1.1.5 Evaluation of the Order of Magnitude of the Quantities in the Equations for Integral Functions

We begin with the estimation of the order of magnitude of the terms in Eq. (1.75). The physical meaning of the terms on the left-hand side of this equation is as follows: VI denotes the lateral friction, V the local time derivative of inertial terms, IV the non-linear inertial terms, III the bottom friction effect, II the bottom relief influence and I the β-effect. On the right-hand side, VII represents the vortex of the tangential wind stress, VIII the wind β-effect and IX the non-stationary wind effect. Other terms on the right-hand side arise due to the sea water inhomogeneity and have been numbered in conformity with the terms on the left-hand side having the same physical meaning. Therefore, the term VI on the right-hand side can be called the "baroclinic effect of lateral friction" and I the baroclinic β-effect and so on.

We shall denote the characteristic values of the corresponding quantities by the same symbols as in Section 1.1.2 but with one difference: in Eq. (1.75) H denotes the actual ocean depth. Therefore, in place of the characteristic baroclinic layer thickness h_0, it is better to use the characteristic ocean depth H_0. This, however, is unimportant because h_0 and H_0 are quantities of the same order of magnitude. We take the β-effect as the standard measure for Eq. (1.75). On equating the characteristic values of the terms indexed I on the right- and left-hand sides of this equation, we obtain:

$$\zeta_0 = \frac{(\delta\rho)_0}{\rho_0} H_0 \,, \tag{1.81}$$

an expression similar to Eq. (1.41). By virtue of Eq. (1.81), we easily find that the terms of the same index on the left- and right-hand sides of Eq. (1.75) are of the same order of magnitude: there is therefore no need to estimate the order of magnitude of all the terms in this equation.

Now we assume that the inertial terms V and IV are quantities of the same order of magnitude. So we can determine the characteristic time scale:

$$t_0 = \frac{l_0 L_0^2}{g\zeta_0} = \frac{L_0}{v_0} \,. \tag{1.82}$$

Now we convert Eq. (1.75) into a dimensionless form, dividing it by the characteristic value of the reference quantity $\frac{\beta_0\zeta_0}{L_0}$ and using formulas (1.81) and (1.82). Using a tilde to denote dimensionless variables, we obtain:

$$-\varepsilon_6 \bar{\Delta}\bar{\Delta}\bar{\zeta} + \varepsilon_5 \frac{\partial}{\partial \bar{t}} \bar{\Delta}\bar{\zeta} + \frac{\varepsilon_5}{\bar{l}} J(\bar{\zeta}_1 \bar{\Delta}\bar{\zeta}) + \varepsilon_3 \frac{\bar{l}}{2\bar{\alpha}\bar{H}} \bar{\Delta}\bar{\zeta} + a_2 \frac{\bar{l}}{\bar{H}} J(\bar{H}, \bar{\zeta}) + \bar{\beta}\frac{\partial\bar{\zeta}}{\partial\bar{x}}$$

$$= \varepsilon_7 \frac{\bar{l}}{\bar{H}} \overline{\operatorname{rot}} \bar{\tau} + \varepsilon_8 \frac{\bar{\beta}}{\bar{H}} \bar{\tau}_x + \varepsilon_9 \frac{1}{\bar{H}} \frac{\partial}{\partial \bar{t}} \overline{\operatorname{div}} \bar{\tau} + \frac{L_0}{\beta_0 \zeta_0} f_1 \,. \tag{1.83}$$

The subscripts of the small dimensionless parameters correspond to the numbers of the groups of terms: the parameter under group II is not small, so it is denoted by some other symbol. The expressions for the dimensionless parameters are as follows:

$$\varepsilon_6 = \frac{\mu}{\beta_0 L_0^3}\,; \qquad \varepsilon_5 = \frac{1}{t_0 \beta_0 L_0}\,; \qquad \varepsilon_3 = \frac{\sqrt{l_0 \nu}}{H_0 L_0 \beta_0}\,;$$

$$a_2 = \frac{l_0}{L_0 \beta_0}\,; \qquad \varepsilon_7 = \frac{l_0 \tau_0}{\rho_0 H_0 \beta_0 \zeta_0 g}\,; \qquad \varepsilon_8 = \frac{\tau_0 L_0}{\rho_0 g H_0 \zeta_0}\,;$$

$$\varepsilon_9 = \frac{\tau_0}{t_0 \rho_0 H_0 g \beta_0 \zeta_0} \,. \tag{1.84}$$

The specified characteristic quantities have the same numerical values as in Section 1.1.2, namely:

$$l_0 = 10^{-4}\,; \qquad \tau_0 = 1\,; \qquad L_0 = 10^8\,; \qquad A_l = 10^7\,;$$
$$\nu = 10^2\,; \qquad \beta_0 = 2 \times 10^{-13}\,. \tag{1.85}$$

Only H_0 and $(\delta\rho)_0$ suffer small variations because the characteristic depth $H_0 = 1\,\text{km} = 10^5\,\text{cm}$ is no longer the baroclinic layer thickness but represents the characteristic ocean depth and $(\delta\rho)_0 = 0.5 \times 10^{-3}$ now represents the vertically averaged density gradient. With the help of Eqs. (1.81), (1.82), (1.25) and (1.34), we find that:

$$\zeta_0 = 50\,; \qquad v_0 = 5\,; \qquad \mathscr{P} = 5.10^4\,; \qquad t_0 = 2 \times 10^7\,, \tag{1.86}$$

which coincide with the values found earlier in Section 1.1.2. By virtue of Eqs. (1.85) and (1.86), we obtain the following values for the dimensionless parameters:

$$\varepsilon_3 = 5 \times 10^{-2}\,; \qquad \varepsilon_5 = 2.5 \times 10^{-3}\,; \qquad \varepsilon_6 = 0.5 \times 10^{-4}\,;$$
$$\varepsilon_7 = 0.1\,; \qquad \varepsilon_8 = 2 \times 10^{-2}\,; \qquad \varepsilon_9 = 0.5 \times 10^{-4}\,; \qquad a_2 = 5\,. \tag{1.87}$$

These estimates reveal the following. The bottom relief effect on the left-hand side of Eq. (1.83) is the only factor that is comparable with the β-effect. Furthermore, ε_6 is five times the Ekman number E_m and ε_5 is five times the Kibel (Rossby) number (see Sect. 1.1.2). This means that the influence of the inertial terms and the lateral exchange in (1.75) are more important than the influence of these factors in the equations of motion (1.1), (1.2). As the parameters ε_5 and ε_6 are small, while studying large-scale stationary currents or seasonal variations of currents we can neglect these factors in the first approximation even in the equation for ζ. The essential factors on the left-hand side of this equation are the β-effect and the bottom relief effect.

On changing over to dimensionless variables in Eq. (1.76) and using Eq. (1.81), we find that the terms in each group acquires the same dimensionless parameter as on the left-hand side of Eq. (1.75). Consequently, we conclude that the main terms on the right-hand side of Eq. (1.75) are the baroclinic β-effect and the joint effect due to baroclinicity and relief (JEBAR) (groups I and II in Eq. (1.79)]. The direct effect of the tangential wind stress, judging from the value of $\varepsilon_7 = 0.1$, plays a less important role than the density anomaly, which reflects the effect of heat and salt exchange and the indirect effect due to the wind field.

We now evaluate the order of magnitude of ψ by means of Eqs. (1.78) and (1.77). The left-hand sides in these relations and the first two terms on the right-hand sides have the same order of magnitude. And we have:

$$\frac{\zeta_0}{L_0} = \frac{l_0}{gH_0}\frac{\psi_0}{L_0} = \frac{H_0(\delta\rho)_0}{\rho_0 L_0}\,,$$

hence,

$$\psi_0 = \frac{gH_0\zeta_0}{l_0} = \frac{H_0^2 g(\delta\rho)_0}{l_0\rho_0}\,. \tag{1.88}$$

Substituting the characteristic values found above, we find that ψ_0 is equal to 5×10^{13} or 50 Sverdrups (Sverdrup $= 10^{12}$ CGS).

Equation (1.79) is converted into a dimensionless form in the same way as Eq. (1.75); moreover, all the dimensionless parameters, except one, are the same as before. This is natural because Eqs. (1.75) and (1.79) are identical and the same standard, namely, the β-effect, is chosen in changing over to dimensionless quantities. Thus, we have:

$$-\varepsilon_6 \overline{\Delta\Delta\psi} + \varepsilon_5 \frac{\partial}{\partial t}\overline{\Delta\psi} + \varepsilon_5 J(\overline{\psi}, \overline{\Delta\psi}) + \varepsilon_3 \frac{\overline{l}}{2\overline{\alpha}\overline{H}}\overline{\Delta\psi} + a_2 \frac{\overline{l}}{\overline{H}} J(\overline{H}, \overline{\psi}) + \overline{\beta}\frac{\partial\overline{\psi}}{\partial x}$$

$$= \varepsilon_7 \overline{\operatorname{rot}\tau} + \varepsilon_{10}\left(\frac{\partial\overline{H}}{\partial y}\frac{\overline{\tau}_x}{\overline{H}} - \frac{\partial\overline{H}}{\partial x}\frac{\overline{\tau}_y}{\overline{H}}\right) + \frac{L_0}{\beta_0\psi_0} f_2\,, \tag{1.89}$$

where $\varepsilon_{10} = \dfrac{\tau_0}{\rho_0\beta_0\psi_0} = 0.1$ and the expressions of and the values of $a_2, \varepsilon_3, \ldots, \varepsilon_8$ are given above. Since the left-hand sides of Eqs. (1.89) and (1.83) are identical, we will only discuss the difference between the right-hand sides.

The right-hand side of Eq. (1.89) contains an additional term, namely the joint effect of wind and bottom relief, which at best is a quantity of the same order of magnitude as the vortex of tangential wind stress.

Unlike Eq. (1.83), the right-hand side of Eq. (1.89) contains only one main term JEBAR. This is represented by group II in Eq. (1.80). We shall not estimate this term because Eq. (1.88) clearly shows that the terms on the left- and right-hand sides of Eq. (1.79), carrying the same serial numbers, are quantities of the same order of magnitude, i.e. the JEBAR is a quantity of the order a_2. In principle, this main term can be reduced or may vanish with the help of some special technique. Such a case is discussed by Kozlov (1975). But this case is more an exception than the rule. Kozlov has shown in the same work that if the density field at the surface of an ocean with variable depth is not homogeneous (which is the case in reality), the joint effect of baroclinicity and bottom relief is different from zero and consequently gives rise to integral circulation of a thermohaline nature.

This analysis is in disagreement with the concept accepted in the total mass transport theory that *rot* τ is the main factor that forms the field ψ. In the sequel we shall repeatedly return to this point with concrete calculations. Here, we only note that the terms with first derivatives on the left-hand side of Eq. (1.79), like in Eq. (1.75), are the basic ones, i.e. Eq. (1.79) is an equation containing higher derivatives multiplied by a small parameter.

1.1.6 Equations for the Calculation of Sea-Surface Topography at the Basin Boundary in an Off-Equatorial Currents Model

It is clear from the previous sections that to evaluate the hydrological characteristics we have to solve a corresponding differential equation for one of the integral functions ζ or ψ. As a rule, in practice, we deal with elliptic equations.

First, we have to determine the magnitudes of these functions on the basin boundary. If the water flux across the boundary contour is known, then it is a simple matter to find ψ, but ζ has to be determined by solving a corresponding differential equation. In Section 1.2.3 we shall present another approach to the solution of boundary value problems for ζ in more advanced models.

Thus, any problem can be solved with the help of either ζ or ψ which is easily determined on the contour. Consequently, it may at first sight seem that ψ has to be chosen as the integral function. Later, we shall show that ζ is a better choice in many cases, meanwhile we construct an equation for determining ζ at the water basin contour. For this purpose, we use the relations (1.61)–(1.62) and the fact that the water flux across the boundary is specified (flow across the solid part of the contour is zero). Neglecting the inertial terms and lateral exchange in these equations, and after simple transformations, we obtain:

$$\frac{\partial \zeta}{\partial x} = \frac{l}{gH} S_y + \frac{1}{\rho_0 H} \int_0^H \frac{\partial \rho}{\partial x} z dz - \frac{1}{\rho_0} \int_0^H \frac{\partial \rho}{\partial x} dz + \frac{\tau_x}{\rho_0 g H} + \frac{1}{2\alpha H}\left(\frac{\partial \zeta}{\partial x} + \frac{\partial \zeta}{\partial y}\right) + \frac{1}{2\alpha H \rho_0}\left(\int_0^H \frac{\partial \rho}{\partial x} dz + \int_0^H \frac{\partial \rho}{\partial y} dz\right); \tag{1.90}$$

$$\frac{\partial \zeta}{\partial y} = -\frac{l}{gH} S_x + \frac{1}{\rho_0 H} \int_0^H \frac{\partial \rho}{\partial y} z dz - \frac{1}{\rho_0} \int_0^H \frac{\partial \rho}{\partial y} dz + \frac{\tau_y}{\rho_0 g H} + \frac{1}{2\alpha H}\left(\frac{\partial \zeta}{\partial y} - \frac{\partial \zeta}{\partial x}\right) + \frac{1}{2\alpha u \rho_0}\left(\int_0^H \frac{\partial \rho}{\partial y} dz - \int_0^H \frac{\partial \rho}{\partial x} dz\right). \tag{1.91}$$

The physical meanings of the terms on the right-hand sides of Eqs. (1.90) and (1.91) are as follows. The first terms represent the integral water flux specified at the boundary contour. The second and third terms represent the baroclinic part of the vertically averaged gradient flow. The fourth terms represent the effect of tangential wind stress at the basin boundary contour. The fifth group of terms represents the bottom friction effect at the basin boundary contour and the last group is the baroclinic effect of bottom friction at the boundary contour.

In Section 1.1.3 we have derived equations for the integral functions making allowance for the inertial terms and the lateral exchange effect. We could as well have retained these factors in deriving Eqs. (1.90) and (1.91). But in Sections 1.1.2 and 1.1.4 we have demonstrated that in studying large-scale flows these factors can be neglected both in the equations of motion and in the equations of the integral functions. It is a simple matter to prove that these factors can be neglected also for the boundary conditions. For this purpose, in place of Eqs. (1.90) and (1.91) we have to derive more bulky expressions with due regard to the inertial terms and lateral exchange and then prove that these factors are small by means of scale analysis. We shall not, however, dwell on this question. In the next section we shall show how to choose equations and boundary conditions for concrete models.

1.2 Diagnostic Sea Current Models

1.2.1 Quasi-Geostrophic Model for the Calculation of Sea-Surface Topography and Flow Velocity

In the beginning of this century Bjerkness, Sandstöm and Helland-Hansen (Sandström and Helland-Hansen 1903) developed a dynamic method for calculating the gradient flows from a specified density field. Though they meant only the calculation of relative currents, this method is used in practice to calculate the flow in the upper ocean layer 1–1.5 km thick under the assumption that the flow velocity is zero at the lower layer boundary. After the well-known work of Defant (1941), many methods for the determination of the reference level have appeared. Thanks to its exclusive simplicity, the dynamic method finds extensive application in oceanography. However, it suffers from certain serious drawbacks. In a real sea or ocean the vertical profile of the gradient flow velocity may behave quite arbitrarily. At some points the flow velocity at a certain depth changes its sign, while at others it decreases slowly with depth and at 1.5–2.0 km depths the velocity is of a magnitude comparable with the near-surface gradient flow. At certain other points the velocity virtually does not decrease with depth. Consequently, there is no continuous zero-velocity surface in a real ocean that could be taken as the reference level. This is the main drawback of the dynamic method, which does not, in addition, take into account the direct effect of tangential wind stress, bottom relief, etc. Moreover, the vertical component of the flow velocity cannot be calculated by the dynamic method. Without going into the details of other demerits of the dynamic method, only from the foregoing, we can conclude that some other approach is necessary for calculating ocean currents. The method for solving the hydrodynamic problem of calculating flow velocity from a specified density field is of practical and theoretical interest. It may also be an alternative to the dynamic method.

The statement of the problem is as follows. There is a sea or ocean with an arbitrary bottom relief and arbitrary coastal contour. In addition to the natural solid coasts, an ocean may also have a certain conditional boundary, namely a "liquid coast" on which the water flux is specified. The density field in the basin and the tangential wind stress (or atmospheric pressure) at the surface are given. It is required to solve the hydrodynamic problem concerning the calculation of the flow field in this basin from the given density and wind fields under appropriate boundary conditions. To solve this problem we have to apply the system of equations and boundary conditions given in Section 1.1.1. Since the density field is supposed to be given, there is no need to use Eqs. (1.5)–(1.7) and the boundary conditions for temperature and salinity. Therefore, the only problem here is to calculate the currents in a baroclinic ocean. We have already proved that the system can be significantly simplified in studying large-scale flows in an extra-equatorial basin.

Scale analysis shows that in studying large-scale flows we can neglect the inertial terms and the lateral exchange effect not only in the equations of motion but also in the equations for the integral functions ψ and ζ. It is therefore natural to construct the simplest model just on the basis of these simplifications. Thus, from Eq. (1.75) we can obtain:

$$\underbrace{\frac{1}{2\alpha}\Delta\zeta}_{\text{III}} + \underbrace{J(H,\zeta)}_{\text{II}} + \underbrace{\frac{H\beta}{l}\frac{\partial\zeta}{\partial x}}_{\text{I}} = \underbrace{\frac{1}{\rho_0 g}\operatorname{rot}\bar{\tau}}_{\text{VII}}$$

$$- \underbrace{\frac{1}{2\alpha\rho_0}\int_0^H \Delta\rho dz}_{\text{III}} - \underbrace{\frac{1}{\rho_0}\int_0^H J(H,\rho)dz}_{\text{II}} - \underbrace{\frac{\beta}{\rho_0 l}\int_0^H (H-z)\frac{\partial\rho}{\partial x}dz}_{\text{I}} . \qquad (1.92)$$

From the foregoing estimates we find that the terms containing the first derivatives of the fields ρ and ζ are basic in Eq. (1.92). Besides these basic terms, Eq. (1.92) also contains the tangential wind stress and bottom friction effects. The latter term is retained because it permits us to solve the problem for a closed domain.

The normal components of the water flux across the lateral boundaries of the domain are assumed to be known. Generally, the boundary of the water basin is approximated by a broken line, each link being parallel to one of the coordinate axes. The value of S_x is specified on the links parallel to the meridian. Equation (1.91) is used in determining the sea-surface topography on the links parallel to the meridian and Eq. (1.90) for the zonal links. We first assume that the right-hand sides of Eqs. (1.90) and (1.91) are specified. Then, writing one of these equations for each of the boundary parts, we obtain a closed system of equations for determining ζ at the basin boundary. The methods of numerical solution of these equations are presented in Chapter 2. But, as has already been pointed out, this is true only when the right-hand sides of Eqs. (1.90) and (1.91) are specified. In reality, they contain the derivatives of the unknown function ζ. But these terms can be taken into account by successive approximations as they are small (they are neglected in the first approximation). We now take up this question in greater detail. Assuming ν to be equal to 10^2 cm^2 s^{-1} and $H = 0.5$ km, which is the minimum for large-scale flows, we readily find that the dimensionless factor $\frac{1}{2}\alpha H$ is of the order of 10^{-2}. Theoretically, we may assume that when the step of the difference grid tends to zero, the derivatives of ζ along the normal to the coast (more exactly, their difference analogues) may become greater than the derivatives of this function along the tangent to the contour. But we cannot do this since by just reducing the step along the horizontal, we indeed pass over from a large-scale process to meso- or small-scale processes for which this model is hardly satisfactory (as the non-linear terms and the lateral exchange effect have been omitted). Indeed, the successive approximations may converge when the horizontal grid step in the difference scheme is bounded from below. However, this is an unimportant constraint because we have so far no accurate observational data that could be suitable for calculations with small horizontal steps.

Thus, in the framework of the quasi-geostrophic model the sea-surface topography at the boundary contour is determined by Eqs. (1.90) and (1.91), and inside the basin by Eq. (1.92). After determining the sea-surface topography it is a simple matter to calculate the pressure field and the flow velocity components. The pressure field is calculated from the simple formula (1.25). The horizontal flow velocity components can be found with the help of formula (1.56). Incidentally, Eq. (1.56) yields to simplification. In dealing with large-scale flows there is no need for examining their fine structure in the surface and bottom layers insofar as the simple Ekman spiral does not describe the flow fine structure. We therefore apply formula (1.56) to calculate the flows in layers separated by a distance greater than the Ekman friction layer thickness (more than tens of metres for $\nu = 1$ cm^2 s^{-1}). Thus, u and v at the ocean surface are calculated from the formulas:

$$u = \frac{1}{2\rho_0 \nu\alpha}(\tau_x + \tau_y) - \frac{g}{l}\frac{\partial\zeta}{\partial y}; \tag{1.93}$$

$$v = \frac{1}{2\rho_0 \nu\alpha}(\tau_y - \tau_x) + \frac{g}{l}\frac{\partial\zeta}{\partial x} . \tag{1.94}$$

At other layers they are found with the help of the geostrophic flow formulas:

$$u = -\frac{g}{l}\frac{\partial\zeta}{\partial y} - \frac{g}{\rho_0 l}\int_0^z \frac{\partial\rho}{\partial y}dz ; \tag{1.95}$$

$$v = \frac{g}{l}\frac{\partial\zeta}{\partial x} + \frac{g}{\rho_0 l}\int_0^z \frac{\partial\rho}{\partial x}dz . \tag{1.96}$$

It is quite easy to derive a formula for calculating the vertical flow component. Using the boundary condition $w = 0$ for $z = 0$, from Eq. (1.4) we obtain:

$$w = -\int_0^z \left(\frac{\partial u}{\partial x} + \frac{\partial v}{\partial y}\right)dz . \tag{1.97}$$

Substituting the values of u and v found from Eq. (1.56) into (1.97), and using Eq. (1.25), we obtain (for the layers outside the boundaries):

$$w = -\frac{1}{\rho_0 l}\operatorname{rot}\bar{\tau} - \frac{\beta}{\rho_0 l^2}\tau_x + \frac{g\beta z}{l^2}\frac{\partial\zeta}{\partial x} + \frac{g\beta}{\rho_0 l^2}\int_0^z (z-\xi)\frac{\partial\rho}{\partial x}d\xi . \tag{1.98}$$

The components of the tangential wind stress are usually determined from the quadratic dependence between $\bar{\tau}$ and the wind velocity. But the wind field has not been studied in detail and has a large variability. Therefore, it is more convenient to calculate $\bar{\tau}$ in terms of the sea-level atmospheric pressure P_a. A simple version of such a formula is the relation (1.48) easily derived from the Akerblom model. This model is the simplest scheme for solving our problem. It can be generalized but any simplification would lead to contradiction. We now demonstrate that the dynamic method is a particular case of this model. Let us consider a basin with a minimal depth not less than 1 km. In this case we can neglect the density anomalies at the basin bottom, i.e. we can take:

$$\int_0^H \frac{\partial\rho}{\partial x}dz \approx \frac{\partial}{\partial x}\int_0^H \rho dz ; \qquad \int_0^H \frac{\partial\rho}{\partial y}dz \approx \frac{\partial}{\partial y}\int_0^H \rho dz . \tag{1.99}$$

Furthermore, since the density anomaly attenuates with depth, as a rule we have:

$$\left|\int_0^H \frac{\partial\rho}{\partial x}dz\right| > \frac{1}{H}\left|\int_0^H z\frac{\partial\rho}{\partial x}dz\right| . \tag{1.100}$$

Therefore, we assume that:

$$\int_0^H (H-z)\frac{\partial\rho}{\partial x}dz \approx H\int_0^H \frac{\partial\rho}{\partial x}dz . \tag{1.101}$$

Theoretically, condition (1.100) may not be satisfied at the points where $\frac{\partial\rho}{\partial x}$ changes its sign along the vertical, but it is practically satisfied almost everywhere because $\frac{\partial\rho}{\partial x}$ decays quite rapidly with depth. We shall demonstrate the validity of this statement with the help of concrete calculations. Finally, we shall neglect the first term on the right-hand side of Eq. (1.92) which can thus be rewritten as follows:

$$L(\zeta) \equiv \frac{l}{2\alpha H}\Delta\zeta + \frac{l}{H}J(H,\zeta) + \beta\frac{\partial\zeta}{\partial x} = -L\left(\frac{1}{\rho_0}\int_0^H \rho dz\right). \tag{1.102}$$

Its solution is:

$$\zeta_d = -\frac{1}{\rho_0}\int_0^H \rho dz . \tag{1.103}$$

This is the relation that is used in the dynamic method to calculate the sea-surface topography, provided $z = H$ is taken as the reference level. This solution also holds for the boundary conditions. Indeed, let the flux through the lateral boundaries be specified by the dynamic method, i.e. S_x and S_y at the contour are determined by the formulas:

$$S_x = \frac{g}{\rho_0 l}\int_0^H z\frac{\partial\rho}{\partial y}dz ; \qquad S_y = -\frac{g}{\rho_0 l}\int_0^H z\frac{\partial\rho}{\partial x}dz . \tag{1.104}$$

Then, it is a simple matter to verify that Eq. (1.103) satisfies the boundary conditions (1.90) and (1.91) under the simplification made above. For example, let us consider Eq. (1.90). The ratio of $\frac{\tau_x}{\rho_0 gH}$ to $\frac{lS_y}{gH}$ has a characteristic value of $\frac{\tau_0}{\rho_0 H_0 l_0 v_0} = 0.02$, therefore, we can neglect the term containing the tangential wind stress. If, in addition, we substitute S_y from Eq. (1.104) into (1.90) and use the simplified correlations (1.99), we obtain:

$$\left(1 - \frac{1}{2\alpha H}\right)\frac{\partial}{\partial x}\left[\zeta + \frac{1}{\rho_0}\int_0^H \rho dz\right] - \frac{1}{2_\alpha H}\frac{\partial}{\partial y}\left[\zeta + \frac{1}{\rho_0}\int_0^H \rho dz\right] = 0 , \tag{1.105}$$

which is satisfied by expression (1.103). Equation (1.91) can also be transformed in a similar way. The approximate formula (1.103) holds even if the more general equation (1.75) and more general equations and conditions are used instead of Eqs. (1.90), (1.91) and (1.92) i.e. when the non-linear terms, effects of lateral exchange and bottom friction are approximately taken into account. Thus, the dynamic method may serve as a first approximation for determining the surface topography of a deep baroclinic ocean or sea. Unlike any version of the dynamic method in which the reference level is chosen with certain arbitrariness, integration in Eq. (1.103) is taken over the whole of the ocean depth, i.e. the real ocean bottom relief is the reference level. Therefore the method in which the sea-surface topography is determined from Eq. (1.103) may be called a quasi-dynamic technique.

These results naturally suggest that the solution of the linear Eq. (1.92) should be sought in the form of a sum:

$$\zeta = \zeta_d + \zeta^1 . \tag{1.106}$$

In this case ζ' satisfies the equation:

$$\frac{1}{2\alpha}\Delta\zeta' + J(H,\zeta') + \frac{H\beta}{l}\frac{\partial\zeta'}{\partial x} = \frac{1}{\rho_0 g} rot\,\bar{\tau} + \frac{\beta}{\rho_0 l}\int_0^H z\frac{\partial\rho}{\partial x}dz\,, \tag{1.107}$$

whose right-hand side is much simpler than in Eq. (1.92). The simplest version of the boundary condition for ζ' is the condition $\zeta' = 0$ at the basin boundary contour. Let us call the function ζ' a non-dynamic correction. The function $\rho_0 g\zeta'$ represents the pressure anomaly of the ocean bottom.

Let us note that if the sea-surface topography is calculated with the help of Eq. (1.103) the results might strongly depend on the mean ρ_0. To eliminate this drawback we have to choose $\rho = \rho_1 - \rho_0 - \bar{\rho}(z)$ as the density anomaly, where $\rho(z)$ is the anomaly distribution of the standard mean vertical density. This model is useful in calculating the flow velocity from a specified density field (diagnostic calculation) only outside the equator. In order to apply this model to large-scale currents of the World Ocean, the computational grid has to be constructed so that there are no grid points at the equator. For instance, in calculating with a 5° step, the points nearest to the equator must be located at 2.5° *N* and *S*. But even at these latitudes the flow velocity proves to be overestimated as the Coriolis parameter is small. This can be eliminated by substituting sin 5° or sin 7.5° for sin 2.5° in formulas (1.93)–(1.96) and (1.98).

In this procedure it is assumed that the specific pecularities of the fine structure of the equatorial flows are purely local and do not influence (in the framework of the diagnostic formulation) the large-scale circulation of the World Ocean. The validity of this statement is proved in Chapter 5.

The calculation for a multi-connected domain (e.g., the presence of large islands like the Antarctic and Australia) can be performed by the method developed in Kamenkovitch (1961), and Bryan (1969a). But this procedure is not convenient in the case of many relatively small-scale islands (New Zealand, Madagascar, etc.) because the computer memory may be overloaded. Small islands can be considered as shallow regions of the ocean so that water density is constant there and no tangential wind stress exists over them. In Section 1.2.3 a more sophisticated model is presented which is valid for a multi-connected domain.

1.2.2 Quasi-Geostrophic Model for the Calculation of the Total Mass Transport Stream Function and Flow Velocity

Using the model described in the previous section we can easily calculate all the hydrodynamic characteristics of a quasi-stationary flow. In particular, using Eqs. (1.77) and (1.78), we can also determine the field ψ. But another method, i.e. the calculation of the circulation characteristics from the field ψ is more common in the theory of ocean currents. Moreover, several specific features of the flow dynamics are easily demonstrated by calculating the fields ζ and ψ independently. Therefore, we present a model for calculating the currents with the help of the integral function ψ. Let us represent Eq. (1.79) in a form convenient for illustrating many models:

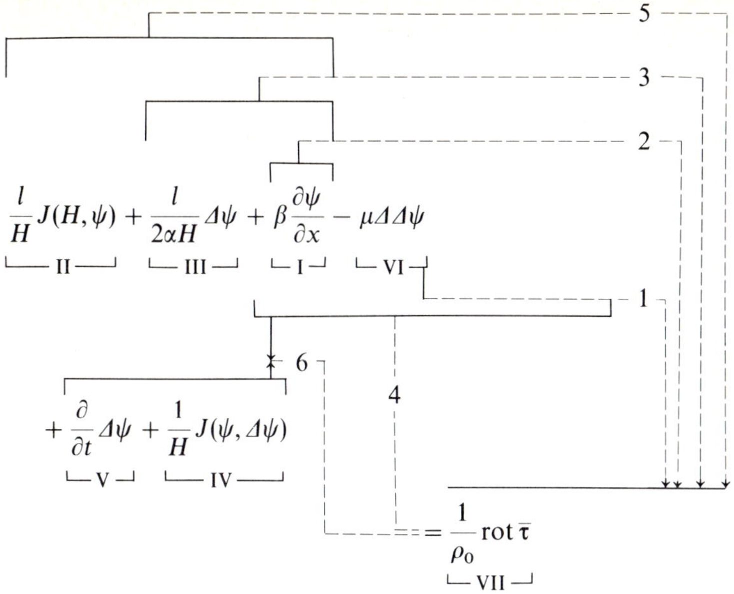

$$\underbrace{\frac{l}{H}J(H,\psi)}_{\text{II}} + \underbrace{\frac{l}{2\alpha H}\Delta\psi}_{\text{III}} + \underbrace{\beta\frac{\partial\psi}{\partial x}}_{\text{I}} - \underbrace{\mu\Delta\Delta\psi}_{\text{VI}} + \underbrace{\frac{\partial}{\partial t}\Delta\psi}_{\text{V}} + \underbrace{\frac{1}{H}J(\psi,\Delta\psi)}_{\text{IV}} = \underbrace{\frac{1}{\rho_0}\operatorname{rot}\bar{\tau}}_{\text{VII}}$$

$$\underbrace{-\frac{g}{\rho_0 H}\int_0^H zJ(H,\rho)\,dz}_{\text{II}} + f_3\,; \tag{1.108}$$

where

$$f_3 = \underbrace{\frac{1}{\rho_0 H}\left(\frac{\partial H}{\partial y}\tau_x - \frac{\partial H}{\partial x}\tau_y\right)}_{\text{X}} - \left(\frac{g}{\rho_0 l}\right)^2 \int_0^H J\left(\int_0^z \rho dz, \Delta\int_0^z \rho dz\right) dz$$

$$\underbrace{+\left(\frac{g}{\rho_0 l}\right)^2 J\left(\int_0^H (H-z)\rho dz, \int_0^H (H-z)\Delta\rho dz\right)}_{\text{IV}}$$

$$\underbrace{-\frac{g}{2\alpha H\rho_0}\int_0^H z\Delta\rho dz}_{\text{III}}\,. \tag{1.109}$$

Note that in Eqs. (1.75), (1.79) and (1.108) the terms generated by the same physical factors are denoted by the same Roman numeral. The physical meanings of these terms are interpreted in Section 1.1.5.

By placing $\rho \equiv 0$ on the right-hand side of Eq. (1.108), we find that it is transformed into the fundamental equation of the total mass transport theory. Therefore, we expressed it in a form readily yielding the basic equations of this approach. Arabic numerals on the lines denote the following:

1. The first work dealing with this theory, according to Shtokman (1946), presupposes that there is a balance between the lateral exchange effect and rot τ. The broken line joining the terms VI and VII illustrates the basic equation of the Shtokman theory.

2. A year later Sverdrup (1947) postulated a relation which assumes a balance between the β-effect and rot τ. Subsequently, this relation became the underlying principle for all the works concerned with the study of integral circulation in the ocean in the framework of the total mass transport theory. The broken line 2 shows the Sverdrup relation.
3. After about a year Stommel (1948) proposed his model which demonstrates the role of the β-effect in the intensification of currents along the western coastline of the ocean. The broken line 3 represents this model.
4. The main equation proposed by Munk (1950).
5. The main relation in the Neuman model (1955, 1958) and Ivanov and Kamenkovich (1959) etc.
6. The main equation used in the works of Carrier and Robinson (1962).

This is not a complete list of all the works on the total mass transport theory, but it is quite sufficient for discussing its fundamental drawbacks. Historically, the total mass transport theory emerged as a countertheory to the homogeneous ocean model. The founders of this theory, Shtokman, Sverdrup, Munk and others, believed that they were studying the integral circulation in a baroclinic ocean. Unfortunately, there is no baroclinicity in the main equation of this theory. All versions of this theory follow only from Eq. (1.108) if the density anomaly is disregarded, in other words, if $\rho \equiv 0$. But no such assumption was made in the works of Shtokman, Sverdrup and Munk. But they made other simplifying assumptions which were responsible for the loss of the baroclinicity in the equation for ψ. Previously, while deriving Eq. (1.75), we have already noted that many terms on the right- and left-hand sides of this equation have the same physical nature and are therefore denoted by the same Roman numerals. The same can also be said of Eq. (1.79) or (1.108). In order to derive Eq. (1.79) from (1.75) we have to apply the relations (1.77) and (1.78). The left-hand side of Eq. (1.108) was obtained due to the first terms on the right-hand sides of Eqs. (1.77) and (1.78), while the terms on the right-hand side of Eq. (1.108) containing ρ due to the second terms on the right-hand sides of these equations. Therefore, the terms having the same physical nature on the right- and left-hand sides of Eq. (1.108) are denoted by the same Roman numerals. For instance, the term III on the left-hand side of Eq. (1.108) is the bottom stress effect, while the term III on the right-hand side is the baroclinic bottom friction effect. Similarly, the term II on the left-hand side is the bottom topography effect, while on the right-hand side it represents the joint effect of baroclinicity and bottom relief, JEBAR, etc. If one neglects the effect of bottom topography, then not only term II of the left-hand side diminishes but the main term of the right-hand side of the equation, i.e. JEBAR, also. Further, if one simplifies the non-linear terms, when constructing Eq. (1.108), by assuming the integral of the product of two functions to be equal to the products of two integrals (as it was done in many works on the total mass transport theory), then term IV still remains on the left-hand side of Eq. (1.108), but all the terms of the same group on the right-hand side of the equation diminish. Finally, if one neglects bottom frictions or supposes that $\frac{\partial u}{\partial z} = \frac{\partial v}{\partial z} = 0$ at some depth H_1, then term III on the left-hand side diminishes, and also "the last of the mohicans" — the terms of the same group III — on the right-hand side. For these three reasons the baroclinicity was entirely lost in the theory of total mass transport, although the assumption of $\rho \equiv 0$ was not made in this theory, due to the belief that it was a theory of the integral circulation of a baroclinic ocean.

A few words on the works of Neumann (1958, 1955). Neumann, on the basis of physical considerations, supposed the JEBAR to be essential, but in concrete calculations he did not take this factor into account. In Neumann's works H is not bottom topography but the depth of the ocean's baroclinic layer. The terms of group II on the right-hand side of Eq. (1.108) existed in Welander's work (1968) too, but Welander supposed that the effects of baroclinicity and bottom topography compensate each other, so that in the vertical average the behaviour of a baroclinic ocean with variable depth is the same as the barotropic ocean with constant depth. For these reasons the works of Neumann and Welander are located in a series of works concerning the total mass transport theory of the barotropic ocean. According to the evaluation of the orders of magnitude in Section 1.1.5 it was shown that the JEBAR is the main term on the right-hand side of Eq. (1.108), and that it is five- to ten-fold larger than rot $\boldsymbol{\tau}$. In models of the total mass transport theory, as it is shown in the scheme of Eq. (1.108), all the arrows converge around rot $\boldsymbol{\tau}$. Some of the authors took one or two terms on the left-hand side of Eq. (1.108) into account, others considered the left-hand side in a more general form, however, the main drawback of all of them is that they all neglected the main source, i.e. JEBAR, and kept the less sufficient factor, rot $\boldsymbol{\tau}$. The Sverdrup correlation is not valid for a baroclinic ocean with variable depth. This was shown not only by the evaluation, but also by concrete diagnostic calculations in Sarkisyan (1969). The intercomparison between the dynamic method and the total mass transport theory is an other factor considered in Sarkisyan (1969). As it is known the sea-surface topography, gradient currents, and consequently, the integral meridional transport in a baroclinic ocean are determined exclusively by the density field in the dynamic method. At the same time the integral meridional transport is determined exclusively by the wind field in the total mass transport theory. Let us present the simple formulas of these two methods. According to the Sverdrup correlation:

$$S_y^{sv} = \frac{1}{\rho_0 \beta} \operatorname{rot} \boldsymbol{\tau} \,. \tag{1.110}$$

According to the dynamic method, as it is easy to see from Eq. (1.13):

$$S_y^d = -\frac{g}{\rho_0 l_1} \int_0^{H_1} z \frac{\partial \rho}{\partial x} dz \,, \tag{1.111}$$

where $l_1 = \text{const}$ is some average value of the Coriolis parameter. These two formulas are absolutely different from each other. Variability of the wind field and the Coriolis parameter is necessary for the calculation of meridional transport according to Eq. (1.110), as for the water density, it may be constant. In contrast, for the calculation of the same quantity according to formula (1.111), the Coriolis parameter and the wind field may be constant and the water should be inhomogeneous. Of course, the wind field is present in formula (1.111), implicitly taking part in the redistribution of the density field. But this indirect effect is absent in formula (1.110) as well as the influence of heat and salt exchange processes. In formula (1.111), on the other hand, we have only an indirect influence of wind, and the influence of heat and salt exchange and other factors through field ρ. From the very beginning of this century calculations according to the dynamic method have been performed and it has been proved that the results agree well with observed data. Since 1947 calculations according to Sverdrup's correlation have been performed, and the results apparently also agree with the observed data. So, for more than a quarter of a century two contradictory correlations have existed and used many times in the sea currents theory, on the one hand, and in the practice of oceanographic calculations, on the other. All the factors are present in the equations for the integral functions (1.92) and (1.108), from which both the above mentioned correlations have been derived, thus there is a possibility to compare them and elucidate the question, which of these two correlations reflects the dynamics of currents better. Let us return to the initial correlation (1.68) for this purpose. From the evaluations, given in Section 1.1.5, it is known that there are only two main factors in Eq. (1.68), the left-hand side and the first term of the right-hand side. For the analysis it is necessary to also preserve the second term of the right-hand side (rot $\boldsymbol{\tau}$), so let us formulate the following simplified version of Eq. (1.68):

$$\beta S'_y = -\frac{1}{\rho_0} J(H, P)_{z=H} + \frac{1}{\rho_0} \operatorname{rot} \boldsymbol{\tau} \, . \tag{1.112}$$

Now, let us examine two extreme cases.

1. $H = \text{const}$. In this case the Jacobian disappears from Eq. (1.112) and we obtain the Sverdrup correlation (1.110).

2. $H \neq \text{const}$, $l = \text{const}$ $(\beta = 0)$. Let us neglect also the second term on the right-hand side of Eq. (1.112), which is small in comparison with the first one. Then, we obtain approximately:

$$J(H, P)_{z=H} = 0 \, . \tag{1.113}$$

Correlation (1.113) is satisfied, when assuming:

$$\left(\frac{\partial P}{\partial x}\right)_{z=H} = \left(\frac{\partial P}{\partial y}\right)_{z=H} = 0 \, , \tag{1.114}$$

i.e. we obtain the dynamic method, in which the actual ocean bottom serves as a reference level. In this case the sea-surface topography is determined by formula (1.103), and the total mass transport stream function, by the formula:

$$\psi_d = -\frac{g}{\rho_0 l_1} \int_0^H z\rho dz \, . \tag{1.115}$$

This formula is obtained from correlations (1.77), (1.78) and taking into account Eq. (1.103) and neglecting small terms caused by the wind stress; Eq. (1.111) follows automatically from Eq. (1.115) according to the definition of the total mass transport stream function. So, Eqs. (1.110) and (1.111) are two extreme cases of correlation (1.112). As these two formulas are physically contradictory, then either both of them are wrong or only one of them is applicable. On the basis of diagnostic calculations, it has been shown that S_y^d is more or less close to the value of S_y, obtained from the general equation for ψ or ζ, and S_y^{sv} is several times less.

Thus, the qualitative analysis and preliminary calculations show the basic drawbacks of the total mass transport theory, but a series of calculations is necessary for reliable conclusions. For this purpose we proceed from Eq. (1.79). Estimations, given in Section 1.1.4, show that this equation can be greatly simplified for large-scale currents in a deep baroclinic ocean. The simplified variant of this equation has the following form:

$$\underbrace{L(\psi) \equiv \frac{1}{2\alpha}\Delta\psi}_{\text{III}} + \underbrace{J(H,\psi)}_{\text{II}} + \underbrace{\frac{H\beta}{l}\frac{\partial\psi}{\partial x}}_{\text{I}} = \underbrace{\frac{H}{\rho_0 l}\operatorname{rot}\tau}_{\text{VII}} - \underbrace{\frac{g}{2\rho_0\alpha l}\int_0^H z\Delta\rho dz}_{\text{III}}$$

$$- \underbrace{\frac{g}{\rho_0 l}\int_0^H zJ(H,\rho)dz}_{\text{II}} \,. \tag{1.116}$$

The differential operator of the left-hand side of this equation is identical to the operator of Eq. (1.92). Both of these equations are obtained on the basis of the same estimations of characteristic values. The only difference between them is in the following: there are two principal terms on the right-hand side of Eq. (1.92), i.e. groups I and II, and there is only one principal term, i.e. group II (JEBAR), on the right-hand side of Eq. (1.116). It leads to the essential distinction in the behaviour of functions ψ and ζ. The Sverdrup correlation is a particular case of Eq. (1.116). Therefore, we can check this correlation by concrete diagnostic calculations. The boundary condition for the solution of Eq. (1.116) is trivial: ψ is specified on the contour of the examined basin.

When ψ is calculated, it is easy to determine u, v, w. The formulas for them are easily obtained from the corresponding formulas of the model with the help of correlations (1.77) and (1.78)[1]. The velocity of the horizontal current is calculated according to the following formulas:

a) at the sea surface

$$u = -\frac{1}{\rho_0 l}\sqrt{\frac{v'}{v}}\frac{\partial P_a}{\partial y} - \frac{1}{H}\frac{\partial\psi}{\partial y} + \frac{g}{\rho_0 lH}\int_0^H (H-z)\frac{\partial\rho}{\partial y}dz\,; \tag{1.117}$$

$$v = \frac{1}{\rho_0 l}\sqrt{\frac{v'}{v}}\frac{\partial P_a}{\partial x} + \frac{1}{H}\frac{\partial\psi}{\partial x} - \frac{g}{\rho_0 lH}\int_0^H (H-z)\frac{\partial\rho}{\partial x}dz\,. \tag{1.118}$$

b) at all other levels

$$u = -\frac{1}{H}\frac{\partial\psi}{\partial y} + \frac{g}{\rho_0 l}\int_0^H \frac{\partial\rho}{\partial y}dz - \frac{g}{\rho_0 lH}\int_0^H z\frac{\partial\rho}{\partial y}dz\,; \tag{1.119}$$

$$v = \frac{1}{H}\frac{\partial\psi}{\partial x} - \frac{g}{\rho_0 l}\int_0^H \frac{\partial\rho}{\partial x}dz + \frac{g}{\rho_0 lH}\int_0^H z\frac{\partial\rho}{\partial x}dz\,. \tag{1.120}$$

---

[1] The terms in the formulas, caused by the wind, are essential only in shallow-water areas.

When passing from formulas (1.93), (1.94) to (1.117), (1.118) we substituted tangential wind stress through P_a using formulas (1.48).

The vertical component of the current velocity is zero at the sea surface, and at the other calculate horizons it is calculated according to the formula:

$$w = -\frac{1}{2\rho_0 l\alpha'}\Delta P_a + \frac{3\beta}{4\rho_0 l^2\alpha'}\left(\frac{\partial P_a}{\partial x} + \frac{\partial P_a}{\partial y}\right) + \frac{\beta z}{lH}\frac{\partial\psi}{\partial x}$$

$$-\frac{g\beta z}{H\rho_0 l^2}\int_0^H (H-z)\frac{\partial\rho}{\partial x}dz + \frac{g\beta}{\rho_0 l^2}\int_0^H (z-\xi)\frac{\partial P}{\partial x}d\xi\,. \qquad (1.121)$$

Equation (1.116) and formulas (1.117)–(1.127) represent the simplest diagnostic model for the calculation of currents by the given density field with the help of the total mass transport streams function Ψ.

1.2.3 A Non-Linear Model for the Calculation of Sea-Surface Topography and Flow Velocity

The previously described models are useful for the calculation of the large-scale circulation of the World Ocean waters without taking into account the local peculiarities of the coastal and equatorial currents. The difference grid in this case is constructed so, that there will not be any grid points at the equator. The grid points nearest to the equator are situated at such a distance to it, where the geostrophic balance is a predominant factor (according to Demin and Sarkisyan 1977; Sarkisyan 1977b; Holland and Hirschman 1972), approximately 2°–3° to the north and south).

At the present point another problem is considered. At the suggestion that at some distances north and south of the equator we know all the necessary characteristic features of a large-scale circulation, the problem of the calculation of currents at the equatorial belt occurs. Then difference grids of both models, which overlap each other, are formed and the circulation of all ocean waters, including the equator is calculated by successive approximations. For instance, the calculation of the World Ocean currents according to the quasi-geostrophic model on the grid with a 5° step, which is constructed so, that the points closest to the equator are at 2.5° N and S latitude is carried out; then an equatorial belt with a 15° width from 7.5° S latitude to 7.5° N latitude, in which the calculation according to a non-linear model is performed; then, the recalculation of currents in the tropical zone is carried out according to the quasi-geostrophic model, taking into account the results obtained according to a non-linear model, etc. Thus, in two strips, located to the north and to the south of the equator, calculations are being made alternately by both models until reaching the steady state of a process. At the very equator and in the narrow belt near it, calculations are performed according to a non-linear model, and in most of the ocean, according to the simple quasi-geostrophic model. The mutual adjustment of solutions for subregions are determined by boundary conditions. Such a method is based on the following considerations.

The simplifications have been done in a quasi-geostrophic model with respect to the fact that non-linear inertial terms are small. Thus, we manage to construct a simple, economic scheme of the calculation of the World Ocean currents by avoiding the narrow equatorial belt with a 4°–5° width.

In a non-linear model we have to take inertial terms into account, but then the dimensions of the domain, where the application of this model is necessary, are not large.

Before presenting a non-linear model let us do a qualitative analysis of the peculiarities of the equatorial zone currents on the basis of a linear model. Let us assume $A = B = 0$ in Eqs. (1.57), (1.58) and solve them with respect to u, v, taking boundary conditions (1.9), (1.15) into account. We obtain (Demin and Sarkisyan 1977; Sarkisyan 1977b):

$$\begin{aligned} u + iv = {} & \frac{sh[(1+i)\alpha(H-z)]}{(1+i)\alpha\nu\rho_0 ch[(1+i)\alpha H]}(\tau_x + i\tau_y) \\ & - \frac{1}{2(1+i)\alpha}\left[\int_0^z e^{(1+i)\alpha(\xi - z)}\mathscr{F}d\xi + \int_z^H e^{-(1+i)\alpha(\xi - z)}\mathscr{F}d\xi\right] \\ & + \frac{e^{-(1+i)\alpha H}}{2(1+i)\alpha ch[(1+i)\alpha H]}\int_0^H \mathscr{F} ch[(1+i)\alpha(\xi - z)]d\xi \\ & - \frac{1}{2(1+i)\alpha ch[(1+i)\alpha H]}\int_0^H \mathscr{F} sh[(1+i)\alpha(H - \xi - z)]d\xi\,, \end{aligned} \tag{1.122}$$

where

$$\alpha = \sqrt{\pm\frac{l}{2\nu}}; \qquad \mathscr{F} = \frac{1}{\rho_0 \nu}\left(\frac{\partial P}{\partial x} + i\frac{\partial P}{\partial y}\right). \tag{1.123}$$

In the last formula the sign "+" refers to the Northern Hemisphere and "−" to the Southern Hemisphere. If we simplify the right-hand side of the formula for the extra-equatorial region of the deep basin ($l \neq 0$, $H > 100$ m), then we obtain the above mentioned approximate formula (1.56). However, here we are interested only in the equatorial zone, in which $|\sin\varphi| \ll 1$.

In the equatorial belt all the expressions of the type αH, $\alpha(\xi - z)$, etc. are infinitely small values, approaching zero closer to the equator. Therefore, all the exponents on the right-hand side of formula (1.122) can be decomposed to the powers of corresponding arguments and then we can limit ourselves to the values of the first-order smallness, i.e. the use of approximate correlations according to the form $e^{\pm(1+i)\alpha H} \approx 1 \pm \alpha H(1+i)$. After this substitution and simple transformations a part of expression (1.122) transfers into a definite form, and some terms at the equator lead to the indefiniteness of the form $\frac{0}{0}$. Solving them by the l'Hospital rule and passing to the limit with $l \to 0$ we come finally to the following formula:

$$u + iv = \frac{H - z}{\nu\rho_0}(\tau_x + i\tau_y) + \int_0^z (z - \xi)\mathscr{F}d\xi - \int_0^H (H - \xi)\mathscr{F}d\xi\,; \tag{1.124}$$

or

$$u = \frac{H-z}{\nu\rho_0}\tau_x + \frac{1}{\nu\rho_0}\int_0^z (z-\xi)\frac{\partial P}{\partial x}d\xi - \frac{1}{\rho_0 \nu}\int_0^H (H-\xi)\frac{\partial P}{\partial x}d\xi\,; \tag{1.125}$$

$$v = \frac{H-z}{\nu\rho_0}\tau_y + \frac{1}{\nu\rho_0}\int_0^z (z-\xi)\frac{\partial P}{\partial y}d\xi - \frac{1}{\rho_0 \nu}\int_0^H (H-\xi)\frac{\partial P}{\partial y}d\xi\,. \tag{1.126}$$

Formulas (1.125), (1.126) can serve for the approximate calculations of u and v in the equatorial belt. At the equator they satisfy exactly Eqs. (1.57), (1.58) (with $A \equiv B \equiv 0$) and boundary conditions (1.9) and (1.15). The accuracy of these formulas deteriorates when moving away from the equator, and at some critical latitude they will become unfit and "give up place" to Eq. (1.56).

It can be seen from correlations (1.125), (1.126) that there are neither boundary layers nor Ekman's spiral at the equator. Here, the wind drift currents, as well as the effect of bottom stress friction are spread from the sea surface to the bottom, furthermore, the wind drift currents are directed exactly downwind.

Let us consider the directions of gradient currents at the equator. For the simplicity of the analysis let us assume:

$$\frac{\partial P}{\partial y} = 0\,, \qquad \frac{\partial P}{\partial x} = \left(\frac{\partial P}{\partial x}\right)_0 e^{-\nu z}\,,$$

where $\nu = \text{const} > 0$. Then we will obtain from formulas (1.125), (1.126) for the components of the gradient part of the current velocity, the formulas:

$$u_g^{(0)} = -\frac{1}{\nu\rho_0\gamma^2}[\gamma(H-z) + e^{-\gamma H} - e^{-\gamma z}]\left(\frac{\partial P}{\partial x}\right)_0\,, \qquad v_g^{(0)} = 0\,. \tag{1.127}$$

Let us assume that the adopted regulation for the pressure gradient is fulfilled not only at the equator, but in the examined equatorial zone. Then, from the usual geostrophic correlations, we will obtain for the Northern Hemisphere strip:

$$u_g^{(N)} = 0\,; \qquad v_g^{(N)} = \frac{e^{-\gamma z}}{2\omega\rho_0 \sin\varphi}\left(\frac{\partial P}{\partial x}\right)_0\,, \tag{1.128}$$

and for the Southern Hemisphere:

$$u_g^{(s)} = 0\,,$$

$$v_g^{(s)} = \frac{e^{-\gamma z}}{2\omega\rho_0 \sin(-\varphi)}\left(\frac{\partial P}{\partial x}\right)_0 = -\frac{e^{-\gamma z}}{2\omega\rho_0 \sin\varphi}\left(\frac{\partial P}{\partial x}\right)_0\,. \tag{1.129}$$

Thus, in the Northern Hemisphere the gradient current deviates to the right of the pressure gradient $= -\text{grad}\,P$, and in the Southern Hemisphere to the left; at the equator it coincides in the direction with the pressure gradient vector.

It is known from the practice of diagnostic calculations that the sea-surface topography in the equatorial belt is higher near the western coast than near the eastern one, i.e. vector **G** and consequently the gradient part of the current is directed at the equator from the west to east. On the ocean surface the current is directed downwind, from east to west. With depth, the drift part of the current diminishes much faster than the gradient one, thus the total current is directed to the east, forming the analogue of the equatorial undercurrent.

So, the considered simple linear model of the equatorial circulation explains qualitatively the possibility of the appearance of the equatorial undercurrent (similar to the currents of Cromwell, Lomonosov, Tareev). But in reality equatorial currents have a far more complicated character, thus demanding a three-dimensional analysis with regards to non-linear terms in the equations of motion.

Before considering the preparation of such a non-linear model of equatorial currents, let us first derive the equations for ζ. Let us integrate Eqs. (1.1), (1.2) with respect to 2 from 0 to H, then differentiate the first of these equations with respect to x and add the second differentiated equation with respect to y. Then, after simple transformations, taking into account correlations (1.25) and (1.63) we obtain:

$$
\begin{aligned}
\underbrace{\Delta\zeta}_{\text{II}} + \underbrace{\frac{1}{H}\frac{\partial H}{\partial x}\frac{\partial \zeta}{\partial x} + \frac{1}{H}\frac{\partial H}{\partial y}\frac{\partial \zeta}{\partial y}}_{\text{I}} &= \underbrace{-\frac{1}{\rho_0 H}\int_0^H (H-z)\delta\rho dz}_{\text{II}} \\
&\quad \underbrace{-\frac{1}{\rho_0 H}\left[\frac{\partial H}{\partial x}\int_0^H \frac{\partial P}{\partial x}dz + \frac{\partial H}{\partial y}\int_0^H \frac{\partial P}{\partial y}dz\right]}_{\text{I}} \underbrace{-\frac{\beta}{gH}\int_0^H udz}_{\text{III}} \\
&\quad \underbrace{+\frac{l}{gH}\int_0^H\left(\frac{\partial v}{\partial x} - \frac{\partial u}{\partial y}\right)dz}_{\text{V}} \underbrace{-\frac{1}{gH}\int_0^H\left(\frac{\partial A_1'}{\partial x} + \frac{\partial B_1'}{\partial y}\right)dz}_{\text{IV}} \\
&\quad \underbrace{+\frac{1}{gH\rho_0}\left(\frac{\partial \tau_x}{\partial x} + \frac{\partial \tau_y}{\partial y}\right)}_{\text{VI}} \underbrace{+\frac{1}{gH\rho_0}\left(\frac{\partial \tau_x^H}{\partial x} + \frac{\partial \tau_y^H}{\partial y}\right)}_{\text{VII}} \\
&\quad \underbrace{+\frac{\mu}{gH}\left(\frac{\partial H}{\partial x}\Delta u + \frac{\partial H}{\partial y}\Delta v - \Delta w\right)_{z=H}}_{\text{VIII}}, \qquad (1.130)
\end{aligned}
$$

where

$$
\tau_x^H = \rho_0 \nu \left(\frac{\partial u}{\partial z}\right)_{z=H}, \qquad \tau_y^H = \rho_0 \nu \left(\frac{\partial v}{\partial z}\right)_{z=H}, \qquad (1.131)
$$

$$
A_1' = \frac{\partial u^2}{\partial x} + \frac{\partial uv}{\partial y}; \qquad B_1' = \frac{\partial uv}{\partial x} + \frac{\partial v^2}{\partial y}. \qquad (1.132)
$$

Some terms of Eqs. (1.130) and (1.75) are formally similar to each other. The terms on the right- and left-hand sides of Eq. (1.130), having the same nature, are denoted by the same number, as well as in Eq. (1.75). But in reality these equations differ radically in a physical sense, as well as in orders of magnitude of the separate terms, therefore in Eq. (1.130) another numbering system is adopted. For instance, the physical sense of the terms of group II are the following: on the left, the effect of the pressure gradient at the sea surface; on the right, an integral of a purely baroclinic part of the pressure gradient. The meaning of the remaining terms is obvious.

With the help of the methodology given in Section 1.1.4 in evaluating the orders of magnitude of Eq. (1.130), it is easy to see that the terms of group II are principal in this equation. Hence, unlike Eq. (1.75), Eq. (1.130) is not the equation, in which the highest derivative has a small parameter. Unlike the vorticity equation (1.75), Eq. (1.130) is an equation of divergency. Such a method of construction leads to the equation for ζ, which does not have a singularity at $l = 0$.

The system of equations for the diagnostic calculation of currents is composed in the following way. For the definition of ζ we start from Eq. (1.130), where unknown terms on the right-hand side are taken into account by successive approximations. The components of the horizontal current velocity are defined directly from Eqs. (1.1), (1.2) without any simplifications. The vertical component of the velocity is determined from formula (1.97), and the pressure field, according to formula (1.25). The method of definition of ζ at the side boundaries is given below.

The constructed model is useful not only for the equatorial zone, but also for the regions of middle latitudes, i.e. it is universal. However, since it is more time-consuming and makes higher demands to the operative memory of a computer, unlike the above considered quasi-geostrophic models, it is reasonable to use it only for the calculation of currents in dynamically complicated areas (the equatorial zone, regions of coastal upwelling, jetlike currents, etc.) For most of the ocean, the extra-equatorial zone where the geostrophic balance predominates, a simpler quasi-geostrophic model is quite applicable. The use of these two models (which are based on the same integral function, the SST) allows one to perform economical calculations of currents of the whole World Ocean. It should be noted that it is possible to construct a non-linear model (see Sect. 1.1.4), in which the equation for the SST would be a traditional vorticity equation, but not a divergency equation. But such a model is applicable only for the dynamically complicated basins of mid-latitudes, but not for the equatorial zone.

This model can be used for the short-term forecast of currents, too. However, first it is necessary to know the initial fields of u, v. At present it is practically impossible to construct initial velocity fields of currents for the ocean on the basis of observed data. It is possible, however, to construct an initial field of ζ according to satellite data. It will simplify the construction of an initial velocity field of a horizontal current, and particularly the velocity field of the sea-surface current. With the help of equations of motion, initial fields of u, v can be constructed according to the known fields of ρ, ζ and P_a. As we have no exact satellite data on the fields of ζ, it is necessary to solve the whole system of the model equations. But the main problem by the construction of initial fields is the inaccuracy of specifying field ρ. Below we will see that the field of ρ is the main indicator of currents in oceans and deep seas. The existing atlases give very approximate density fields for this or that season, and for the short-term forecast these fields are necessary at least for each pentade. Below, in chapter 5 methods of adaptation of the fields of ρ will be examined and the problem of the short-term forecast of currents on the basis of the known fields of ρ, regardless of whether they are constructed according to the observed data or by calculation, will be discussed briefly.

As the density field is far more conservative than the field of currents, it can be considered as stationary for the time of the short-term forecast. Then by the solution of the model equations, according to the method of initialization, one can define the initial fields of u, v, adjusted to the given fields of ρ, P_a. In this case the spin-up time has no special physical meaning. It only shows the time interval in which the currents' field adjusts to the given fields of ρ, P_a. After the initial fields of u, v, ζ have been constructed we can consider the short-term forecast of the currents, the changes of which are caused by the variability of the atmospheric pressure field P_a. For this purpose we start from the same model equations, use constructed initial fields of u, v, ζ, substitute new fields of P_a into the right-hand sides of the motion equations, considering ρ to be stationary and calculate prognostic fields of u, v, w, ζ by the time steps. This is a principal scheme of the short-term forecast of currents, based on a non-linear model.

Let us return to the problem of the SST calculation. For the solution of Eq. (1.130) boundary conditions are necessary. These conditions should not be set up arbitrarily, they must be obtained from the initial system of equations and boundary conditions for the velocity. In a quasi-geostrophic model the equations of motion, integrated by the vertical, served as boundary conditions for the SST (with the use of the Akerblom formulas to express the bottom friction directly through function ζ). As a result, a tangential derivative of the SST was used at the contour of the basin (in a more general statement the problem developed into a case with an arbitrarily directed derivative). Such an approach is based on the assumption of the adopted sea currents theory that on the boundaries of the basin the equations of motion are valid as well as boundary conditions for the velocity (if, of course, the boundaries are vertical walls, as usual). By using this approach in the considered model and integrating equations of motion (1.1), (1.2) with respect to z from 0 to H we will obtain:

$$\frac{\partial \zeta}{\partial x} = -\frac{1}{\rho_0 H}\int_0^H (H-z)\frac{\partial \rho}{\partial x}dz - \frac{1}{gH}\frac{\partial}{\partial t}\int_0^H u dz + \frac{l}{gH}\int_0^H v dz + \frac{\tau_x}{\rho_0 gH} + \frac{\tau_x^H}{\rho_0 gH} + \frac{\mu}{gH}\int_0^H \Delta u dz\,; \tag{1.133}$$

$$\frac{\partial \zeta}{\partial y} = -\frac{1}{\rho_0 H}\int_0^H (H-z)\frac{\partial \rho}{\partial y}dz - \frac{1}{gH}\frac{\partial}{\partial t}\int_0^H v dz - \frac{l}{gH}\int_0^H u dz + \frac{\tau_y}{\rho_0 gH} + \frac{\tau_y^H}{\rho_0 gH} + \frac{\mu}{gH}\int_0^H \Delta v dz\,. \tag{1.134}$$

In a quasi-geostrophic model, where lateral exchange and non-linear terms are not taken into account and for the flow velocity softened boundary conditions are set up (the integral mass flux normal to the contour is specified), at each part of the boundary, only one of the correlations of this type is correct: at the meridional part for $\frac{\partial \zeta}{\partial x}$; at the zonal parts, for $\frac{\partial \zeta}{\partial y}$. Thus, at the contour a tangential derivative of ζ is specified. In the considered model at the boundary both correlations (1.133), (1.134) are generally correct. For the solution of Eq. (1.130), however, only the specification of one of these conditions is adequate: either a tangential or a normal derivative of the sea level. Then each of these problems is solvable only under definite conditions. The solution of the problem with a tangential derivative is given in Section 2.2. For the problem with the normal derivative (Neumann's problems) of the form:

$$\Delta\zeta + \frac{1}{H}\frac{\partial H}{\partial x}\frac{\partial \zeta}{\partial x} + \frac{1}{H}\frac{\partial H}{\partial y}\frac{\partial \zeta}{\partial y} = R(x, y), \qquad (x, y) \in G_2 \tag{1.135}$$

$$\frac{\partial \zeta}{\partial n} = r(x, y), \qquad (x, y) \in L_2, \tag{1.136}$$

the necessary condition for the solvability is the coordination of the right-hand sides $R(x, y)$ and $r(x, y)$ of the form: (Samarsky and Andreyev 1976)

$$E = \int_{G_2} R(x, y) dG_2 - \int_{L_2} Hr(x, y) dL_2 = 0, \tag{1.137}$$

where G_2 is a two-dimensional domain of the basin surface, L_2 is the contour of the G_2 domain.

The fulfillment of condition (1.137) follows from the derivation of the equation for the sea-surface topography (SST). While solving the problem numerically, condition (1.137) is fulfilled only approximately due to the errors of difference approximation. The numerical methods of solving both boundary problems for ζ will be presented in the following chapter. Here, we will dwell only on the main peculiarities and comparative advantages of the Neumann problem.

Boundary condition (1.136) is placed at the internal contour, to be more exact, in the grid units nearest to the external contour of the basin, where boundary conditions for the velocity are given. Then the boundaries of the domains for the SST and velocity are spaced one grid step apart, whereby the equations of motion are not necessarily satisfied at the basin boundary as in the problem with the tangential derivative. The essence of the approach is that at the internal contour, in addition to the equation of the second order for ζ, a boundary equation is used, i.e. the correlation for the normal derivative of ζ. This allows one to exclude from the principal equation the normal derivative of ζ, expressing it through other functions (density, velocity, etc). Thus, the approach does not demand knowledge of the SST on the basin boundary, though one can define it, for instance, at the end of calculation, using Eq. (1.136). The construction of the boundary problem for the SST as Neumann's problem allows one to avoid a number of drawbacks of a more traditional approach, based on a boundary equation with a tangential derivative: (1) the necessity in numerical boundary conditions for the velocity, caused by the presence of the second-order derivatives of velocity in correlations for the tangential derivative; (2) traditional complications in multi-connected domains (the same as in the problems for ψ (Kamenkovitch 1961); (3) the complexity of calculations, caused by the necessity of recalculations at the contours at every time step (which is especially essential by calculations with a computer having a small operative memory).

1.2.4 The List of the Main Correlations in the Spherical System of Coordinates and for the Southern Hemisphere

All the above presented equations and formulas were written for convenience in a rectangular system of coordinates and are applicable to the Northern Hemisphere. But many calculations of large-scale circulation are carried out for the whole World

Ocean, taking into account the Earth's sphericity. Here, we present the most important correlations in a spherical system of coordinates for which the polar angle θ, longitude λ and vertical coordinate z are taken. Many formulas and terms are the same for both hemispheres. A number of terms have two signs, from which the upper corresponds to the Northern Hemisphere, and the lower to the Southern Hemisphere.

The equations of motion are:

$$\frac{\partial v_\lambda}{\partial t} + \frac{v_\theta}{a}\frac{\partial v_\lambda}{\partial \theta} + \frac{v_\lambda}{a\sin\theta}\frac{\partial v_\lambda}{\partial \lambda} + v_z\frac{\partial v_\lambda}{\partial_z} + \frac{v_\theta v_\lambda}{a}\operatorname{ctg}\theta$$

$$+ 2\omega\cos\theta\cdot v_\theta = -\frac{1}{\rho_0 a\sin\theta}\frac{\partial P}{\partial\lambda} + \nu\frac{\partial^2 v_\lambda}{\partial z^2} + \mu\frac{\Delta v_\lambda}{a^2}\,; \tag{1.138}$$

$$\frac{\partial v_\theta}{\partial t} + \frac{v_\theta}{a}\frac{\partial v_\theta}{\partial \theta} + \frac{v_\lambda}{a\sin\theta}\frac{\partial v_\theta}{\partial \lambda} + v_z\frac{\partial v_\theta}{\partial z} - \frac{v_\lambda^2}{a}\operatorname{ctg}\theta$$

$$- 2\omega\cos\theta\cdot v_\lambda = -\frac{1}{\rho_0 a}\frac{\partial P}{\partial\theta} + \nu\frac{\partial^2 v_\theta}{\partial z^2} + \mu\frac{\Delta v_\theta}{a^2}\,. \tag{1.139}$$

The equation of imcompressible liquid continuity is:

$$\frac{\partial v_z}{\partial z} + \frac{1}{a\sin\theta}\left(\frac{\partial}{\partial\theta}\sin\theta v_\theta + \frac{\partial v_\lambda}{\partial\lambda}\right) = 0 \tag{1.140}$$

The equation of heat transport is:

$$\frac{\partial T}{\partial t} + \frac{v_\lambda}{a\sin\theta}\frac{\partial T}{\partial\lambda} + \frac{v_\theta}{a}\frac{\partial T}{\partial\theta} + v_z\frac{\partial T}{\partial z} = \nu_1\frac{\partial^2 T}{\partial z^2} + \frac{\mu_1}{a^2}\Delta T\,. \tag{1.141}$$

The equation of salinity transport should be written quite analogously. The boundary condition of Eq. (1.17) in the spherical system of coordinates is of the form:

$$v_z = \frac{v_\theta}{a}\frac{\partial H}{\partial\theta} + \frac{v_\lambda}{a\sin\theta}\frac{\partial H}{\partial\lambda}\,; \qquad z = H(\theta,\lambda)\,. \tag{1.142}$$

The Laplas operator on the sphere of a unit radius is:

$$\Delta = \frac{1}{\sin\theta}\left(\frac{\partial}{\partial\theta}\sin\theta\frac{\partial}{\partial\theta} + \frac{1}{\sin\theta}\frac{\partial^2}{\partial\lambda^2}\right)\,. \tag{1.143}$$

Correlations (1.90) and (1.91) for the determination of ζ at the basin contour in the spherical system of coordinates are of the form:

$$\left(1 - \frac{1}{2\alpha H}\right)\frac{\partial\zeta}{\partial\theta} = -\frac{1}{\rho_0 H}\int_0^H (H-z)\frac{\partial\rho}{\partial\theta}dz$$

$$+ \frac{1}{2\alpha H\rho_0}\int_0^H\left(\frac{\partial\rho}{\partial\theta} \pm \frac{1}{\sin\theta}\frac{\partial\rho}{\partial\lambda}\right)dz \pm \frac{1}{2\alpha H\sin\theta}\frac{\partial\zeta}{\partial\lambda}$$

$$-\frac{1}{2\alpha'\rho_0 gH}\left(\frac{\partial P_a}{\partial\theta} \pm \frac{1}{\sin\theta}\frac{\partial P_a}{\partial\lambda}\right) + \frac{2\omega a\cos\theta}{g}V_\lambda\,; \tag{1.144}$$

$$\left(1-\frac{1}{2\alpha H}\right)\frac{\partial\zeta}{\partial\lambda}=-\frac{1}{\rho_0 H}\int_0^H (H-z)\frac{\partial\rho}{\partial\lambda}dz$$

$$+\frac{1}{2\alpha H\rho_0}\int_0^H\left(\frac{\partial\rho}{\partial\lambda}\mp\sin\theta\frac{\partial\rho}{\partial\theta}\right)dz\mp\frac{\sin\theta}{2\alpha H}\frac{\partial\zeta}{\partial\theta}$$

$$-\frac{1}{2\alpha'\rho_0 gH}\left(\frac{\partial P_a}{\partial\lambda}\mp\sin\theta\frac{\partial P_a}{\partial\theta}\right)-\frac{2\omega a\sin\theta\cos\theta}{g}V_\theta\,, \tag{1.145}$$

where

$$\alpha=\sqrt{\frac{\omega|\cos\theta|}{\nu}}\,;\qquad \alpha'=\sqrt{\frac{\omega|\cos\theta|}{\nu'}}\,, \tag{1.146}$$

and V_λ, V_θ are the components of the vertical average velocity of a current.

Equations (1.92) and (1.116) for the determination of ζ and ψ inside the basin in a spherical system of coordinates are of the form:

$$\frac{1}{2\alpha}\Delta\zeta\pm\frac{1}{\sin\theta}J(H,\zeta)+\frac{H}{|\cos\theta|}\frac{\partial\zeta}{\partial\lambda}=\frac{1}{2\alpha'\rho_0 g}\Delta P_a-\frac{1}{2\alpha\rho_0}\int_0^H\Delta\rho dz$$

$$\mp\frac{1}{\rho_0\sin\theta}\int_0^H J(H,\rho)dz-\frac{1}{\rho_0|\cos\theta|}\int_0^H(H-z)\frac{\partial\rho}{\partial\lambda}dz\,; \tag{1.147}$$

$$\frac{1}{2\alpha}\Delta\psi\pm\frac{1}{\sin\theta}J(H,\psi)+\frac{H}{|\cos\theta|}\frac{\partial\psi}{\partial\lambda}=\frac{H}{4\omega\alpha'\rho_0\cos\theta}\Delta P_a$$

$$-\frac{g}{4\omega\alpha\rho_0\cos\theta}\int_0^H z\Delta\rho dz-\frac{g}{2\omega\rho_0\sin\theta|\cos\theta|}\int_0^H zJ(H,\rho)dz\,. \tag{1.148}$$

Here,

$$J(b,c)=\frac{\partial b}{\partial\theta}\frac{\partial c}{\partial\lambda}-\frac{\partial b}{\partial\lambda}\frac{\partial c}{\partial\theta}\,. \tag{1.149}$$

The components of the tangential wind stress are expressed through atmospheric pressure at sea level by formulas:

$$\tau_\lambda=\frac{1}{2\alpha' a}\left(\mp\frac{\partial P_a}{\partial\theta}+\frac{1}{\sin\theta}\frac{\partial P_a}{\partial\lambda}\right),$$

$$\tau_\theta=-\frac{1}{2\alpha' a}\left(\frac{\partial P_a}{\partial\theta}\pm\frac{1}{\sin\theta}\frac{\partial P_a}{\partial\lambda}\right), \tag{1.150}$$

which are the analogues of correlations (1.48).

The formulas for the calculation of horizontal current velocity components through the total mass transport stream function are of the form:

$$v_\theta=v_\theta^d-\frac{1}{aH\sin\theta}\frac{\partial\psi}{\partial\lambda}+\frac{g}{2\omega\rho_0 a\sin\theta\cos\theta}\left[\int_z^H\frac{\partial\rho}{\partial\lambda}dz-\frac{1}{H}\int_0^H z\frac{\partial\rho}{\partial\lambda}dz\right]$$

$$v_\lambda=v_\lambda^d+\frac{1}{aH}\frac{\partial\psi}{\partial\theta}-\frac{g}{2\omega\rho_0 a\cos\theta}\left[\int_z^H\frac{\partial\rho}{\partial\theta}dz-\frac{1}{H}\int_0^H z\frac{\partial\rho}{\partial\theta}dz\right], \tag{1.151}$$

where v_θ^d, v_λ^d are the wind drift components of the current velocity:

$$v_\theta^d = \frac{1}{2\omega a \rho_0 \cos\theta}\sqrt{\frac{v'}{v}}\left(\pm\frac{\partial P_a}{\partial\theta}\sin\alpha z - \frac{1}{\sin\theta}\frac{\partial P_a}{\partial\lambda}\cos\alpha z\right)e^{-\alpha z};$$

$$v_\lambda^d = -\frac{1}{2\omega a \rho_0 \cos\theta}\sqrt{\frac{v'}{v}}\left(\frac{\partial P_a}{\partial\theta}\cos\alpha z \pm \frac{1}{\sin\theta}\frac{\partial P_a}{\partial\lambda}\sin\alpha z\right)e^{-\alpha z}. \tag{1.152}$$

At the sea surface the vertical component of the current velocity is assumed to be zero, and at other calculating horizons it is calculated according to the formula:

$$\begin{aligned} v_z = {} & \mp\frac{1}{4\omega\rho_0 a^2\cos\theta\cdot\alpha'}\Delta P_a + \frac{\sin\theta}{4\alpha'\omega a^2\rho_0\cos^2\theta}\left(\mp\frac{\partial P_a}{\partial\theta} + \frac{1}{\sin\theta}\frac{\partial P_a}{\partial\lambda}\right) \\ & + \frac{z}{Ha^2\cos\theta}\frac{\partial\psi}{\partial\lambda} - \frac{gz}{2\omega\rho_0 Ha^2\cos^2\theta}\int_0^H (H-z)\frac{\partial\rho}{\partial\lambda}dz \\ & + \frac{g}{2\omega\rho_0 a^2\cos^2\theta}\int_0^z (z-\xi)\frac{\partial\rho}{\partial\lambda}d\xi. \end{aligned} \tag{1.153}$$

When calculating with the help of the integral function ζ, the formula for the vertical velocity is the following:

$$\begin{aligned} v_z = {} & \mp\frac{1}{4\omega a^2\rho_0\alpha'\cos\theta}\Delta P_a + \frac{\sin\theta}{4\omega\alpha' a^2\rho_0\cos^2\theta}\times\left(\mp\frac{\partial P_a}{\partial\theta} + \frac{1}{\sin\theta}\frac{\partial P_a}{\partial\lambda}\right) \\ & + \frac{gz}{2\omega a^2\cos^2\theta}\frac{\partial\zeta}{\partial\lambda} + \frac{g}{2\omega a^2\rho_0\cos^2\theta}\int_0^z (z-\xi)\frac{\partial\rho}{\partial\lambda}d\xi. \end{aligned} \tag{1.154}$$

The equation of the SST in a non-linear model is:

$$\begin{aligned} \Delta\zeta & + \frac{1}{H}\left(\frac{\partial H}{\partial\theta}\frac{\partial\zeta}{\partial\theta} + \frac{1}{\sin^2\theta}\frac{\partial H}{\partial\lambda}\frac{\partial\zeta}{\partial\lambda}\right) = \frac{1}{H\rho_0}\int_0^H (H-z)\Delta\rho dz - \frac{1}{H\rho_0}\frac{\partial H}{\partial\theta}\int_0^H\frac{\partial\rho}{\partial\theta}dz \\ & - \frac{1}{H\rho_0\sin^2\theta}\frac{\partial H}{\partial\lambda}\int_0^H\frac{\partial\rho}{\partial\lambda}dz + \frac{2\omega a}{gH}\operatorname{ctg}\theta\left(\frac{\partial}{\partial\theta}\sin\theta S_\lambda - \frac{\partial S_\theta}{\partial\lambda}\right) \\ & - \frac{2\omega a\sin\theta}{gH}S_\lambda - \frac{a}{gH\sin\theta}\int_0^H\left(\frac{\partial}{\partial\theta}\sin\theta B_1 + \frac{\partial A_1}{\partial\lambda}\right)dz \\ & + \frac{a}{\rho_0 gH\sin\theta}\left(\frac{\partial}{\partial\theta}\sin\theta\cdot\tau_\theta^H + \frac{\partial\tau_\lambda^H}{\partial\lambda}\right) + \frac{1}{2\alpha' gH}\Delta P_a, \end{aligned} \tag{1.155}$$

where

$$A_1 = \frac{v_\theta}{a}\frac{\partial v_\theta}{\partial\theta} + \frac{v_\lambda}{a\sin\theta}\frac{\partial v_\theta}{\partial\lambda} + v_z\frac{\partial v_\theta}{\partial z} - \frac{v_\lambda^2\operatorname{ctg}\theta}{a};$$

$$B_1 = \frac{v_\theta}{a}\frac{\partial v_\lambda}{\partial\theta} + \frac{v_\lambda}{a\sin\theta}\frac{\partial v_\lambda}{\partial\lambda} + v_z\frac{\partial v_\lambda}{\partial z} + \frac{v_\theta v_\lambda}{a}\operatorname{ctg}\theta. \tag{1.156}$$

1.3 Some Numerical Methods of Solving Simplified Equations of Hydrodynamics

The governing equation is an important component of the hydrodynamic process, whereby particles transfer along the trajectory, etc. The great attention which researchers pay to the solution of these equations is explained by the importance of applications.

The development of numerical methods for solving hydrodynamic equations was stimulated by necessity in hydrodynamic and gasodynamic calculations. For the past 20–30 years numerical methods of solving hydrodynamic problems have been enriched by the number of interesting ideas, suggested by many authors. Recently, the intensive study of weather forecast and ocean dynamics has given a new impulse to the establishment of efficient numerical methods of solving hydrodynamics equations, which, thanks to works of Kurihara (1965), Arakawa (1966), Bryan (1969a), Murchuk (1974) and others led to the creation of universal and effective algorithms for solving such problems. In the present section some approaches, used in the following chapters, will be discussed.

The equation of substance transfer along the particle trajectory is one of the simplest problems of mathematical physics, which can be presented as:

$$\frac{d\varphi}{dt} = 0\,,$$

where

$$\frac{d}{dt} = \frac{\partial}{\partial t} + u\frac{\partial}{\partial x} + v\frac{\partial}{\partial y} + w\frac{\partial}{\partial z}$$

is the full derivative of the function $\varphi(x, y, z, t)$ by time, u, v and w are the components of the velocity vector $\mathbf{u} = u\mathbf{i} + v\mathbf{j} + w\mathbf{k}$ and

$$u = \frac{dx}{dt}\,, \qquad v = \frac{dy}{dt}\,, \qquad w = \frac{dz}{dt}\,.$$

The considered equation is solved under additional conditions, the simplest of which for the limited area will be:

$$\varphi(x, y, z, 0) = f(x, y, z)\,.$$

An analogous problem appears as an element of the general algorithm of the numerical solution of hydrodynamic equations, of the radiation transfer theory and of many others. Taking this situation into consideration, let us discuss in detail all the possible ways of numerically solving such problems. First, let us examine the simplest equations of motion.

While solving the problems of hydrodynamics, hydrothermodynamics, weather forecast, ocean dynamics and others, we have to deal with the equations of substance transfer along the trajectories. The simplest equation of this kind is the equation:

$$\frac{\partial\varphi}{\partial t} + u\frac{\partial\varphi}{\partial x} = 0\,,$$

$$\varphi = f(x) \qquad \text{when } t = 0\,, \tag{1.157}$$

where u is the specified velocity, and $f(x)$ is the initial distribution of φ. We assume that functions $\varphi(x,t)$ and $f(x)$ are periodical functions of x with a 2π period. If $u = \text{const}$, then the problem (1.157) has the obvious solution:

$$\varphi(x,t) = f(x - ut)\,, \tag{1.158}$$

if $f(x)$ is a differentiable function. This solution (1.158) describes the process of initial propagation along the characteristic:

$$x - ut = \text{const}\,.$$

This means that $\varphi(x,t) = \text{const}$ on any straight line $x - ut = \text{const}$.

Thus, problem (1.157) with $u > 0$ determines the process of disturbance propagation in the x direction. These well-known statements should be taken into consideration when constructing the difference analogues of problem (1.157). If the velocity $u = u(x,t)$ is variable, then it is impossible to solve problem (1.157) in an analytical form. In these cases it is necessary to apply numerical methods, based on the finite difference approximations.

Let us examine the simplest difference schemes with $u = \text{const}$. For definiteness, let us assume $u > 0$. Then we have an explicit scheme:

$$\frac{\varphi_k^{j+1} - \varphi_k^j}{\tau} + u\frac{\varphi_k^j - \varphi_{k-1}^j}{\Delta x} = 0 \tag{1.159}$$

and the implicit one

$$\frac{\varphi_k^{j+1} - \varphi_k^j}{\tau} + u\frac{\varphi_k^{j+1} - \varphi_{k-1}^{j+1}}{\Delta x} = 0\,. \tag{1.160}$$

Both of these schemes have the first-order approximation by Δx and τ. Thus, let us assume that the initial values $f(x)$ and $\varphi(x,t)$ are rather smooth functions. Then let us expand the solution of Eq. (1.157) into the Tailor series in the vicinity of point $x = x_k$, $t = t_j$:

$$\varphi(x,t) = (\varphi)_k^j + (\varphi_t)_k^j(t - t_j) + (\varphi_x)_k^j(x - x_k) + \cdots \tag{1.161}$$

Substituting series (1.161) in Eqs. (1.159), (1.160) we obtain:

$$\frac{\partial\varphi}{\partial t} + u\frac{\partial\varphi}{\partial x} = \frac{u\Delta x}{2}\frac{\partial^2\varphi}{\partial x^2} - \frac{\tau}{2}\frac{\partial^2\varphi}{\partial t^2}\,. \tag{1.162}$$

The neglected terms in Eq. (1.162) have a higher order of smallness. It follows from Eq. (1.157) that:

$$\frac{\partial^2\varphi}{\partial t^2} = u^2\frac{\partial^2\varphi}{\partial x^2}\,. \tag{1.163}$$

Then expression (1.162) will have the form:

$$\frac{\partial\varphi}{\partial t} + u\frac{\partial\varphi}{\partial x} = \frac{u\Delta x - \tau u^2}{2}\frac{\partial^2\varphi}{\partial x^2}\,. \tag{1.164}$$

Such an analysis of difference schemes was suggested by A.I. Zhukov (see Rozhdestvensky and Yanenko 1968). The analysis of correlation (1.164) shows that when:

$$\frac{u\tau}{\Delta x} < 1 \, ,$$

correlation (1.164) can be interpreted as the equation with the solution determination domain $x_{k-1} \leqslant x \leqslant x_k$, $t_j \leqslant t \leqslant t_{j+1}$.

If we assume that the neglected terms in Eq. (1.164) are small, we obtain the heat conductivity equation:

$$\frac{\partial \varphi}{\partial t} + u\frac{\partial \varphi}{\partial x} = \mu\frac{\partial^2 \varphi}{\partial x^2} \, ,$$

where

$$\mu = \frac{u\Delta x - u^2\tau}{2}$$

is the so-called coefficient of an artificial or "numerical" viscosity. It should be noted, moreover, that if:

$$\frac{u\tau}{\Delta x} = 1 \, ,$$

then $\mu = 0$ and all other neglected terms will be zero and the explicit scheme (1.159) will develop into the scheme of an infinite order of approximation by Δx and τ.

In particular, we should note the case when:

$$\frac{u\tau}{\Delta x} > 1 \, .$$

In this case we obtain the equation:

$$\frac{\partial \varphi}{\partial t} + u\frac{\partial \varphi}{\partial x} = -|\mu|\frac{\partial^2 \varphi}{\partial x^2} \, . \tag{1.165}$$

It can be easily seen that Eq. (1.165) with the initial condition:

$$\varphi = \varphi^0(x) \, , \qquad \text{when } t = 0 \tag{1.166}$$

leads to the problem, incorrectly constructed according to Adamar. The solution of this problem is unstable in relation to small variations of the initial data. While constructing difference equations for the problems of form (1.157), it is always necessary to take into consideration the condition of the correctness of the problem:

$$\mu = \frac{u\tau}{\Delta x} \leqslant 1 \, .$$

Now let us investigate the problem of numerical stability of scheme (1.159). For this purpose let us first examine the spectral problem:

$$(A^h\omega)_k \equiv u\frac{\omega_k - \omega_{k-1}}{\Delta x} = \lambda\omega_k \tag{1.167}$$

on an infinite grid interval $D_h = (-\infty < x_k < \infty)$. The solution to Eq. (1.167), limited in D_h, has the form:

$$\omega_k = e^{ikP\Delta x}, \tag{1.168}$$

where P is an arbitrary integer. Substituting (1.168) into (1.167), we obtain the expression of the eigenvalue:

$$\lambda_P = \frac{u}{\Delta x}\left(2\sin^2\frac{P\Delta x}{2} + i\sin P\Delta x\right). \tag{1.169}$$

Thus, Eq. (1.159) in the operator form is:

$$\frac{\varphi^{j+1} - \varphi^j}{\tau} + A^h\varphi^j = 0\,. \tag{1.170}$$

The solution to Eq. (1.170) can be obtained in the form:

$$\varphi^j = \sum_{P=-\infty}^{\infty} \varphi_P^j e^{ikP\Delta x}, \tag{1.171}$$

where φ_P^j is the Fourier coefficient of function φ^j. For the coefficients we obtain the equation:

$$\frac{\varphi_P^{j+1} - \varphi_P^j}{\tau} + \lambda_P\varphi_P^j = 0\,.$$

Hence,

$$\varphi_P^{j+1} = T_P\varphi_P^j\,, \tag{1.172}$$

where $T_P = 1 - \tau\lambda_P$ is the factor of transfer for the Fourier coefficients.

Let us then find the condition under which the modulus of all the components does not increase. Such a condition will be:

$$|1 - \tau\lambda_P| \leqslant 1\,. \tag{1.173}$$

This inequality takes place, if:

$$\frac{u\tau}{\Delta x} \leqslant 1\,.$$

Indeed

$$|1 - \tau\lambda_P|^2 = \left(1 - \frac{2u\tau}{\Delta x}\sin^2\frac{P\Delta x}{2}\right)^2 + \left(\frac{u\tau}{\Delta x}\right)^2\sin^2 P\Delta x$$

$$= 1 - 4\frac{u\tau}{\Delta x}\sin^2\frac{P\Delta x}{2} + \left(\frac{u\tau}{\Delta x}\right)^2\left(4\sin^4\frac{P\Delta x}{2} + \sin^2 P\Delta x\right)$$

$$= 1 - 4\frac{u\tau}{\Delta x}\sin^2\frac{P\Delta x}{2} + 4\left(\frac{u\tau}{\Delta x}\right)^2\sin^2\frac{P\Delta x}{2}\left(\sin^2\frac{P\Delta x}{2}\right.$$

$$\left. + \cos^2\frac{P\Delta x}{2}\right) = 1 - 4\sin^2\frac{P\Delta x}{2}\left(\frac{u\tau}{\Delta x}\right)\left(1 - \frac{u\tau}{\Delta x}\right).$$

If $u\tau/\Delta x > 1$, then $|1 - \tau\lambda_P| > 1$. So, we have to examine the domain of the variation of $|1 - \tau\lambda_P|$ with $0 < u\tau/\Delta x \leq 1$ (it should be kept in mind that we have assumed $u > 0$). The function $u\tau/\Delta x(1 - u\tau/\Delta x)$ in this interval represents the largest value, which is equal to $\frac{1}{4}$, if

$$\frac{u\tau}{\Delta x} = \frac{1}{2} .$$

And in this case $|1 - \tau\lambda_P| = 1 - \sin^2 P\Delta x/2 = \cos^2 P\Delta x/2$. At other values of $\frac{u\tau}{\Delta x}$ from interval $0 \leqslant \frac{u\tau}{\Delta x} \leqslant 1$, this positive function, decreasing, tends to zero at the ends, i.e. at $\frac{u\tau}{\Delta x} = 0$ and $\frac{u\tau}{\Delta x} = 1$. This means that the inequality

$$4\sin^2\frac{P\Delta x}{2}\left(\frac{u\tau}{\Delta x}\right)\left(1 - \frac{u\tau}{\Delta x}\right) \leqslant 1$$

takes place. From this the correctness of the above formulated statement follows. Thus, we come to the condition of the numerical stability of the difference scheme. It is easy to see that in this case the construction of the stability criteria coincides with the correctness of Eq. (1.164).

Now the implicit difference scheme (1.160) will be discussed. According to the above formulated method it is easy to show that scheme (1.160) also has the first-order approximation by Δx and τ. With the help of expansion in the Taylor series at $x = x_k$ and $t = t_j$, we obtain the following equation, analogous to (1.164):

$$\frac{\partial\varphi}{\partial t} + u\frac{\partial\varphi}{\partial x} = \mu\frac{\partial^2\varphi}{\partial x^2} + \cdots , \tag{1.174}$$

where

$$\mu = \frac{u\Delta x + u^2\tau}{2} .$$

Already here we can state the principle difference between correlations (1.164) and (1.174). In the latter equation the coefficient of numerical viscosity is always positive. So, Eq. (1.174) with the rather smooth corresponding initial data is always correct. It is not difficult to show that the difference equation (1.160) is stable at any grid step correlation, i.e. it is absolutely stable, as the transfer factor for each Fourier coefficient is equal to:

$$T_P = \frac{1}{1 + \tau\lambda_P} .$$

Thus, it can be deduced that:

$$|T_P| \leqslant 1.$$

From the interesting and rather frequently used difference schemes, with the exception of Eqs. (1.159) and (1.160), we obtain, for example, the following:

$$\frac{\varphi_k^{j+1} - \varphi_k^j}{\tau} + u\frac{\varphi_{k+1}^{j+1} - \varphi_{k-1}^{j+1}}{2\Delta x} = 0 \tag{1.175}$$

and

$$\frac{\varphi_k^{j+1} - \varphi_k^j}{\tau} + \frac{u(\varphi_{k+1}^{j+1/2} - \varphi_{k-1}^{j+1/2})}{2\Delta x} = 0\,, \tag{1.176}$$

where

$$\varphi_k^{j+1/2} = \tfrac{1}{2}(\varphi_k^{j+1} + \varphi_k^j)\,.$$

It is easy to show that scheme (1.175) is a scheme of the first-order approximation by τ and second, by Δx. Corresponding to this scheme, the differential equation will take the form of Eq. (1.174), where

$$\mu = \frac{u^2\tau}{2}\,.$$

As far as the stability is concerned, it is determined by the transfer operators for the Fourier coefficients. Thus, the spectral problem occurs:

$$(A^h\omega)_k \equiv u\frac{\omega_{k+1} - \omega_{k-1}}{2\Delta x} = \lambda\omega_k\,. \tag{1.177}$$

The solution to Eq. (1.177) can be expressed in the form of Eq. (1.168). As a result we will have for λ_P:

$$\lambda_P = i\frac{u}{\Delta x}\sin P\Delta x\,, \tag{1.178}$$

from which it follows directly that for Fourier coefficients we obtain the equations:

$$\frac{\varphi_P^{j+1} - \varphi_P^j}{\tau} + \lambda_P\varphi_P^{j+1} = 0$$

and

$$\varphi_P^{j+1} = T_P\varphi_P^j\,, \tag{1.179}$$

where

$$T_P = \frac{1}{1 + \tau\lambda_P}\,.$$

With respect to Eq. (1.178):

$$T_P = \frac{1}{1 + i\dfrac{u\tau}{\Delta x}\sin P\Delta x}$$

and so,

$$|T_P| = \frac{1}{\sqrt{1+\left(\dfrac{u\tau}{\Delta x}\right)^2 \sin^2 P\Delta x}} \leqslant 1\,.$$

Thus, the absolute numerical stability of scheme (1.175) is confirmed.

The Crank-Nickolson scheme (1.176) is the most interesting to apply. It is not difficult to see that it is a second-order approximation scheme by τ, Δx and it is not dissipative. This means that in the differential equation (1.174) we have $\mu = 0$, and the neglected terms have the order $\tau\Delta x$, τ^2 and Δx^2. As far as the numerical stability is concerned, in the present case we have:

$$T_P = \frac{1 - i\dfrac{u\tau}{2\Delta x}\sin P\Delta x}{1 + i\dfrac{u\tau}{2\Delta x}\sin P\Delta x}$$

and, so

$$|T_P| = 1\,.$$

Thus, this scheme is absolutely stable. It should be noted, that if in scheme (1.159) the difference expression for $u\dfrac{\partial\varphi}{\partial x}$, in the form:

$$u\frac{\varphi_k^j - \varphi_{k-1}^j}{\Delta x}$$

is to be replace by:

$$u\frac{\varphi_{k+1}^j - \varphi_k^j}{\Delta x}\,,$$

then the obtained difference scheme, when $u > 0$, will be unstable at any step correlation. The proof is obvious.

In conclusion, let us examine one more interesting numerical solution of problem (1.157) on the basis of the so-called scheme of a running calculation. Suggested by L.D. Landau, N.N. Neuman and I.M. Khalatnikov (1956) this scheme has the form:

$$\frac{\varphi_k^{j+1} - \varphi_k^j}{\tau} + \frac{u}{\Delta x}\left[\left(\frac{\varphi_k^{j+1} + \varphi_k^j}{2}\right) - \left(\frac{\varphi_{k-1}^{j+1} + \varphi_{k-1}^j}{2}\right)\right] = 0\,. \tag{1.180}$$

It is easy to show that it is a second-order approximation scheme by τ and a first-order one, by Δx. It is realized by the recurrent correlation:

$$\varphi_k^{j+1} = \frac{1 - \dfrac{u\tau}{2\Delta x}}{1 + \dfrac{u\tau}{2\Delta x}}\varphi_k^j + \frac{\dfrac{u\tau}{2\Delta x}}{1 + \dfrac{u\tau}{2\Delta x}}(\varphi_{k-1}^{j+1} + \varphi_{k-1}^j)\,. \tag{1.181}$$

On the basis of the stability analysis, according to Neuman and with the help of the Fourier method, it is not difficult to prove that scheme (1.180) is absolutely stable.

Likewise we can construct an absolutely stable scheme of the "running calculation" for the multi-dimensional problem of motion and prove its absolute stability in the case of an equation with constant coefficients.

Let us draw attention to the fact that previously it was assumed that u is constant and positive. If u is negative, then by replacing x by $-x$ we obtain the equation examined above. But, the case when $u = u(x, t)$ is especially interesting for applications. The simplest analysis shows that in this case, even by using implicit dissipative difference schemes, the violation of numerical stability is possible. It becomes especially apparent in non-linear problems, which is elucidated in the following. If we expand the solution of the difference problem and coefficient $u(x_k, t_j)$ in the Fourier series, then the products of the Fourier Series will lead to the harmonics, longer or shorter than interacting ones.

As a result of such a process, in a number of cases, the "energy" transfer, due to rounding-off errors from long waves to shorter ones, can take place, and the calculating process will be unstable, despite the fact that the present difference scheme with the constant coefficient is numerically stable. Such an instability is usually called a non-linear instability. It sometimes appears also when solving the linear problems with variable coefficients. Therefore, the construction of difference schemes for the non-linear equations with variable coefficients, stable to any perturbations, is a very actual problem. In the majority of cases the suppression of the numerical instability is possible with the help of dissipative difference schemes, corresponding to the particular choice of the numerical viscosity coefficient μ. However, such schemes, as a rule, are the schemes of the first-order approximation, either by τ or by Δx, or both by τ and Δx.

Of special interest in practical applications are the equations in the form:

$$\frac{\partial \varphi}{\partial t} + \frac{\partial u \varphi}{\partial x} = 0 , \tag{1.182}$$

where

$$u = u(x, t) .$$

The difference schemes for the equations of this kind, i.e. absolutely stable and of the first- or even second-order approximation with some classes of coefficients, will be studied while examining multi-dimensional equations of the form (1.182).

Let us consider on the plane (x, y) the problem of ensemble motion of particles according to specified trajectories. Within the framework of the continuous medium, mechanics method we have the following problem:

$$\frac{\partial \varphi}{\partial t} + u\frac{\partial \varphi}{\partial x} + v\frac{\partial \varphi}{\partial y} = 0 , \qquad \text{in } D \times D_t ,$$

$$\varphi(x, y, 0) = g , \qquad \text{in } D \tag{1.183}$$

here,

$$u = u(x, y, t)\,, \qquad v = v(x, y, t)\,.$$

Let us assume further that the components of the velocity vector u, v at each moment of the time satisfy the continuity equation:

$$\frac{\partial u}{\partial x} + \frac{\partial v}{\partial y} = 0\,. \tag{1.184}$$

Let the rectangle $\{0 \leqslant x \leqslant a, 0 \leqslant y \leqslant b\}$ be the domain of D and the solution to problem (1.183), together with the coefficients u and v, is periodic and has the same values at the opposite boundaries of the rectangle.

Thus, the evolutional equations from (1.183) take the form of an operator:

$$\frac{\partial \varphi}{\partial t} + A\varphi = 0\,,$$

$$\varphi = g\,, \qquad t = 0\,, \tag{1.185}$$

where

$$A = u\frac{\partial}{\partial x} + v\frac{\partial}{\partial y}\,.$$

It is not difficult to show that in this problem, operator A satisfies the condition $(A\varphi, \varphi) = 0$. Moreover, introducing the scalar product in the Gilbert space, let us formulate the equation:

$$(A\varphi, \varphi) = \int_0^a dx \int_0^b dy \left(u\frac{\partial \varphi}{\partial x} + v\frac{\partial \varphi}{\partial y}\right)\varphi\,. \tag{1.186}$$

The subintegral expression with respect to Eq. is transformed into the form:

$$\left(u\frac{\partial \varphi}{\partial x} + v\frac{\partial \varphi}{\partial y}\right)\varphi = \frac{\partial u\frac{\varphi^2}{2}}{\partial x} + \frac{\partial v\frac{\varphi^2}{2}}{\partial y}\,.$$

Therefore,

$$(A\varphi, \varphi) = \int_0^a dx \int_0^b dy \left(\frac{\partial u\frac{\varphi^2}{2}}{\partial x} + \frac{\partial v\frac{\varphi^2}{2}}{\partial y}\right)\,. \tag{1.187}$$

From this, with the help of the periodicity of the solution at the boundaries we obtain:

$$(A\varphi, \varphi) = 0\,. \tag{1.188}$$

Thus, operator A is skew-symmetric and so, allows the construction of the absolutely stable difference schemes.

Now let us try to split operator A in such a way, that each of the elementary operators will also satisfy the condition:

$$(A\alpha\varphi, \varphi) = 0\,. \tag{1.189}$$

In this case the difference scheme of the componentwise splitting allows one to obtain an absolutely stable difference scheme of second-order accuracy.

The formal decomposition of the operator A into components:

$$A_1 = u\frac{\partial}{\partial x}, \qquad A_2 = v\frac{\partial}{\partial y} \tag{1.190}$$

does not satisfy condition (1.189). It is not difficult to see that correlations:

$$(A_1\varphi, \varphi) = -\frac{1}{2}\int_0^a dx \int_0^b \varphi^2 \frac{\partial u}{\partial x} dy\,,$$

$$(A_2\varphi, \varphi) = -\frac{1}{2}\int_0^a dx \int_0^b \varphi^2 \frac{\partial v}{\partial y} dy$$

take place. Therefore, the operators A_1 and A_2 cannot be taken to be elementary for the construction of the successive splitting scheme.

Now let us choose the operators A_1 and A_2 in a more complicated form:

$$A_1\varphi = u\frac{\partial \varphi}{\partial x} + \frac{\varphi}{2}\frac{\partial u}{\partial x}\,,$$

$$A_2\varphi = v\frac{\partial \varphi}{\partial y} + \frac{\varphi}{2}\frac{\partial v}{\partial y}\,. \tag{1.191}$$

It is easy to see that each of such operators now satisfy the demand (1.189), and the sum of these operators is exactly equal to A. Thus, the sum:

$$(A_1 + A_2)\varphi = u\frac{\partial \varphi}{\partial x} + v\frac{\partial \varphi}{\partial y} + \frac{\varphi}{2}\left(\frac{\partial u}{\partial x} + \frac{\partial v}{\partial y}\right) = u\frac{\partial \varphi}{\partial x} + v\frac{\partial \varphi}{\partial y} = A\varphi\,.$$

Here, we have made use of the fact that coefficients u and v satisfy Eq. (1.184).

Thus, all the necessary conditions for the application of the splitting method are now fulfilled and we come to the splitting scheme at the interval $t_{j-1} \leqslant t \leqslant t_{j+1}$:

$$\frac{\varphi^{j-1/2} - \varphi^{j-1}}{\tau} + \left(u^j\frac{\partial}{\partial x} + \frac{1}{2}\frac{\partial u^j}{\partial x}\right)\frac{\varphi^{j-1/2} + \varphi^{j-1}}{2} = 0\,,$$

$$\frac{\varphi^{j} - \varphi^{j-1/2}}{\tau} + \left(v^j\frac{\partial}{\partial y} + \frac{1}{2}\frac{\partial v^j}{\partial y}\right)\frac{\varphi^{j} + \varphi^{j-1/2}}{2} = 0\,,$$

$$\frac{\varphi^{j+1/2} - \varphi^{j}}{\tau} + \left(v^j\frac{\partial}{\partial y} + \frac{1}{2}\frac{\partial v^j}{\partial y}\right)\frac{\varphi^{j+1/2} + \varphi^{j}}{2} = 0\,,$$

$$\frac{\varphi^{j+1} - \varphi^{j+1/2}}{\tau} + \left(u^j\frac{\partial}{\partial x} + \frac{1}{2}\frac{\partial u^j}{\partial x}\right)\frac{\varphi^{j+1} + \varphi^{j+1/2}}{2} = 0\,. \tag{1.192}$$

If the functions u and v and the solution φ are smooth enough by all variables, then scheme (1.192) has a second-order approximation and will be absolutely stable according to:

$$\|\varphi^{j+1}\| = \|\varphi^{j-1}\| = \cdots = \|g\|\,. \tag{1.193}$$

This is an instructive example of how the formal splitting into operators (1.190) can compromise the very idea of the splitting method, and only supplementary considerations lead to the schemes, theoretically justified and efficient in applications.

After such a preliminary introduction the construction of difference schemes for problem (1.183) by spacial variables and time will be considered. For this purpose let us first discuss the rational methods of approximating operator A by spacial variables x and y. As it has been noted, the convenient method of approximating mathematical physics problems with the conservation of the operator's additive properties and its qualitative peculiarities is the method of coordinatewise approximation, of which we will make use for the construction of difference schemes.

Let us assume that the coefficients u and v are smooth enough and consider Eq. (1.183) in a divergent form:

$$\frac{\partial \varphi}{\partial t} + \frac{\partial u\varphi}{\partial x} + \frac{\partial v\varphi}{\partial y} = 0 \quad \text{in } D \times D_t ,$$

$$\varphi = g \quad \text{in } D \text{ at } t = 0 . \tag{1.194}$$

To construct a difference scheme let us take as a basis operator A, determined by the expression:

$$A\varphi = \frac{\partial u\varphi}{\partial x} + \frac{\partial v\varphi}{\partial y} - \frac{\varphi}{2}\left(\frac{\partial u}{\partial x} + \frac{\partial v}{\partial y}\right) \tag{1.195}$$

whereby the difference analogue of this correlation is considered in the form:

$$(A^h\varphi)_{k,l} = \frac{u_{k+1,l}\cdot\varphi_{k+1,l} - u_{k-1,l}\varphi_{k-1,l}}{2\Delta x} + \frac{v_{k,l+1}\varphi_{k,k+1} - v_{k,l-1}\varphi_{k,l-1}}{2\Delta y}$$
$$- \frac{\varphi_{k,l}}{2}\left(\frac{u_{k+1,l} - u_{k-1,l}}{2\Delta x} + \frac{v_{k,l+1} - v_{k,l-1}}{2\Delta y}\right) . \tag{1.196}$$

It is obvious that the difference expression (1.196) approximates (1.195) with second-order accuracy with respect to Δx and Δy with the smooth enough functions u, v and φ. But the expression (1.196) has an essential drawback, since in such a form, the operator A^h loses its skew-symmetric structure, i.e.

$$(A^h\varphi, \varphi) \neq 0 . \tag{1.197}$$

Therefore, the approximation commonly used is not satisfactory for the construction of the numerical algorithm to solve problem (1.183).

Let us show that the approximation of expression (1.195) in the form:

$$(A^h\varphi)_{k,l} = \frac{u_{k+1/2,l}\varphi_{k+1,l} - u_{k-1/2,l}\varphi_{k-1,l}}{2\Delta x} + \frac{v_{k,l+1/2}\varphi_{k,l+1} - v_{k,l-1/2}\varphi_{k,l-1}}{2\Delta y} \tag{1.198}$$

satisfies the principal correlation:

$$(A^h\varphi, \varphi) = 0 , \tag{1.199}$$

and approximates expression (1.195) with second-order accuracy by Δx and Δy. To show this, let us use the following approximation of the coefficients:

$$u_{k+1/2,l} = u_{k+1,l} - \frac{u_{k+1,l} - u_{k,l}}{2}\,;\; v_{k,l+1/2} = v_{k,l+1} - \frac{v_{k,l+1} - v_{k,l}}{2}\,;$$

$$u_{k-1/2,l} = u_{k-1,l} + \frac{u_{k,l} - u_{k-1,l}}{2}\,;\; v_{k,l-1/2} = v_{k,l-1} + \frac{v_{k,l} - v_{k,l-1}}{2}\,. \tag{1.200}$$

Let us substitute expressions (1.200) into (1.198). Then, after simple transformations, we will obtain:

$$\begin{aligned}(A^h\varphi)_{k,l} = {} & \frac{u_{k+1,l}\varphi_{k+1,l} - u_{k-1,l}\varphi_{k-1,l}}{2\Delta x} + \frac{v_{k,l+1}\varphi_{k,l+1} - v_{k,l-1}\varphi_{k,l-1}}{2\Delta y} \\ & - \frac{\varphi_{k,l}}{2}\left(\frac{u_{k+1,l} - u_{k-1,l}}{2\Delta x} + \frac{v_{k,l+1} - v_{k,l-1}}{2\Delta y}\right) \\ & - (\Delta x^2 R_{k,l} + \Delta y^2 Q_{k,l})\,,\end{aligned} \tag{1.201}$$

where the values $R_{k,l}$ and $Q_{k,l}$ with $\Delta x \to 0$ and $\Delta y \to 0$ tend to the following:

$$R_{k,l} \to \frac{1}{4}\frac{\partial}{\partial x}\left(\frac{\partial u}{\partial x}\frac{\partial \varphi}{\partial x}\right);$$

$$Q_{k,l} \to \frac{1}{4}\frac{\partial}{\partial y}\left(\frac{\partial v}{\partial y}\frac{\partial \varphi}{\partial y}\right).$$

Now let us assume that the coefficients $u_{k,l}$, $v_{k,l}$ satisfy the difference analogue of the continuity equation:

$$\frac{u_{k+1,l} - u_{k-1,l}}{2\Delta x} + \frac{v_{k,l+1} - v_{k,l-1}}{2\Delta y} = 0(h^2)\,. \tag{1.202}$$

If the coefficients u, v and the solution φ have a limited second-order derivation by x and y, then we can see that the expression (1.201) with condition (1.202) differs from Eq. (1.196) by the same second-order of smallness as expression (1.196) from (1.195). Thus, we have shown that expression (1.198) approximates (1.195) with the second-order accuracy relative to Δx and Δy.

Now, let us show that the operator A^h, constructed in this way, satisfies condition (1.199) and, moreover, that each of the operators A_1^h and A_2^h, determined by the expressions:

$$A_1^h\varphi = \frac{u_{k+1/2,l}\cdot\varphi_{k+1,l} - u_{k-1/2,l}\varphi_{k-1,l}}{2\Delta x}\,,$$

$$A_2^h\varphi = \frac{v_{k,l+1/2}\varphi_{k,l+1} - v_{k,l-1/2}\varphi_{k,l-1}}{2\Delta y}\,, \tag{1.203}$$

also satisfies the conditions:

$$(A^h\alpha\varphi, \varphi) = 0\,. \tag{1.204}$$

For this purpose let us introduce the scalar product for the vector values a, b:

$$(a, b) = \sum_k \sum_l a_{k,l} b_{k,l} \Delta x \Delta y\,,$$

so,

$$(A_1^h \varphi, \varphi) = \frac{1}{2} \sum_k \sum_l \Delta y (u_{k+1/2,l} \varphi_{k+1,l} - u_{k-1/2,l} \varphi_{k-1,l}) \varphi_{k,l}\,,$$

$$(A_2^h \varphi, \varphi) = \frac{1}{2} \sum_k \sum_l \Delta x (v_{k,l+1/2} \varphi_{k,l+1} - v_{k,l-1/2} \varphi_{k,l-1}) \varphi_{k,l}\,. \tag{1.205}$$

Grouping similar terms in (1.205), we come to the equality (1.204). From condition (1.204), condition (1.199) follows immediately. So, the necessary spacial approximations are made. Now, the problem is to make a reduction by time of the ordinary differential equations system:

$$\frac{d\varphi^h}{dt} + A^h \varphi^h = 0 \quad \text{in } D_h \times T\,,$$

$$\varphi^h = g \quad \text{in } D_h \text{ at } t = 0\,, \tag{1.206}$$

where φ^h is a vector function, whereby the components $\varphi_{k,l}$ and A_h^n satisfy condition (1.204). This means that problem (1.206) can be solved with the help of the splitting method. Neglecting the index h in the functions and operators as non-essential, at the interval $t_{j-1} \leqslant t \leqslant t_{j+1}$ we come to the system:

$$\frac{\varphi^{j-1/2} - \varphi^{j-1}}{\tau} + A_1^j \frac{\varphi^{j-1/2} + \varphi^{j-1}}{2} = 0\,;$$

$$\frac{\varphi^{j} - \varphi^{j-1/2}}{\tau} + A_2^j \frac{\varphi^{j} + \varphi^{j-1/2}}{2} = 0\,;$$

$$\frac{\varphi^{j+1/2} - \varphi^{j}}{\tau} + A_2^j \frac{\varphi^{j+1/2} + \varphi^{j}}{2} = 0\,;$$

$$\frac{\varphi^{j+1} - \varphi^{j+1/2}}{\tau} + A_1^j \frac{\varphi^{j+1} + \varphi^{j+1/2}}{2} = 0\,. \tag{1.207}$$

Thus, problem (1.183) is reduced to the system of the simplest one-dimensional difference equations, the solution of which is possible with the help of the factorization method of the three-point difference equations. The equation of hydromechanics in three-dimensional space may be solved similarly, when $A = A_1 + A_2 + A_3$.

Chapter 2
The Simplest Methods of Difference Approximation and Constructed Equations Solution

2.1 The Construction of Difference Grids

When solving hydrodynamical problems by finite-difference methods, the examined basin can be approximated by some grid domain, in the grid points of which the values of the initial fields are given and the values of the sought characteristics are determined (excluding the calculations on so-called imbedded and shifted grids, of which we will speak later). The grids with rectangle nets are most frequently used. The basin boundary in this case is a broken line, the sequences of which are parallel to corresponding coordinate axes. Other, more accurate boundary approximation methods are seldom used in practice because of the essential complication of calculations. Here, we will consider only the grids with rectangle nets, both inside the domain and in the boundary region.

When constructing difference grids it is necessary to take into account a number of aspects. First, at least one interior calculating point should be next to one of each pair of neighbouring boundary points. This condition should be fulfilled especially precisely during the approximation of the complex parts of the basin's real boundary, for instance, caps, gulfs, straits. Otherwise, the grid domain boundary at these parts can become indefinite, besides the loss potential calculating points in this case. The analogous demand is made also to the representation of the basin's bottom topography. Here, if such a condition is not fulfilled, bottom hollows actually become excluded from the examination, which is equal to the smoothing of the basin's bottom irregularities.

Second, the accuracy of the basin's bottom topography should be coordinated with the accuracy of the grid resolution by the vertical near the bottom, i.e. the specified fields of density and bottom topography should be coordinated in the difference sense. The most frequent case is that in which the basin's depth at each point is taken equal to the horizon depth, nearest to the real bottom. In this case in the basin's bottom points the density should be specified. If these conditions are not fulfilled for some reason and bottom depths differ considerably from the specified horizons' depths, then we have to specify at the basin bottom (where the boundary conditions for u, v, w are taken) some density values (in an explicit or implicit form), usually mean densities for the given depth or anomaly. These values are generally equal to zero, which are necessary for the calculation of integrals from density. Further, due to the discrepancy in the grids of density and bottom topography fields, additional complications arise when approximating the density derivatives, which lead to the rough representation of density fields. Moreover, we should take into account the fact that in this case the step by the vertical in the near-bottom domain will be variable in the basin area. Since in practical calculations the steps by the vertical are often too large (in the deep layers they are usually equal to 500 or 1000 m and seldom less), then the unsuccessful adjustment of the grid set of the bottom depths with the grid set of the depths at the calculating levels can lead to the case, when the nearest to the bottom grid step by the vertical will decrease to a few metres, i.e. it will be less than the width of the bottom friction layer (in the limits of which, naturally, the calculating horizon closest to the bottom will be located). When applying the quasi-geostrophic model, with the calculation of u, v according to the explicit correlations, it will be necessary to add coordinating boundary layer corrections to these correlations. A more complicated situation in this case arises in the models with feedback, when the velocity components can be found directly from equations of motion. Because of the sharp irregularity of the grid resolution by the vertical (when neighbouring steps can differ over a hundred times from each other), the formation of the calculating wave, "swinging" the solution of the difference problem is possible.

Third, when composing the initial grid sets, it is always recommended to check whether the density is specified in all internal and boundary grid points, situated above and at the basin bottom. Due to such errors, either the values, left in the grid nets from the previous calculations, or zeroes, if corresponding memory sets are substituted by zero before the calculation (which is usually foreseen in the corresponding calcuation programs), get into the corresponding computer memory grid nets. Such errors may be overseen when analyzing the results of calculations.

Fourth, if the initial density, bottom topography and wind fields (especially the first one) are characterized by sharp, anomalous gradient values in the basin region (which is often marked in the frontier zones, by the coasts, near the underwater heights), then it is reasonable to construct the grid with a smaller spacial step for the given region. Otherwise, the chosen regular grid step, quite satisfactory for the adequate description of the circulation in the "calm" part of the basin will be too large in the regions with the large gradients, which will result either in smoothing of the difference solution in them, or in the loss of the essential local peculiarities of the circulation between the grid points. Thus, in this case it is necessary to construct an irregular grid. Of course, a sufficiently fine grid can be constructed for the whole domain, but is usually non-economical, and when the power of the computer is insufficient, it is difficult to carry out.

Finally, while constructing the equatorial basin grid one should take into account the possibilities of the applied model. Thus, the above considered quasi-geostrophic model is not applicable at the very equator (the formal reason is that the Coriolis parameter is in the denominator, and in the equatorial belt it gives, as well as the dynamic method, very rough results, as the quasi-geostrophic balance of forces in this belt is poorly fulfilled). Therefore, the grid for the quasi-geostrophic model should be constructed in such a way, to avoid the very equator and one should keep in mind the essential inaccuracy of the calculating results in the equatorial belt, in particular, the commonly marked overestimation of the velocities. The assumption $\varphi = \text{const}$ (for instance, equal to 2.5° or 5°) in the equatorial zone slightly improves the possibilities of the quasi-geostrophic model. This assumption results in the smoothing of the solution in the equatorial zone. The best method, however, will be the application of the non-linear models (Bryan 1969a, b; Marchuk and Sarkisyan 1980; Gill 1975; Philander and Pacanowski 1980; Semtner and Holland 1980).

2.2 The Methods of Approximation and Equation Solutions

2.2.1 The Methods of Calculating the Sea-Surface Topography (SST) at the Basin Boundary

In the previous chapter it was shown that for the all considered currents models the definition of the SST ζ or the total mass transport stream function ψ is necessary for solving the problem. For the definition of ζ at the basin boundary all the corresponding equations are obtained. In this section only the problems with the tangential derivative of the SST will be considered. Formally, the form of the boundary equations in a quasi-geostrophic and non-linear model is the same, however, the following difference can be observed. In the equations of the quasi-geostrophic model we have neglected non-linear terms and the lateral exchange effect. It does not allow one to satisfy accurately the boundary conditions for the velocity at all levels, therefore the condition is constructed only for the average current velocity by height. The function ζ is found according to this method, thus satisfying this condition. As far as the non-linear model is concerned, the current velocity components at each level here satisfy the boundary conditions.

Let us see what the equations for ζ at the basin boundary in these models have in common. We will assume briefly that the right-hand sides of correlations (1.90), (1.91) and (1.133), (1.134) are specified. As in a quasi-geostrophic model, as well as in a non-linear one, at the meridional parts of the boundary $\partial\zeta/\partial y$ is specified, and at the zonal ones, $\partial\zeta/\partial x$. In other words, the tangential to the contour derivative of the unknown function is specified. Therefore, the method for determining ζ at the boundary is the same for both models. For simplicity, we will give the scheme for calculating ζ at the boundary of the closed, singly-connected domain in the example of a quasi-geostrophic model. We will consider some part of the boundary (the form of which is clear from the difference correlations presented below, in which i is an index by the x-axis and j by the y-axis) and apply Eqs. (1.90) and (1.91) to the middle sections, of the corresponding links of the broken line. The right-hand sides of these equations consist not only of the specified values, but also the unknown functions. The possibility of considering the latter ones by the method of successive approximations was discussed in Chapter 1. Here, we will assume that the right-hand sides are specified and are equal to f and f^1 respectively. For the middle of the interval $(i, j; i+1, j)$ we will apply Eq. (1.90). Substituting the derivative of ζ with the central difference relation, we have:

$$\zeta_{i+1,j} - \zeta_{i,j} = \delta x f_{i+1/2,j} \,. \tag{2.1}$$

In an analogous way with the help of Eq. (1.91) for the middle of the interval

$(i, j; i, j+1)$ we obtain:

$$\zeta_{i,j+1} - \zeta_{i,j} = \delta y f^1_{i,j+1/2} \; . \tag{2.2}$$

If we write the equations of type (2.1), (2.2) for all parts of the closed contour in this way, we will obtain the system of n equations with n unknown values $\zeta_{i,j}$. But the determinant of this system is equal to zero, so it cannot be resolved if the right-hand side is arbitrary. For simplicity, let us demonstrate the situation on a very simple example, when the contour is a rectangle and the number of the calculating points is equal to four. Denoting the right-hand sides of the equations of type (2.1), (2.2) through $b_k (k = 1, \dots, 4)$, we will obtain:

$$\zeta_{0,1} - \zeta_{0,0} = b_1 \; ; \qquad \zeta_{1,1} - \zeta_{0,1} = b_2 \; ;$$

$$\zeta_{1,0} - \zeta_{1,1} = b_3 \; ; \qquad \zeta_{0,0} - \zeta_{1,0} = b_4 \; . \tag{2.3}$$

Adding together all four equations of system (2.3) we will find the solvability condition:

$$\sum_{k=1}^{4} b_k = 0 \; , \tag{2.4}$$

at the fulfillment of which the unknown values $\zeta_{i,j}$ are determined with the accuracy of the additive constant. We assume that one of the values ζ is known and is equal to zero. Then, neglecting one of the equations of system (2.3), containing the specified value $\zeta_{i,j}$ at a chosen point, we will define the left values from the other equations.

So, in order to solve the equations system (2.1), (2.2), first it is necessary to check whether the condition of type (2.3) is fulfilled, and then we can assume that one of the unknown values $\zeta_{i,j}$ is equal to zero and can neglect one of the equations, containing this value. The system, obtained afterwards from $n - 1$ equations with $n - 1$ unknown quantities, has a unique solution. Now we have only to discuss the possibility of fulfilling condition (2.4).

It was shown in Chapter 1 that with some simplications, the equations for the SST at the basin boundary (as well as inside) have an exact solution [see Eq. (1.103)]. In this case the condition of type (2.4) is naturally fulfilled. If we neglect on the right-hand sides of Eqs. (1.90) and (1.91) the terms containing the derivatives of ζ, or take them into account by successive approximations, the condition of type (2.4) will be violated. Generally when investigating large-scale currents of a deep baroclinic ocean, when the depth at the contour is not less that 100 m, the formed residual, as the calculations have shown, is very small. But when the basin is not very deep, or when the baroclinicity of the sea water is not taken into account, the residual may become noticeable. In this case one should correct the right-hand sides of the equations of type (2.1), (2.3) so as to fulfil conditions (2.4). This can be done in the following way: to calculate the arithmetic mean value of the right-hand sides of the equations of type (2.1), (2.2), and then to subtract it from each of them. The resulting right-hand sides will satisfy condition (2.4).

Let us consider systems (2.1), (2.2) once more. Adding together these equations, we obtain:

$$\zeta_{i+1,j} - 2\zeta_{i,j} + \zeta_{i,j+1} = \delta y f^1_{i,j+1/2} + \delta x f_{i+1/2,j} \; . \tag{2.5}$$

In a number of works the SST at the boundary is determined from the system of type (2.5). Having chosen any point as a reference point, we can solve system (2.5) by the factorization method (Marchuk 1972, 1973, 1975a, b, 1980b). The simple method of the solution is also the Gauss-Zheidel procedure. If, for instance, we go round the contour clockwise, then the iterative formula has the form:

$$\zeta_{i,j}^{n} = \tfrac{1}{2}(\zeta_{i+1,j}^{n} + \zeta_{i,j+1}^{n-1}) - \tfrac{1}{2}(\delta y f_{i,j+1/2}^{1} + \delta x f_{i+1/2,j}) \,. \tag{2.6}$$

The calculation continues until the difference between two successive approximations becomes less than the a priori specified ε. It is practically sufficient if $\varepsilon = 10^{-2} \div 10^{-3}$ cm. The process converges very quickly if we take ζ, calculated by formula (1.103) as the first approximation. The advantage of this method is in its stability to rounding off errors, but here it is also necessary to fulfil the condition of type (2.4), which guarantees the independence of the result from the choice of a reference point.

When applying the described procedure to Eqs. (1.90), (1.91) the components of the right-hand sides of these equations, containing ζ, are usually taken into account by the method of successive approximations. The calculations at the contour and inside the basin in this case can be fulfilled simultaneously. In the numerical modelling of large- or intermediate-scale currents (the grid step is more than 50 km), the terms of the right-hand sides of Eqs. (1.90), (1.91), containing ζ, are small as they possess the small factor $1/2\alpha H$.

The calculation of ζ at the boundary in the case of a non-linear model is carried out with the help of Eqs. (1.133), (1.134). The concrete calculations can be performed by the above described method, taking into account the terms, depending on velocities, by successive approximations. So, in both models the definition of ζ is reduced to the solution of the Dirichlet problem.

Presenting the above described algorithms, we have started with the assumption that all iterative processes converge. The convergency of the successive approximations method is established each time experimentally. Let us also note that at each part of the boundary only one of the correlations (1.133), (1.134) is used, then the second one, taking into account the lateral exchange in the model, can serve as a criterion of the accuracy of the obtained solution.

Let us mention one more method, suggested by Marchuk (1969) for solving the problem with a skew derivative. The essence of the method is the following. The differential equation of type (2.1) is used not only for the internal points of the basin, but also for the boundary points. Together with the conditions of types (1.90), (1.91) it allows one by the difference approximation to exclude fictitious points, lying outside the boundary and to obtain the closed system of equations for the determination of the values ζ at the contour and inside the basin.

For the calculations at large grid meshes (with a step of 1° and more) in the first approximation, the specification of $\zeta|_r = \zeta_d$ at the boundary is valid. This boundary condition is often used in diagnostic calculations and corresponds to the case of "liquid" boundaries. In this case the water fluxes across the boundaries are specified by the quasi-dynamic method and the direct wind effect and the secondary effect of the bottom relief are neglected. Let us note here that a distinct criterion for the determination of the boundary as a "liquid" one, i.e. with some non-zero water flux through it, or as a "solid" one, i.e. with zero water flux, does not exist. Principally this is connected with the fact that at the boundary, i.e. a vertical wall, the density is always specified and the possibility of specifying zero fluxes through it is determined, in fact, only by the relative height of the wall in comparison to the basin's characteristic depth. We believe that in the diagnostic calculation of currents with large grid steps, it is usually reasonable to assume the boundary as a "liquid" one. The experience of such calculations, having accumulated by now, proves it.

When solving the boundary problem for the SST with non-simplified boundary conditions, information on the density field is necessary, but not on density anomalies, because then only the derivatives of density by horizontal coordinates exist in all the correlations of the model. However, when specifying the SST at the boundary by the quasi-dynamic method or when solving the problems for the bottom pressure anomalies, it is also necessary to know the density anomaly. Usually, the density anomaly field at each level is calculated as a field of density deviations from its average value at this level.

2.2.2 Methods of Difference Approximation and Solutions of the Equations of Quasi-Geostrophic Models

For the numerical application of the model it is necessary to construct a stable difference scheme, approximating the initial differential problem. The resulting difference problem is a system of algebraic equations and can be solved by any direct or iterative method, chosen for reasons of economy, calculating stability, simplicity of programming and the experience of numerical calculations. These problems will be considered here in the application to the above described quasi-geostrophic model, based on the equation for the SST.

First, let us dwell on the problems of constructing a stable difference scheme. When solving problems by finite difference methods, all differential derivatives are substituted by difference derivatives (difference relations). Let us formulate here the expressions for difference relations, approximating corresponding differentials, which are included in the correlations of the model. Let us introduce them in the example of the derivatives by x for the case of irregular grid steps.

The first derivative can be approximated either by the directed difference relations (right or left) or by the central one. In the first case we have: when approximating by the right difference:

$$\frac{\partial f}{\partial x} \approx \frac{f_{i+1} - f_i}{\delta x_{i+1}}, \tag{2.7}$$

when approximating by the left difference:

$$\frac{\partial f}{\partial x} \approx \frac{f_i - f_{i-1}}{\delta x_i}, \tag{2.8}$$

where δx_i is an irregular step of the grid by x, f_i is a value of the grid function in the i point.

The directed difference derivatives have the first-order approximation. The approximation of the first derivatives by the central difference relations is more accurate. On a regular grid set they give the second-order approximation.

The central difference derivative has the form:

$$\frac{\partial f}{\partial x} \approx \frac{\delta x_i \dfrac{f_{i+1} - f_i}{\delta x_{i+1}} + \delta x_{i+1} \dfrac{f_i - f_{i-1}}{\delta x_i}}{\delta x_i + \delta x_{i+1}}$$

$$= \frac{\delta x_i}{\delta x_{i+1}(\delta x_{i+1} + \delta x_i)} f_{i+1} + \frac{\delta x_{i+1} - \delta x_i}{\delta x_i \delta x_{i+1}} f_i - \frac{\delta x_{i+1}}{\delta x_i(\delta x_{i+1} + \delta x_i)} f_{i-1} \,. \tag{2.9}$$

It is a half-sum of the left- and right-directed difference derivations with the weights inversely proportional to the corresponding grid steps. On a regular grid the intermediate term becomes zero, and the correlation becomes the very simple form:

$$\frac{\partial f}{\partial x} \approx \frac{f_{i+1} - f_{i-1}}{2\delta x} \,. \tag{2.10}$$

The second derivative can be approximated in the following way:

$$\frac{\partial^2 f}{\partial x^2} \approx \frac{\dfrac{f_{i+1} - f_i}{\delta x_{i+1}} - \dfrac{f_i - f_{i-1}}{\delta x_i}}{\dfrac{\delta x_i + \delta x_{i+1}}{2}}$$

$$= 2\left[\frac{f_{i+1}}{\delta x_{i+1}(\delta x_i + \delta x_{i+1})} - \frac{f_i}{\delta x_i \delta x_{i+1}} + \frac{f_{i-1}}{\delta x_i(\delta x_i + \delta x_{i+1})}\right] . \tag{2.11}$$

It is the difference between right- and left-directed difference derivatives, divided by the half-sum of the corresponding grid steps. On a regular grid, it takes the form:

$$\frac{\partial^2 f}{\partial x^2} \approx \frac{f_{i+1} - 2f_i + f_{i-1}}{\delta x^2} \,. \tag{2.12}$$

The written set of approximation correlations is quite sufficient for constructing simple difference schemes for our models. At the same time the difference formulas for all three components of the velocity are easily obtained by the simple substitution of differential derivations by the corresponding difference ones. To increase the order of approximation of the first derivatives they are substituted by central difference relations. Taking this into account, the difference formula for u, for example, on a regular grid will have the form:

$$u_{i,j,k} = \frac{e^{-\alpha_j z_k}}{2\alpha_j \rho_0 \nu} [(\tau_{xi,j} \pm \tau_{yi,j}) \cos \alpha_j z_k - (\tau_{xi,j} \mp \tau_{yi,j}) \sin \alpha_j z_k]$$

$$- \frac{g}{l}\left[\frac{\delta y_j}{\delta y_{j+1}(\delta y_j + \delta y_{j+1})} \zeta_{i,j+1} + \frac{\delta y_{j+1} - \delta y_j}{\delta y_j \delta y_{j+1}} \zeta_{i,j}\right.$$

$$\left. - \frac{\delta y_{j+1}}{\delta y_j(\delta y_j + \delta y_{j+1})} \zeta_{i,j-1}\right] - \delta_k \frac{g}{\rho_0 l_j} \Phi_{i,j,k} \,, \tag{2.13}$$

where i, j, k are the numbers of the grid points by x, y, z; δyj is an irregular grid step by y; z_k is the depth of some calculating horizon ($z_1 = 0$ corresponds to the ocean surface);

$$\delta_k = \begin{cases} 0 & \text{for } k = 1 \\ 1 & \text{for } k > 1 \, ; \end{cases}$$

$\Phi_{i,j,k}$ is an integral of the grid function, approximating the density integral in the formula for u.

The integral approximation can be carried out by a number of methods. In practice the trapezoid method is usually used. In this case for $\Phi_{i,j,k}$ we will obtain:

$$\Phi_{i,j,k} = \sum_{\eta=2}^{k} \left[\frac{(c_{j+1}^{(y)} \rho_{i,j+1,\eta} + c_j^{(y)} \rho_{i,j,\eta} - c_{j-1}^{(y)} \rho_{i,j-1,\eta})}{2} + \frac{(c_{j+1}^{(y)} \rho_{i,j+1,\eta-1} + c_j^{(y)} \rho_{i,j,\eta-1} - c_{j-1}^{(y)} \rho_{i,j-1,\eta-1})}{2} \right] (z_\eta - z_{\eta-1}) \, , \tag{2.14}$$

where k is the number of the calculating horizon; η is an integration variable (the variable number of the horizon); $C_{j+1}^{(y)}$, $C_j^{(y)}$, $C_{j-1}^{(y)}$ are the grid coefficients, located before the corresponding terms of the central difference derivative of the first order by y [from left to right in the correlations for $\partial\zeta/\partial y$ in Eq. (2.13)].

Likewise it is not difficult to obtain the difference formulas for v, w. The main problem of the numerical application of quasi-geostrophic models is in the solution of boundary problems for the integral function: the SST, total mass transport stream function or bottom pressure anomaly. The equations for them are the equations of the elliptic type with the small parameter at the highest derivative. The latter means that the terms with the highest derivative (in this case it is the derivative of the second order) play a relatively small role in the balance of forces. Physically, these terms of the equations under consideration express the effect of the bottom friction and are relatively important only if the grid steps are sufficiently small and only in the boundary layers. However, it is necessary to take into account these terms in the model, as it allows one to correctly construct the problem for a closed domain.

In the majority of diagnostic investigations the difference scheme of the boundary problem for an integral function can be constructed either by the method of directed differences or by the Ilyin method (Ilyin 1969). Other methods are seldom applied for the quasi-geostrophic and linear diagnostic models. For non-linear models one of the most well-known schemes is the Bryan scheme (Bryan 1969a). It is applied, for instance, for the diagnostic calculation of large-scale currents of the North Atlantic on a 1° grid in Holland and Hirschman (1972). The scheme satisfies the integral laws of conservation and has a second-order approximation. Its drawback is that for the stability of the numerical solution one has to take overestimated values of the coefficient of the horizontal turbulent exchange (in Holland and Hirschman 1972), for example, $\mu = 4 \times 10^8 \text{ cm}^2 \text{ s}^{-1}$ leading to the extreme smoothness of the solution.

The Ilyin scheme, as well as the Bryan scheme, has the second-order approximation (at least for the case of constant coefficients). Besides, it is simple enough. However, the numerous calculations, which have been carried out by now, showed that it leads practically to the same results as the scheme of directed differences, which has the first-order approximation. This can be explained by the fact that monotonous difference schemes (the scheme of directed differences and the Ilyin scheme also belong to this class), having the first as well as the second order of approximation, give similar results, if "difference Reynolds numbers" are much larger than 1 (Kozlov 1977).

For the equation under consideration these numbers have the form:

$$r^{(x)} = \delta x \alpha \left| \frac{H\beta}{l} - \frac{\partial H}{\partial y} \right| , \qquad r^{(y)} = \delta y \alpha \left| \frac{\partial H}{\partial x} \right| . \tag{2.15}$$

It is clear that their quantity depends essentially on a grid step and the degree of irregularity of the bottom relief. The calculations of these numbers for the North Atlantic on the grid with a 1° step, carried out in Sarkisyan et al. (1980), showed that they are relatively small (smaller than ten) only in a very limited ocean area with small gradients of the bottom relief. The estimations show that for real oceanic basins, difference Reynolds numbers, as a rule, are greater than 1 also on a half-degree grid.

Due to these reasons and also because the method of directed differences leads to simpler and at the same time very stable difference schemes, this very method is more frequently used in diagnostic problems. Besides, on its basis, if it is necessary, it is not difficult to also construct a scheme of the second-order of approximation, for example, by the methods presented in Marchuk (1980b); Kochergin (1978). We will dwell on one of these methods of increasing the approximation order of difference schemes later.

The largest number of numerical experiments are fulfilled with the help of the integal function ζ, therefore we turn to Eq. (1.92). We denote the right-hand side of this equation through Φ and write it in the form:

$$\frac{1}{2\alpha} \Delta\zeta + A^{(0)} \frac{\partial \zeta}{\partial x} + B^{(0)} \frac{\partial \zeta}{\partial y} = \Phi , \tag{2.16}$$

where

$$A^{(0)} = \frac{\beta H}{l} - \frac{\partial H}{\partial y} ; \qquad B^{(0)} = \frac{\partial H}{\partial x} . \tag{2.17}$$

As the equations for ζ and ψ are the equations with small parameters at the highest derivatives, it is not difficult to work out stable difference schemes of the second order of accuracy for them. The drawbacks of such schemes are discussed, for example, in Ilyin (1969).

a) The Method of Directed Differences. The method of directed differences is the simplest of the schemes with the first order of accuracy. Many works on the diagnostic investigations of currents were supported with the help of these very schemes, therefore we will begin our presentation with this method. Its essence is extremely simple: the derivatives of the first order are substituted by the differences, directed forward or backward, depending on the coefficients' signs, in such a way that the diagonal terms of the coefficients' matrix possess the maximal weights. For instance, let us substitute the derivative by x in Eq. (2.16) by the directed difference relation in the following way:

$$\delta x\left(\frac{\partial\zeta}{\partial x}\right)_{i,j} = \delta_1\zeta_{i+1,j} + (1-2\delta_1)\zeta_{i,j} + (\delta_1 - 1)\zeta_{i-1,j}\,, \tag{2.18}$$

where

$$\delta_1 = \begin{cases} 0 & \text{at } A^{(0)}_{i,j} < 0 \\ 1 & \text{at } A^{(0)}_{i,j} > 0\,, \end{cases} \tag{2.19}$$

i.e. at

$$A^{(0)}_{i,j} > 0$$

$$\delta x\left(\frac{\partial\zeta}{\partial x}\right)_{i,j} = \zeta_{i+1,j} - \zeta_{i,j}\,, \tag{2.20}$$

and at

$$A^{(0)}_{i,j} < 0$$

$$\delta x\left(\frac{\partial\zeta}{\partial x}\right)_{i,j} = \zeta_{i,j} - \zeta_{i-1,j}\,. \tag{2.21}$$

The same applies for:

$$\delta y\left(\frac{\partial\zeta}{\partial y}\right)_{i,j} = \delta_2\zeta_{i,j+1} + (1-2\delta_2)\zeta_{i,j} + (\delta_2 - 1)\zeta_{i,j-1}\,, \tag{2.22}$$

where

$$\delta_2 = \begin{cases} 0 & \text{at } B^{(0)}_{i,j} < 0 \\ 1 & \text{at } B^{(0)}_{i,j} > 0\,. \end{cases} \tag{2.23}$$

If we now write the finite-difference analogue of the sum $A^{(0)}\dfrac{\partial\zeta}{\partial x} + B^{(0)}\dfrac{\partial\zeta}{\partial y}$, then in this sum $\zeta_{i,j}$ will have the coefficient $[|A^{(0)}_{i,j}| + |B^{(0)}_{i,j}|]$, i.e. regardless of the signs of coefficients $A^{(0)}$ and $B^{(0)}$, the diagonal predominance will take place in the obtained system of algebraic equations. The Laplacian operator is substituted by central difference relations in the usual way. After the indicated transformations we will obtain the following difference approximation of Eq. (2.16) on the grid G^h with the steps δx, δy:

$$\frac{1}{2\alpha_j}\left\{\frac{\zeta_{i-1,j} + \zeta_{i+1,j} - 2\zeta_{i,j}}{(\delta x)^2} + \frac{\zeta_{i,j-1} + \zeta_{i,j+1} - 2\zeta_{i,j}}{(\delta y)^2}\right\}$$
$$+ A^{(0)}_{i,j}\frac{\delta_1\zeta_{i+1,j} + (1-2\delta_1)\zeta_{i,j} + (\delta_1 - 1)\zeta_{i-1,j}}{\delta x\delta y}$$
$$+ B^{(0)}_{i,j}\frac{\delta_2\zeta_{i,j+1} + (1-2\delta_2)\zeta_{i,j} + (\delta_2 - 1)\zeta_{i,j-1}}{\delta y\delta x} = \Phi_{i,j}\,. \tag{2.24}$$

If, at last, we substitute $A^{(0)}$, $B^{(0)}$ by the central difference relations and resolve the obtained equation relative to $\zeta_{i,j}$, we will obtain the formula, with the help of which we can calculate $\zeta_{i,j}$ by the method of successive approximations. For simplicity, this formula is given for the case, when the grid steps are identical in both directions ($\delta x = \delta y$):

$$\zeta_{i,j} = \frac{1}{c_{i,j}}\left\{\left[\frac{1}{2\alpha_j} + A^{(0)}_{i,j}\delta_1\right]\zeta_{i+1,j} + \left[\frac{1}{2\alpha_j} + (\delta_1 - 1)A^{(0)}_{i,j}\right]\zeta_{i-1,j}\right.$$
$$\left. + \left[\frac{1}{2\alpha_j} + B^{(0)}_{i,j}\delta_2\right]\zeta_{i,j+1} + \left[\frac{1}{2\alpha_j} + (\delta_2 - 1)B^{(0)}_{i,j}\right]\zeta_{i,j-1} - \frac{(\delta x)^2}{c_{ij}}\Phi_{i,j}\right., \quad (2.25)$$

where

$$A^{(0)}_{i,j} = \frac{\delta x \beta j H_{i,j}}{lj} - (H_{i,j+1} - H_{i,j-1}),$$

$$B^{(0)}_{i,j} = H_{i+1,j} - H_{i-1,j}. \quad (2.26)$$

The obtained system of algebraic equations can be solved by either a direct or an iterative method. The Gauss-Zheidel method and relaxation methods are used most frequently. The required accuracy of the iterative process is determined experimentally, for example, by the comparison of the results of the calculation, having two different initial approximation, and/or by the analysis of the behaviour of the maximal residual (maximal for all the grid points of the absolute difference of the function values at two neighbouring iterations), the tentative determination of the process, reaching the steady regime, and further by controlling the calculation on a number of iterations. The parameters of the iterative process convergency (the number of iterations and the maximal residual) depend essentially on the grid step, the number of the points of the grid domain, the smoothness of the initial data, etc. Usually, the number of iterations, necessary for obtaining the spin-up solution of the problem for ζ, extends from a few tens to a few hundred in a quasi-geostrophic model and the maximal residual is from 0.1 cm to 0.001 cm.

The main drawback of the method of directed differences is the numerical viscosity, which is characteristic of it. We will dwell on the question of how it influences the solution when we discuss the results of calculations.

b) The Ilyin Difference Scheme. Ilyin (1969) has offered another method of the difference approximation of such equations. The author's aim was to work out the scheme of the second order of accuracy, in such a way, that it will be exact for the equation with constant coefficients.

The Ilyin method can be applied to the equation of form (2.16) with variable coefficients $A^{(0)}$ and $B^{(0)}$. But this scheme has the second order of accuracy only at very small grid step values by the horizontal and at small values of the coefficients $A^{(0)}$ and $B^{(0)}$. With the values $A^{(0)}$ and $B^{(0)}$, corresponding to the real bottom relief, and at a grid size practical for the large-scale circulation, the Ilyin scheme possesses the numerical viscosity, comparable with the viscosity of the method of directed differences (Sarkisyan 1977a, b). Not dwelling on simple transformations, we will present the approximation of Eq. (2.16) by the Ilyin method:

$$A_{i,j}^{(0)} \operatorname{cth}(\alpha_{i,j} A_{i,j}^{(0)} \delta x)(\zeta_{i+1,j} + \zeta_{i-1,j} - 2\zeta_{i,j})$$
$$+ B_{i,j}^{(0)} \operatorname{cth}(\alpha_{i,j} B_{i,j}^{(0)} \delta y)(\zeta_{i,j+1} + \zeta_{i,j-1} - 2\zeta_{i,j})$$
$$+ A_{i,j}^{(0)}(\zeta_{i+1,j} - \zeta_{i-1,j}) + B_{i,j}^{(0)}(\zeta_{i,j+1} - \zeta_{i,j-1}) = 2\Phi_{i,j}\delta x \,. \tag{2.27}$$

In order to obtain an iterative formula from Eq. (2.27) it is necessary to resolve this equation relative to $\zeta_{i,j}$. However, one should keep in mind that unlike the method of directed differences, while using formula (2.27) practically, it is necessary to separate the extreme cases, when the arguments of hyperbolic cotangents are too small ($<10^{-3}$) or too large (>10). In the first case one should multiply and divide the terms with cotangents by α_j and make use of the asymptotic value $\lim_{x\to 0} x \operatorname{cth} x = \mp 1$.

Now we will dwell briefly on the obtained difference equations. Correlation (2.25) is already a calculating formula for the calculation of $\zeta_{i,j}$ by the method of successive approximations. The Gauss-Zheidel procedure is very convenient for this purpose. For instance, when the count begins with the upper right-point of the domain and finishes in the lower left one, the iterative formula has the form:

$$\zeta_{i,j}^{(n+1)} = \frac{1}{c_{i,j}} \left\{ \left[\frac{1}{2\alpha_j} + A_{i,j}^{(0)} \delta_1 \right] \zeta_{i+1,j}^{(n+1)} + \left[\frac{1}{2\alpha_j} + (\delta_1 - 1) A_{i,j}^{(0)} \right] \zeta_{i,j}^{(n)} \right.$$
$$\left. + \left[\frac{1}{2\alpha_j} + B_{i,j}^{(0)} \delta_2 \right] \zeta_{i,j+1}^{(n+1)} + \left[\frac{1}{2\alpha_j} + (\delta_2 - 1) B_{i,j}^{(0)} \right] \zeta_{i,j-1}^{(n)} \right\} - \frac{(\delta x)^2}{c_{i,j}} \Phi_{i,j} \,, \tag{2.28}$$

i.e. as soon as $(n+1)$-th approximation is calculated in the given point, it substitutes the n-th approximation of the unknown function, which is in the computer's memory. At the domain boundary the values of ζ should be calculated beforehand according to the above method. Further, after the analysis of the results of concrete diagnostic calculations, we will present additional methodological instructions on the choice of the first approximation, the choice of the most suitable integral function, etc. After resolving Eq. (2.27) relative to $\zeta_{i,j}$ we also obtain the formula of type (2.25), on the basis of which the calculations can be carried out by the Gauss-Zheidel method.

c) *The Methods of Increasing the Approximation Order of Difference Schemes.* Both presented approximation methods of Eq. (1.92) possess numerical viscosity. Thus, the necessity arises to increase the approximation order and to reduce the numerical viscosity to a minimum. The general method of increasing the approximation of difference schemes is presented in the works of Marchuk (1975a, 1980b). In Kochergin (1978) the application of this method to the equation of type (2.16) is given. The method, very simple and convenient for calculation on the computer, is based on using the solutions of this equation, obtained by the schemes of directed differences, and is applied consequently on difference grids.

Let us assume that we have the difference grid C^h with the steps δx by the x-axis and δy by the y-axis. We approximate Eq. (1.92) at the grid C^h by the method of directed differences. Then we will denote the solution of the obtained difference Eq. (2.24) by ζ^h. Now we will consider another difference scheme C^{2h} with the steps $2\delta x$ and $2\delta y$ all the points of which belong to the grid C^h. The solution of the problem with the grid C^{2h} will be denoted by ζ^{2h}. Now we will construct for the grid C^{2h} the corrector:

$$\bar{\zeta} = 2\zeta^h - \zeta^{2h} . \tag{2.29}$$

It is easy to show that $\bar{\zeta}$ solves Eq. (1.92) with second-order accuracy. For this purpose we will expand $\zeta^h_{i\pm1,j}, \zeta^h_{i,j\pm1}, \zeta^{2h}_{i,j\pm1}, \zeta^{2h}_{i\pm1,j}$ by the Taylor series in the vicinity of the point $(2\delta x \cdot i, 2\delta y \cdot j)$. We will substitute expansions $\zeta^h_{i\pm1,j}, \zeta^h_{i,j\pm1}$ in Eq. (2.24) and carry out the analogous operation for the difference approximation with the grid C^{2h}. Now, taking into account formula (2.29), we will obtain for the corrector $\bar{\zeta}$:

$$\frac{1}{2\alpha} \Delta\bar{\zeta} + A^{(0)} \frac{\partial \bar{\zeta}}{\partial x} + B^{(0)} \frac{\partial \bar{\zeta}}{\partial y} - A^{(0)} \frac{(\delta x)^2}{3} \frac{\partial^3 \bar{\zeta}}{\partial x^3} - B^{(0)} \frac{(\delta y)^2}{3} \frac{\partial^3 \bar{\zeta}}{\partial y^3} + \cdots = \Phi . \tag{2.30}$$

Thus, on the basis of the method of directed differences with the help of the simple corrector (2.29) we can solve Eq. (1.92) with second-order accuracy with the grid C^{2h}. The obtained solution can also be interpolated on the grid C^h, for example, linearly or with the help of spline polynomials. Kochergin and Sherbakov (Kochergin 1978) noted that the corrector can be constructed in the same way, which gives the third-order accuracy. But in practice it results only in a rather slight increase of the solution's accuracy. For instance, the addition, obtained due to the third-order corrector, is on average four times less than the addition given by the second-order corrector. The accuracy of the solution of a concrete problem in the ocean, obtained by the method of directed differences with the grid step width 2.5°, is enchanced by the second-order corrector only by 10% (Kochergin (1978). The presented method is rather complicated. Besides, the interpolation results in the additional smoothing of the solution. In Kochergin (1978) one more method of increasing the approximation is given, which is considerably simpler. The essence of the method can be shown on the example of the difference approximation of the term $A^{(0)} \frac{\partial \zeta}{\partial x}$, when $A^{(0)} > 0$. Let us expand $\zeta_{i+1,j}$ by the Taylor series expansion and conserve the first three terms of the series:

$$\zeta_{i+1,j} \approx \zeta_{i,j} + \delta x \left(\frac{\partial \zeta}{\partial x}\right)_{i,j} + \frac{(\delta x)^2}{2} \left(\frac{\partial^2 \zeta}{\partial x^2}\right)_{i,j} , \tag{2.31}$$

hence,

$$\delta x \left(\frac{\partial \zeta}{\partial x}\right)_{i,j} = \zeta_{i+1,j} - \zeta_{i,j} - \frac{(\delta x)^2}{2} \left(\frac{\partial^2 \zeta}{\partial x^2}\right)_{i,j} . \tag{2.32}$$

Analogously, the term $B^{(0)} \frac{\partial \zeta}{\partial y}$ of Eq. (1.92) can be transformed. Now we have to repeat the procedure of directed differences, to transfer the second derivatives of the expressions of type (2.31) to the right-hand side of Eq. (2.24), to substitute these derivatives by the central difference relations and calculate them by the method of successive approximations. In particular, in the case of $\delta x = \delta y$ on the right-hand side of Eq. (2.25) an additional term appears:

$$-\frac{1}{2c_{i,j}}[A^{(0)}_{i,j}(\zeta_{i+2,j} - 2\zeta_{i,j} + \zeta_{i-2,j}) + B^{(0)}_{i,j}(\zeta_{i,j+2} - 2\zeta_{i,j} + \zeta_{i,j-2})] \,. \tag{2.33}$$

The scheme of the successive approximations is the following. First, the calculations according to formula (2.25) are carried out, not taking into account expression (2.33). As a result, the solution of Eq. (1.92) with the first-order approximation can be found. The determined value ζ is substituted into expression (2.33), which is added to the right-hand side of Eq. (2.25), and then the second approximation of ζ is calculated. Expression (2.33) is a numerical viscosity of the method of directed differences. Calculating this error, we not only enhance the approximation order, but also determine the error of the method of directed differences. For example, with the help of Eq. (2.33) we can determine the value of the optimal horizontal grid step, which makes the numerical viscosity less than the physical one.

The application of imbedded and shifted grids is very important also in the calculation of currents velocities, especially in the regions of narrow meandering streams. The shifted grid, for example, is used when calculating the World Ocean currents (Chap. 5). In applying the velocity components to the calculation according to the formulas of a quasi-geostrophic model, the essence of this approach lies in the fact that the unknown functions are calculated in the centres of each grid cell by the specified or calculated beforehand in the four closest grid points SST values, tangential wind stresses and integrals of density. This procedure enhances the accuracy of the difference expressions of the SST derivatives and density (which are approximated by some average for a grid cell central difference relation, i.e. central relative to the centre of the cell). As a result, it enhances the accuracy of the velocity calculation.

The considered schemes are valid for the calculation of the SST as well as for the calculation of the bottom pressure anomaly ζ'. When solving the problem for ζ' all the derivatives of the right-hand side, except the derivatives of density, are also approximated by central difference relations. The derivatives of density are also approximated by central difference relations when the density is specified in all three points of the difference grid. If the density is specified only in two points, then the corresponding directed difference is taken. Such a situation is usually found near the boundaries and close to the basin bottom with an irregular relief. Besides, it can be artificially produced because of the rather inaccurate adjustment of the density fields and bottom relief, when even the calculating point (which is inside the domain) has neither on the right or on the left (or sometimes both sides) neighbouring points on the basin bottom, nor is the density specified there. This situation was discussed in Section 2.1.

While approximating the right-hand side of the equation for the SST all the derivatives, except for the density ones, are also approximated by the central differences. The derivatives of density can be approximated in the same way, as in the case with ζ' . However, in order to enhance the accuracy of the numerical solution, it is reasonable to substitute the first derivatives of density by the differences, directed in the same way as the corresponding derivatives of the SST (Sarkisyan 1977a, b). It allows one to obtain rather accurately the main part of the solution $\zeta d = -\frac{1}{\rho_0}\int_0^H \rho dz$ in the implicit form without the preliminary division of the solution into the sum ζd and ζ'. But for this purpose it is necessary that the density in two corresponding points be specified. If this condition is not fulfilled, the difference is to be taken in the direction, where the density is specified, but this, of course, increases the error of the solution.

2.2.3 The Methods of Difference Approximation and Solution of the Equations of a Non-Linear Model

Before constructing difference schemes of a non-linear model, we will make some preliminary difference transformations. In the equations of these models the derivatives of the first and second order by the vertical are present. The simple substitution of these derivatives by difference relations can result in essential errors in surface and bottom boundary layers. The transformations performed below are intended to decrease these errors.

Let us denote the components of the current velocity by the x-axis on the ocean surface by u_0. The corresponding value of this function at the horizon $z = h_1$ closest to the surface, we will denote by u_1. We will expand u_1 by the Taylor series, and limit ourselves to three terms of the series and take into account the boundary condition (1.9):

$$\begin{aligned} u_1 &= u_0 + h_1\left(\frac{\partial u}{\partial z}\right)_{z=0} + \frac{y_1^2}{2}\left(\frac{\partial^2 u}{\partial z^2}\right)_{z=0} \\ &= u_0 + h_1\left(\frac{\partial u}{\partial z}\right)_{z=0} + \frac{h_1^2}{2\nu}\left(\frac{\partial}{\partial z}\nu\frac{\partial u}{\partial z}\right)_{z=0} - \frac{h_1^2}{2\nu}\left(\frac{\partial u}{\partial z}\frac{\partial \nu}{\partial z}\right)_{z=0} . \end{aligned} \tag{2.34}$$

Hence,

$$\left(\frac{\partial}{\partial z}\nu\frac{\partial u}{\partial z}\right)_{z=0} = \frac{2\nu_0}{h_1^2}(u_1 - u_0) + \frac{2}{h_1\rho_0}\tau_x - \frac{\tau_x}{\rho_0\nu}\left(\frac{\partial \nu}{\partial z}\right)_{z=0} . \tag{2.35}$$

Likewise we obtain the correlation for $\frac{\partial}{\partial z}\nu\frac{\partial v}{\partial x}$. We transform the term $\frac{\operatorname{div}\tau_x^{(H)}}{\rho_0 H}$ of the equation for the SST, which expresses the bottom friction effect. In the case $\nu =$ Const using the approximation of $\left.\frac{\partial u}{\partial z}\right|_H, \left.\frac{\partial v}{\partial z}\right|_H$ by directed differences and the non-slip condition, we will obtain:

$$\frac{\operatorname{div}\tau_x^{(H)}}{\rho_0 H} \sim \frac{\nu_H}{H}\left(\frac{\partial u_m}{\partial x} + \frac{\partial v_m}{\partial y}\right) = -\frac{\nu_H}{H}\frac{\partial w_m}{\partial z} , \tag{2.36}$$

where $m(x, y)$ is the number of the horizon closest to the bottom. In the case $\nu \neq$ Const and the approximation of the first derivatives by z according to the central differences, the correlation becomes more complicated, but in essence, i.e. the use of the continuity equation, it remains the same.

All the calculations according to this model are based on linear difference schemes. In Eq. (1.1) we will substitute the derivative by time according to the difference relation, directed forward, assuming $\mu = \text{Const}$ and rewriting it in the form:

$$\mu \Delta u^{(n+1)} + \frac{\partial}{\partial z} v \frac{\partial u^{(n+1)}}{\partial z} - u^{(n)} \frac{\partial u^{(n+1)}}{\partial x} - v^{(n)} \frac{\partial u^{(n+1)}}{\partial y}$$

$$- w^{(n)} \frac{\partial u^{(n+1)}}{\partial z} - \frac{u^{(n+1)}}{\delta t} = -\frac{u^{(n)}}{\delta t} - l v^{(n)} + \frac{1}{\rho_0} \frac{\partial p^{(n)}}{\partial x} = f^{(n)} . \tag{2.37}$$

Equation (2.37) will serve for the determination of $u^{(n+1)}$, the current velocity in the time moment $(n+1)\delta t$. We referred the coefficients of this equation to the previous moment of time $n\delta t$ and, so, linearized it. As a result, the equation has taken the same form as Eq. (1.92). The only difference is that in Eq. (2.37) we deal with the function of three independent variables.

The solution methods of Eq. (1.92) presented previously, can be generalized for a three-dimensional equation and applied to the solution of Eq. (2.37). First, we will present the difference approximation of the model equations for the constant μ, μ_1 on the example of an implicit linear scheme of directed differences. Let us formulate the difference analogues by time of the equations of motion, equations for the SST, temperature and salinity[2]:

$$-\frac{u^{(n+1)} - u^{(n)}}{\delta t} + \mu \Delta u^{(n+1)} + \frac{\partial}{\partial z} v \frac{\partial u^{(n+1)}}{\partial z} - u^{(n)} \frac{\partial u^{(n+1)}}{\partial x} - v^{(n)} \frac{\partial u^{(n+1)}}{\partial y} - w^{(n)} \frac{\partial u^{(n+1)}}{\partial z}$$

$$= g \frac{\partial \zeta}{\partial x} + \frac{g}{\rho_0} \int_0^z \frac{\partial \rho^{(n)}}{\partial x} d\xi - l v^{(n)} ; \tag{2.38}$$

$$-\frac{v^{(n+1)} - v^{(n)}}{\delta t} + \mu \Delta v^{(n+1)} + \frac{\partial}{\partial z} v \frac{\partial v^{(n+1)}}{\partial z} - u^{(n)} \frac{\partial v^{(n+1)}}{\partial x} - v^{(n)} \frac{\partial v^{(n+1)}}{\partial y} - w^{(n)} \frac{\partial v^{(n+1)}}{\partial z}$$

$$= g \frac{\partial \zeta}{\partial y} + \frac{g}{\rho_0} \int_0^z \frac{\partial \rho^{(n)}}{\partial y} d\xi + l u^{(n)} ; \tag{2.39}$$

$$\Delta \zeta^{(n+1)} + \frac{1}{H} \frac{\partial H}{\partial x} \frac{\partial \zeta^{(n+1)}}{\partial x} + \frac{1}{H} \frac{\partial H}{\partial y} \frac{\partial \zeta^{(n+1)}}{\partial y} = F_\zeta^{(n)} ; \tag{2.40}$$

$$-\frac{T^{(n+1)} - T^{(n)}}{\delta t} + \mu_1 \Delta T^{(n+1)} + \frac{\partial}{\partial z} v_1 \frac{\partial T^{(n+1)}}{\partial z} - u^{(n)} \frac{\partial T^{(n+1)}}{\partial x}$$

$$- v^{(n)} \frac{\partial T^{(n+1)}}{\partial y} - w^{(n)} \frac{\partial T^{(n+1)}}{\partial z} = 0 ; \tag{2.41}$$

$$-\frac{S^{(n+1)} - S^{(n)}}{\delta t} + \mu_1 \Delta S^{(n+1)} + \frac{\partial}{\partial z} v_1 \frac{\partial S^{(n+1)}}{\partial z} - u^{(n)} \frac{\partial S^{(n+1)}}{\partial x}$$

$$- v^{(n)} \frac{\partial S^{(n+1)}}{\partial y} - w^{(n)} \frac{\partial S^{(n+1)}}{\partial z} = 0 . \tag{2.42}$$

Let us transform the equations of the system into a general form. It will allow one to use for their solution the same difference algorithm, and by calculation with

[2] In high latitudes the semi-implicit representation of the Coriolis force of the type $\frac{1}{2} l (u^{(n+1)} + u^{(n)})$ is reasonable.

a computer, the same subprogram. Taking into account the fact that some equations, containing derivatives by the vertical, change their form at $z = 0$, and denoting $u^{(n+1)}$, $v^{(n+1)}$, $T^{(n+1)}$, $\zeta^{(n+1)}$, $S^{(n+1)}$ by $\varphi^{(n+1)}$, and the right-hand sides of the equations $F_u^{(z)}, F_v^{(z)}, F_T^{(z)}, F_S^{(z)}, F_u^{(0)}, F_v^{(0)}, F_T^{(0)}, F_S^{(0)}, F_\zeta$ by $\Phi^{(n)}$ ($F_f^{(0)}$ corresponds to $z = 0$, $F_f^{(z)} - z > 0$) we will obtain:

$$-a_1 \frac{\varphi^{(n+1)} - \varphi^{(n)}}{\delta t} + a_2 \Delta \varphi^{(n+1)} + a_3 \frac{\partial}{\partial z} v \frac{\partial \varphi^{(n+1)}}{\partial z} + a_4 \frac{\partial \varphi^{(n+1)}}{\partial x}$$
$$+ a_5 \frac{\partial \varphi^{(n+1)}}{\partial y} + a_6 \frac{\partial \varphi^{(n+1)}}{\partial z} = \Phi^{(n)} . \tag{2.43}$$

Here

$$a_1, a_2, a_3, a_4, a_5, a_6 = \begin{cases} 1, \mu, \delta_0, -u^{(n)}, -v^{(n)}, -\delta_0 w^{(n)} & \text{for } u, v ; \\ 0, 1, 0, \dfrac{1}{H}\dfrac{\partial u}{\partial x}, \dfrac{1}{H}\dfrac{\partial H}{\partial y}, 0 & \text{for } \zeta ; \\ 1, \mu_1, \delta_0, -u^{(n)}, -v^{(n)}, -\delta_0 w^{(n)} & \text{for } T, S , \end{cases}$$

$$\delta_0 = \begin{cases} 1 , & z \neq 0 ; \\ 0 , & z = 0 , \end{cases}$$

$$a_1 = \begin{cases} 1 & \text{for } u, v, T, S ; \\ 0 & \text{for } \zeta , \end{cases} \qquad a_3 = \begin{cases} \delta_0 & \text{for } u, v, T, S ; \\ 0 & \text{for } \zeta . \end{cases}$$

In order to approximate Eq. (2.43) we will use the method of directed differences, which allows one to construct a stable difference scheme rather easily, guaranteeing the fulfillment of the diagonal predominance. On an irregular grid by x, y, z we will obtain:

$$-a_1 \frac{\varphi^{(n+1)} - \varphi^{(n)}}{\delta t} + a_2 (cxx_{1,i} \varphi_{i+1,j,k} - cxx_{2,i} \varphi_{i,j,k} + cxx_{3,i} \varphi_{i-1,j,k})^{(n+1)}$$
$$+ a_2 (cyy_{1,j} \varphi_{i,j+1,k} - cyy_{2,j} \varphi_{i,j,k} + cyy_{3,j} \varphi_{i,j-1,k})^{(n+1)}$$
$$+ a_3 (czz_{1,k} \varphi_{i,j,k+1} - czz_{2,k} \varphi_{i,j,k} + czz_{3,k} \varphi_{i,j,k-1})^{(n+1)}$$
$$+ a_4 \begin{Bmatrix} (\varphi_{i+1,j,k} - \varphi_{i,j,k})^{(n+1)} / \delta x_{i+1} \\ (\varphi_{i,j,k} - \varphi_{i-1,j,k})^{(n+1)} / \delta x_i \end{Bmatrix} \begin{matrix} k_x \\ (1 - k_x) \end{matrix}$$
$$+ a_5 \begin{Bmatrix} (\varphi_{i,j+1,k} - \varphi_{i,j,k})^{(n+1)} / \delta y_{j+1} \\ (\varphi_{i,j,k} - \varphi_{i,j-1,k})^{(n+1)} / \delta y_j \end{Bmatrix} \begin{matrix} k_y \\ (1 - k_y) \end{matrix}$$
$$+ a_6 \begin{Bmatrix} (\varphi_{i,j,k+1} - \varphi_{i,j,k})^{(n+1)} / \delta z_{k+1} \\ (\varphi_{i,j,k} - \varphi_{i,j,k-1})^{(n+1)} / \delta z_k \end{Bmatrix} \begin{matrix} k_z \\ (1 - k_z) \end{matrix} = \Phi^{(n)}_{i,j,k} . \tag{2.44}$$

Here

$$k_x, k_y, k_z = \begin{cases} 0 , & \text{if } a_4, a_5, a_6 < 0 , \\ 1 , & \text{if } a_4, a_5, a_6 > 0 ; \end{cases}$$

$$cxx_{1,i} = \frac{2}{\delta x_{i+1}(\delta x_{i+1} + \delta x_i)} ; \quad cCxx_{2,i} = \frac{2}{\delta x_i \delta x_{i+1}} ;$$

$$cxx_{3,i} = \frac{2}{\delta x_i(\delta x_i + \delta x_{i+1})} ;$$

$$cyy_{1,j} = \frac{2}{\delta y_{j+1}(\delta y_{j+1} + \delta y_j)}; \quad cyy_{2,j} = \frac{2}{\delta yy_j \delta y_{j+1}};$$

$$cyy_{3,j} = \frac{2}{\delta y_j(\delta y_j + \delta y_{j+1})};$$

$$czz_{1,k} = \frac{2}{\delta z_{k+1}(\delta z_{k+1} + \delta z_k)}; \quad czz_{2,k} = \frac{2}{\delta z_k \delta z_{k+1}};$$

$$czz_{3,k} = \frac{2}{\delta z_k(\delta z_k + \delta z_{k+1})}.$$

The obtained system of algebraic equations is solved at each step by time either by the Gauss-Zheidel iterative method or by the method of overrelaxation.

The vertical component of the current velocity can be determined in difference ways. The simplest one is the calculation of w directly from the integrated continuity equation (1.97) with the use of the "rigid lid" condition. Such a calculation of the vertical velocity possesses, however, the following drawback: according to the numerical performance of Eq. (1.97) the solution will have an error, which accumulates at the basin bottom. In order to improve the certainty of calculations when using one boundary condition for w and to decrease the accumulation of the error when using the limited computer accuracy, we apply both boundary conditions by z for w. For this purpose we will differentiate the continuity equation by z:

$$\frac{\partial^2 w}{\partial z^2} = -\frac{\partial^2 u}{\partial x\, \partial z} - \frac{\partial^2 v}{\partial y\, \partial z}. \tag{2.45}$$

The vertical velocities, obtained from Eq. (2.45) and the continuity equation, will be identical, if in the numerical model the integral continuity equation is fulfilled (i.e. the integral law of mass conservation on the grid). We will discuss the calculation technique of w in detail in Section 5.7.

Let us formulate the difference analogue (2.45), taking into account the fact that the steps by the vertical are irregular:

$$\frac{2}{\delta z_{k+1}(\delta z_k + \delta z_{k+1})} w_{i,j,k+1} - \frac{2}{\delta z_k \delta z_{k+1}} w_{i,j,k} + \frac{2}{\delta z_k(\delta z_k + \delta z_{k+1})} w_{i,j,k-1} = F_{i,j,k}, \tag{2.46}$$

where $\delta z_k = z_k - z_{u-1}$. Denoting $\Delta x_i = \Delta y_j = h$, we will obtain the expression for $F_{i,j,k}$:

$$\begin{aligned} F_{i,j,k} = &-\Bigg\{\frac{1}{(\delta z_k + \delta z_{k+1})}\left[\frac{\delta z_k}{\delta z_{k+1}}(u_{i,j+1,k+1} - u_{i,j+1,k}) - \frac{\delta z_{k+1}}{\delta z_k}(u_{i,j+1,j+1} - u_{i,k+1,k})\right] \\ &- \frac{1}{(\delta z_k + \delta z_{k+1})}\left[\frac{\delta z_k}{\delta z_{k+1}}(u_{i,j-1,k+1} - u_{i,j-1,k}) - \frac{\delta z_{k+1}}{\delta z_k}(u_{i,j-1,k} - u_{i,j-1,k-1})\right]\Bigg\}\frac{1}{2h} \\ &- \Bigg\{\frac{1}{(\delta z_k + \delta z_{k+1})}\left[\frac{\delta z_k}{\delta z_{k+1}}(v_{i+1,j,k+1} - vu_{i+1,j,k}) - \frac{\delta z_{k+1}}{\delta z_k}(v_{i+1,j,k-1} - v_{i+1,j,k})\right] \\ &- \frac{1}{(\delta z_k + \delta z_{k+1})}\left[\frac{\delta z_k}{\delta z_{k+1}}(v_{i-1,j,k+1} - v_{i-1,j,k}) - \frac{\delta z_{k+1}}{\delta z_k}(v_{i-1,j,k-1} - v_{i-1,j,k})\right]\Bigg\}\frac{1}{2h}. \end{aligned} \tag{2.47}$$

Equation (2.46) can be solved efficiently by the factorization method. The Gauss-Zheidel method is even simpler, but the computer time, spent on the solution of Eq. (2.46), increases considerably.

The succession in solving the model equations system is the following: using some initial fields for $u^{(0)}$, $v^{(0)}$, $T^{(0)}$, $w^{(0)}$, $S^{(0)}$, $\zeta^{(0)}$ and specified τ_x, τ_y, we find $\rho^{(0)}$ and then $\zeta^{(1)}$ by making several internal iterations (until the specified accuracy is obtained); then with the help of $\zeta^{(1)}$, $u^{(0)}$, $v^{(0)}$, $w^{(0)}$, $\rho^{(0)}$, and making also several iterations, we determine $u^{(1)}$, $v^{(1)}$ at each horizon. After $u^{(1)}$ and $v^{(1)}$ are found for all the horizons, we calculate $w^{(1)}$; then $T^{(1)}$ and $S^{(1)}$. Afterward we also find $u^{(2)}$, $v^{(2)}$, $w^{(2)}$, $\zeta^{(2)}$, $T^{(2)}$, $S^{(2)}$, etc., moving by time steps until reaching the steady state of the process.

The constructed, implicit, linear scheme possesses a high stability, allows one to carry out the calculations with non-overestimated values of the turbulent exchange coefficients and does not make high demands to the time step. Besides, it allows one to also solve non-stationary problems as well as problems pertaining to the attainment of the steady state of the process. Its relative drawback is the necessity to keep two sets of u, v, T, S simultaneously in the computer operative memory. In practice this limitation can be very essential. In such cases it is reasonable to use a semi-implicit scheme, which differs from the formulated implicit scheme in the following: the computer operative memory keeps only one set of u, v, T, S, at each step by time only one internal iteration is made; the Zheidel process is used so, that from the four function values in the extreme points of the "cross" (in the central point, the old function value is kept, substituted by the new one after the calculation) two values are taken from the old time layer, and two from the new one. In a number of concrete calculations a modified, semi-implicit scheme was also applied, including the internal iterative process for ζ at each time step. For this purpose the Gauss-Zheidel method was used. Furthermore, the equation for ζ in some calculations was approximated by the scheme of central differences (i.e. with the second-order approximation). The equation for ζ in this model, unlike a quasi-geostrophic one, does not possess a small parameter at the higher derivative, on the contrary, the term with the higher derivative is the main term in the balance of forces. The semi-implicit scheme was used most frequently in concrete calculations. It makes almost two-fold less demands on the computer operative memory, but in turn places rather strict limits on the time step. Therefore, it is reasonable to apply the considered semi-implicit scheme only for problems with very large grid sets, but which do not require a long integration by time. We will discuss this scheme in detail in Chapter 5. A number of efficient difference schemes have been presented in Zalesny (1984), Zeng et al. (1985), Marchuk (1967a), Marchuk et al. (1973), Marchuk (1974a,b) and Marchuk et al. (1976b, 1979, 1985).

Diagnostic problems can also be solved according to these schemes, but their application is much simpler, as there is no need to solve three-dimensional equations of heat and salt transfer. In the adaptation problems, where the main aim is to obtain the hydrodynamically adjusted fields of currents and density (but not of temperature and salinity), we can use only one equation, i.e. the equation of Lineykin's density diffusion (Lineykin 1957) instead of two equations for temperature and salinity:

$$\frac{d\rho}{dt} = æ_\rho \frac{\partial^2 \rho}{\partial z^2} + A_\rho \Delta_\rho \,. \tag{2.48}$$

In order to solve a non-linear problem we can also use other methods, for example, the well-known Fiadeiro-Veronis scheme (1977), the scheme of directed differences with the scales (Roache 1976), the methods of imbedded grids (which is possible because the scheme is linear), the Ilyin method, etc. Here, we will formulate the difference formula for u, obtained by the Ilyin method. The derivatives of the first and second order by z are substituted here by central difference relations by the formulas for the irregular grid points:

$$\frac{\partial u^{(n+1)}}{\partial z} \approx \frac{1}{\delta z_{k+1} + \delta z_k}\left[\frac{\delta z_k}{\delta z_{k+1}}(u^{(n+1)}_{i,j,k+1} - u^{(n+1)}_{i,j,k}) + \frac{\delta z_{k+1}}{\delta z_k}(u^{(n+1)}_{i,j,k} - u^{(n+1)}_{i,j,k-1})\right], \tag{2.49}$$

$$\frac{\partial^2 u^{(n+1)}}{\partial z^2} \approx \frac{2}{\delta z_{k+1} + \delta z_k}\left[\frac{u^{(n+1)}_{i,j,k+1} - u^{(n+1)}_{i,j,k}}{\delta z_{k+1}} - \frac{u^{(n+1)}_{i,j,k} - u^{(n+1)}_{i,j,k-1}}{\delta z_k}\right], \tag{2.50}$$

where

$$\delta z_k = z_k - z_{k-1} \,.$$

The general formula for the calculation of $u^{(n+1)}_{i,j,k}$ has the form:

$$\begin{aligned} u^{(n+1)}_{i,j,k} = \frac{1}{c_{i,j,k}}&(u^{(n)}_{i,j,k} + c_{i+1,j,k}u^{(n+1)}_{i+1,j,k} + c_{i-1,j,k}u^{(n+1)}_{i-1,j,k} \\ &+ c_{i,j+1,k}u^{(n+1)}_{i,j,k} + c_{i,j-1,k}u^{(n+1)}_{i,j-1,k} + c_{i,j,k+1}u^{(n+1)}_{i,j,k+1} \\ &+ c_{i,j,k-1}u^{(n+1)}_{i,j,k-1} - \delta t f_{i,j,k}) \,, \end{aligned} \tag{2.51}$$

where

$$c_{i+1,j,k} = \frac{\delta t}{2\delta x}u^{(n)}_{i,j,k}\left(\mathrm{cth}\frac{u^{(n)}_{i,j,k}\delta x}{2A_e} - 1\right); \quad c_{i-1,j,k} = \frac{\delta t}{\delta x}\frac{u^{(n)}_{i,j,k}}{2}\left(\mathrm{cth}\frac{u^{(n)}_{i,j,k}\delta x}{2A_l} + 1\right);$$

$$c_{i,j+1,k} = \frac{\delta t}{\delta y}\frac{v^{(n)}_{i,j,k}}{2}\left(\mathrm{cth}\frac{v^{(n)}_{i,j,k}\delta y}{2A_l} - 1\right); \quad c_{i,j-1,k} = \frac{\delta t}{\delta y}\frac{v^{(n)}_{i,j,k}}{2}\left(\mathrm{cth}\frac{v^{(n)}_{i,j,k}\delta y}{2A_l} + 1\right);$$

$$c_{i,j,k+1} = \frac{\delta t}{\delta z_{k+1}(\delta z_k + \delta z_{k+1})}(2\nu - w^{(n)}_{i,j,k}\delta z_k);$$

$$c_{i,j,k-1} = \frac{\delta t}{\delta z_k(\delta z_k + \delta z_{k+1})}(2\nu + w^{(n)}_{i,j,k}\delta z_{k+1});$$

$$\begin{aligned} c_{i,j,k} = 1 + \delta t&\left[\frac{u^{(n)}_{i,j,k}}{\delta x}\mathrm{cth}\frac{u^{(n)}_{i,j,k}\delta x}{2A_l} + \frac{v^{(n)}_{i,j,k}}{\delta y}\mathrm{cth}\frac{v^{(n)}_{i,j,k}\delta y}{2A_l}\right. \\ &\left. + \frac{2\nu}{\delta z_k \delta z_{k+1}} + \frac{\delta z_{k+1} - \delta z_k}{\delta z_k \delta z_{k+1}} w^{(n)}_{i,j,k}\right] . \end{aligned} \tag{2.52}$$

At the ocean's surface $w = 0$, and besides, we use here formula (2.35) instead of (2.50). Then with constant v the last term of the right-hand side of Eq. (2.51) and some of its coefficients at $k = 0$ take the form:

$$c_{i,j,k} = \frac{2v_0\delta t}{z_1^2}\,; \qquad C_{i,j-1} = 0\,;$$

$$c_{i,j,0} = 1 + \delta t\left[\frac{u_{i,j,0}^{(n)}}{\delta x}\operatorname{cth}\frac{u_{i,j,0}^{(n)}\delta x}{2A_l} + \frac{v_{i,j,0}^{(n)}}{\delta y}\operatorname{cth}\frac{v_{i,j,0}^{(n)}\delta y}{2A_l} + \frac{2v_0}{h_i}\right];$$

$$f_{i,j,0} = \frac{g(\zeta_{i+1,j}^{(n+1)} - \zeta_{i-1,j}^{(n+1)})}{2\delta x} - l_j v_{i,j,0}^{(n)} - \frac{2\tau \times i,j}{h_i\rho_0}\,.$$

The expressions for other coefficients at $k = 0$ remain the same.

Thus, the difference formula for u has been constructed. The corresponding formula for v has an analogous form. The equation for ζ is also approximated easily by the Ilyin scheme. The obtained system of algebraic equations can be solved by the Gauss-Zheidel method or the method of overrelaxation.

The numerical calculation of the SST at the boundary in the boundary-value problem with a tangential derivative was presented in Section 2.2.2. Here, we will dwell on the numerical solution of the Neumann problem for the SST, given in correlations (1.135), (1.136) (Denim and Ibraev 1985). The formulation and peculiarities of the problem, the solvability condition and its advantages in comparison with the problem with a tangential derivative were considered in Section 1.2.3. The Neumann condition for the SST is constructed at the internal contour, next to the external one, where the boundary conditions for the velocity are specified. The use of the boundary condition in the form of a normal derivative, in addition to the equation for the SST at the points of the internal contour, allows one to exclude the normal derivative from the basic equation, expressing it through other functions at each time step. In this case both correlations (1.133), (1.134) are used at the points of external angles of the internal contour, while at the points of the internal angles only the equation for ζ is used.

When applying the schemes of directed differences only those first derivatives are excluded, which are closely connected with the external contour, where the boundary conditions for the velocity are specified (i.e. only at those grid points, where the condition of the diagonal predominance requires that the first derivatives should be directed to the contour, but not inside the domain). This also refers to the first derivatives, which form the Laplacian difference analogue of the SST. But as the equation for ζ in this model does not have a small parameter at the higher derivative, it easily allows more accurate approximations (Denim and Trukhchev 1985). The boundary condition is also easily approximated with second-order accuracy.

Thus, when using this approach, systems (1.135), (1.136) are approximated at the points of the internal contour, and at other internal points, which are more than one grid step apart from the basin boundary, only Eq. (1.135) is approximated. This approach, as we can see, does not require knowledge of the SST at the basin boundary (boundaries). The first derivatives of ζ, which are included in the right-hand sides of the equations of motion, can be calculated at the points of the internal contour as half-sums (with weights in the case of an irregular grid) of the derivatives, directed inside the domain and calculated by the field ζ, and derivatives directed to the external contour and calculated from Eq. (1.136) of through the density, velocity, etc. But the SST at the basin boundary can also be determined, for example, at the end of the calculation, using Eq. (1.136).

From the deduction of the equation for ζ, the fulfillment of the solvability condition (1.137) in the differential formulation follows. In numerical models condition (1.137) can be fulfilled only approximately, because of approximation errors. To eliminate the imbalance, the correction of the right-hand side of the equation for ζ can be carried out. At each time step (or in a certain number of steps) the fulfillment of condition (1.137) is controlled and the value $E \neq 0$ is found, then a new right-hand side of $R' = R - E/S$ is calculated, where S is the surface area of the basin, and the boundary problem for ζ with the previously corrected right-hand side is solved. Let us note that the analogous method of correcting the solution is presented in Roache (1976).

A few remarks should now be made. First, we will deal with the solvability condition of the Neumann boundary problem for the SST. Controlling this condition and the corresponding correction of the right-hand side of the equation for the SST are necessary. In the first numerical experiments, where the correction was not carried out, the difference problem diverged with time. The imbalance value E depends on the quality of the difference scheme. With time it decreases, tending to some constant. In Denim and Ibraev (1985), for example, its value composed 20% of the right-hand side value (in the calculations, the scheme of directed differences was used). Along with the application of conservative and more accurate schemes, the imbalance value can also be decreased by the additional use of the continuity equation when constructing the right-hand side of the equation for ζ. As one can easily see, this method has already been used in non-linear terms. When carrying out calculations at mid-latitudes, it is reasonable to use this method, first of all, in the terms, which express the vorticity of the Coriolis force (i.e. most importantly in the mid-latitude terms of the right-hand side, depending on velocities). By introducing the continuity equation in the term instead of $\frac{l}{gH}\int_0^H \left(\frac{\partial v}{\partial x} - \frac{\partial u}{\partial y}\right) dz$, we will obtain $\frac{l}{gH}\int_0^H \left[\frac{\partial}{\partial x}(u + v) = -\frac{\partial}{\partial y}(u - v)\right] dz$, using the condition of the "rigid lid" and the non-slip condition for the vertical velocity. According to Roache (1976), where such methods are described, they are extremely effective for the numerical solution of the Neumann problem for pressure anomaly.

The considered technique is applicable not only to the divergency equations for the SST or bottom pressure anomaly, but also to the vorticity equations. The exception is, apparently, problems pertaining to softened boundary conditions for velocity. The common drawback of the known vorticity equations for the SST is, however, the fact that they are not applicable to equatorial currents.

Chapter 3

Numerical Methods of Solving Ocean Dynamics Problems

3.1 The Construction and Methods of Solving Simplified Problems of Ocean Dynamics

Let us formulate the series of possible correct statements for simplified problems of circulation in the ocean and present the methods of their solution. If we neglect the effects of turbulent viscosity and diffusion in the system of equations (1.1)–(1.6), we will obtain the simplified equations:

$$\frac{du}{dt} - lv + \frac{1}{\rho_0}\frac{\partial P}{\partial x} = 0\,,$$

$$\frac{dv}{dt} + lu + \frac{1}{\rho_0}\frac{\partial P}{\partial y} = 0\,,$$

$$\frac{dP}{dz} = g\rho\,,$$

$$\frac{\partial u}{\partial x} + \frac{\partial v}{\partial y} + \frac{\partial w}{\partial z} = 0\,,$$

$$\frac{dT}{dt} = 0\,,$$

$$\frac{dS}{dt} = 0\,. \tag{3.1}$$

Let us add the equation of state (1.7) to the system of Eq. (3.1) in the general form:

$$\rho = f(S, T)\,, \tag{3.2}$$

where f is the simplified function of temperature and salinity.

We can slightly simplify the formulated equations system (3.1), (3.2) by passing from the physical values and anomalies to their deviations. For example, if φ is some physical characteristic or the anomaly of the characteristic, then let us assume:

$$\varphi = \bar{\varphi} + \varphi'\,.$$

Let

$$\rho = \bar{\rho}(z) + \rho' ,$$
$$T = \bar{T}(z) + T' ,$$
$$S = \bar{S}(z) + S' , \tag{3.3}$$

where $\bar{\rho}(z)$, $\bar{T}(z)$ and $\bar{S}(z)$ are some average distributions of pressure, temperature and salinity by depth for the considered ocean water area. Let us assume further that:

$$\bar{\rho} \gg \rho' ,$$
$$\bar{T} \gg T' ,$$
$$\bar{S} \gg S' . \tag{3.4}$$

Besides, let us suggest that $\bar{T}$ and $\bar{S}$ satisfy the equation of state and statics:

$$\bar{\rho} = f(\bar{T}, \bar{S}) , \qquad \frac{d\bar{P}}{dz} = g\bar{\rho} . \tag{3.5}$$

Then we will have with the second-order accuracy:

$$\rho' = \alpha_T T' + \alpha_s S' ; \tag{3.6}$$

where $\alpha_T = \dfrac{\partial f}{\partial \bar{T}}$, $\alpha_s = \dfrac{\partial f}{\partial \bar{S}}$.

From Eqs. (3.3), (3.4), (3.5) it follows that:

$$\rho' \ll \bar{\rho} . \tag{3.7}$$

Using correlations (3.3), (3.6) and estimation (3.7), we transform the system of the basic equations into the form:

$$\frac{du}{dt} - lv + \frac{1}{\rho_0} \frac{\partial P'}{\partial x} = 0 ,$$

$$\frac{dv}{dt} + lu + \frac{1}{\rho_0} \frac{\partial P'}{\partial y} = 0 ,$$

$$\frac{\partial P'}{\partial z} = g\rho' ,$$

$$\frac{\partial u}{\partial x} + \frac{\partial v}{\partial y} + \frac{\partial w}{\partial z} = 0 ,$$

$$\frac{dT'}{dt} + \gamma_T w = 0 ;$$

$$\frac{dS'}{dt} + \gamma_s w = 0 ; \qquad \rho' = \alpha_T T' + \alpha_s S' . \tag{3.8}$$

Here,

$$\gamma_T = \frac{d\bar{T}}{dz} \quad \text{and} \quad \gamma_s = \frac{d\bar{S}}{dz} . \tag{3.9}$$

In the layer of the main thermocline $0 \leqslant z \leqslant H$, values γ_T and γ_S can be considered as constant with good accuracy. Below the layer of the main thermocline they become small. However, let us assume that:

$$T = T_0 + \gamma_T z + T' ;$$

$$S = S_0 + \gamma_S z + S' \tag{3.10}$$

can be used for the whole World Ocean.[3]

Of course, it is assumed here that condition (3.5) is fulfilled. Let us assume further that

$$\alpha_T = \text{const} , \qquad \alpha_S = \text{const} . \tag{3.11}$$

To prove the theorem of the uniqueness in solving non-stationary problems of ocean dynamics and to construct the correct conditions on the boundaries, thus ensuring the correctness in the formulation of the problem, we will start from the system of equations (3.8). For this purpose let us fix the small time interval $t_j \leqslant t \leqslant t_{j+1}$ and linearize Eqs. (3.8) at this interval. Then system (3.8) will take the form:

$$\frac{\partial u}{\partial t} + u^j \frac{\partial u}{\partial x} + v^j \frac{\partial u}{\partial y} + w^j \frac{\partial u}{\partial z} - lv + \frac{1}{\rho_0} \frac{\partial P'}{\partial x} = 0 ,$$

$$\frac{\partial v}{\partial t} + u^j \frac{\partial v}{\partial x} + v^j \frac{\partial v}{\partial y} + w^j \frac{\partial v}{\partial z} + lu + \frac{1}{\rho_0} \frac{\partial P'}{\partial y} = 0 ,$$

$$\frac{\partial T'}{\partial t} + u^j \frac{\partial T'}{\partial x} + v^j \frac{\partial T'}{\partial y} + w^j \frac{\partial T'}{\partial z} + \gamma_T w = 0 ,$$

$$\frac{\partial S'}{\partial t} + u^j \frac{\partial S'}{\partial x} + v^j \frac{\partial S'}{\partial y} + w^j \frac{\partial S'}{\partial z} + \gamma_S w = 0 ,$$

$$\frac{\partial u}{\partial x} + \frac{\partial v}{\partial y} + \frac{\partial w}{\partial z} = 0 ,$$

$$\frac{\partial P'}{\partial z} = g\rho' ,$$

$$\rho' = \alpha_T T' + \alpha_s S' , \tag{3.12}$$

where u^j, v^j and w^j are the components of the velocity vector, which are assumed

---

[3] Transformation (3.10) transfers the initial system of equations into an identical one and thus is mathematically formal. Therefore, such a transformation can be used even in those cases, in which the thermocline is absent in the considered water area or partially absent, or when it differs greatly from the adopted "average" one. At the same time the given transformation, not introducing additional errors into the initial system of equations, allows one to perform the theoretical analysis in the simplest way.

to be specified functions with $t = t_j$ and which satisfy the continuity equation:

$$\frac{\partial u^j}{\partial x} + \frac{\partial v^j}{\partial y} + \frac{\partial w^j}{\partial z} = 0 . \tag{3.13}$$

Let us multiply the first five equations of system (3.12) correspondingly by $\rho_0 u$, $\rho_0 v$, $\frac{g\alpha_T}{\gamma_T} T'$, $\frac{g\alpha_s}{\gamma_s} S'$, P' sum the results and integrate over the whole World Ocean. Then taking into account the two last correlations in Eqs. (3.12) and (3.13) we will obtain:

$$\frac{d}{dt} \iiint_D \pi dD + \iiint_D \operatorname{div}(\mathbf{u}^j \pi + \mathbf{u}P') dD = 0 . \tag{3.14}$$

Here,

$$\pi = \frac{1}{2}\left[\rho_0(u^2 + v^2) + \frac{g\alpha_T}{\gamma_T} T'^2 + \frac{g\alpha_s}{\gamma_s} S'^2 \right] . \tag{3.15}$$

Let us make some transformations in Eq. (3.14):

$$\iiint_D \operatorname{div}(\mathbf{u}^j \pi + \mathbf{u}P') dD = \iint_S (u_n^j \pi + u_n P') dS , \tag{3.16}$$

where u_n^j and u_n are the normal components of the velocity vectors $\mathbf{u}^j$ and $\mathbf{u}$ respectively. As a result, expression (3.14) transforms into the form:

$$\frac{d}{dt} \iiint_D \pi dD = -\iint_S (u_n^j \pi + u_n P') dS . \tag{3.17}$$

Now we will analyze the possible formulations of the problems and prove the uniqueness theorem of the solution of the hydrodynamics equations system, linearized at the interval $t_j \leqslant t \leqslant t_{j+1}$.

1. Let us consider the simplest formulation of the problem related to system (3.12). As boundary conditions at the sea surface we will consider the following:

$$u_n = 0 \qquad \text{at } S . \tag{3.18}$$

Further let us assume that the analogous correlation is fulfilled also for the initial conditions:

$$u_n^0 = 0 \qquad \text{at } S . \tag{3.19}$$

At solid boundaries, conditions (3.18)–(3.19) are the conditions of free slip, and at the sea surface they are equivalent to the condition of equality of the vertical component of the velocity vector to zero. It should be noted that no conditions are placed on T' and S' at the boundary.

Let us take as the initial data at $t = t_j$, the condition

$$\begin{aligned} u &= u_0 , & v &= v_0 , \\ T' &= T_0' , & S' &= S_0' . \end{aligned} \tag{3.20}$$

Let us assume that due to condition (3.20) and the domain D, the solutions of problems (3.12), (3.18), (3.20) are rather smooth. Further we will assume that there are at least two solutions with the same boundary conditions, which differ from each other, thus:

$$\{u_1, v_1, w_1, P_1', T_1', S_1'\}, \quad \{u_2, v_2, w_2, P_2', T_2', S_2'\}.$$

A new solution can be formulated in the form of the difference:

$$\{u_1 - u_2, v_1 - v_2, P_1' - P_2', T_1' - T_2', S_1' - S_2'\}.$$

Due to the linearity of the problem at the interval $t_j \leqslant t \leqslant t_{j+1}$, the difference of the solutions, which we will denote through

$$\{u = u_1 - u_2, v = v_1 - v_2, P' = P_1' - P_2', T' = T_1' - T_2', S' = S_1' - S_2'\}$$

will satisfy the equations system (3.12), the homogeneous boundary conditions (3.18), (3.19) and zero initial data:

$$u = 0, \quad v = 0, \quad T' = 0, \quad S' = 0 \qquad \text{at } t = t_j. \tag{3.21}$$

According to the above formulated assumptions and taking into account Eq. (3.17) we obtain:

$$\frac{d}{dt}\iiint_D \pi dD = 0. \tag{3.22}$$

If we integrate correlation (3.22) by t in the limits of the interval $t_j \leqslant t \leqslant t_{j+1}$, then

$$\iiint_D \frac{u^2 + v^2 + \dfrac{g\alpha_T}{\gamma_T}T'^2 + \dfrac{g\alpha_s}{\gamma_s}S'^2}{2} dD = 0. \tag{3.23}$$

As there is a quadratic function under the integral in Eq. (3.23), then all the components:

$$u = 0, \quad v = 0, \quad T' = 0, \quad S' = 0, \tag{3.24}$$

if the function (3.23) equals zero.

As a result of the continuity equation and if

$$w = 0 \qquad \text{at } z = 0, \tag{3.25}$$

it follows immediately that

$$w = 0 \qquad \text{for all } t \in [t_j, t_{j+1}]. \tag{3.26}$$

As far as the pressure P' is concerned, it is not included in the present formulation, in the equations or in the boundary conditions except through the deviations. Therefore, the pressure will be determined by the accuracy of the arbitrary additive constant. This means that for the complete determination of the problem, it is necessary to specify the pressure at one grid point of the solution's determination domain. Then the homogeneous problem will lead us to the zero solution for all the components, including P', i.e. we will obtain:

$$P' = 0\,, \quad t \in [t_j, t_{j+1}]\,. \tag{3.27}$$

Conditions (3.24), (3.26), (3.27) indicate that

$$u_1 = u_2\,, \quad v_1 = v_2\,, \quad w_1 = w_2\,, \quad P_1' = P_2'\,, \quad T_1' = T_2'\,, \quad S_1' = S_2'\,.$$

Thus, the solution of the problem is unique.

2. Now we will examine another formulation of Eq. (3.12), where we should be able to solve the equations system with the boundary conditions:

$$\begin{aligned} &w = 0 \quad \text{at } z = 0\,,\\ &w = 0 \quad \text{at } z = H\,,\\ &\left.\begin{aligned} u &= f_1\,,\\ v &= f_2\,,\\ T' &= f_3\,,\\ S' &= f_4 \end{aligned}\right\} \quad \text{at } \sigma\,, \end{aligned} \tag{3.28}$$

where σ is a side cylinder surface, selected arbitarily from domain D. Functions f_j are assumed to be specified in all the moments of time $t_j \leqslant t \leqslant t_{j+1}$.

Let us assume further that we again have at least two different solutions, which satisfy boundary conditions (3.28) and initial data (3.21). Then, by compounding the solutions' difference and making the transformation to the function of form (3.15), we will obtain correlations (3.14) with the conditions:

$$\begin{aligned} &w = 0 \quad \text{at } z = 0\,, \qquad \left.\begin{aligned} u &= 0\\ v &= 0\\ T' &= 0\\ S' &= 0 \end{aligned}\right\} \quad \text{at } \sigma\,.\\ &w = 0 \quad \text{at } z = H\,, \end{aligned} \tag{3.29}$$

Thus, it follows that

$$\pi = 0 \quad \text{at } \sigma\,, \qquad u_n = 0 \quad \text{at } S\,. \tag{3.30}$$

If we add to conditions (3.30) the assumptions that the coefficients of the equations system (3.12) satisfy conditions

$$wj = 0 \quad \text{at } z = 0\,, \quad \text{at } z = H\,, \tag{3.31}$$

then it follows that the surface integral on the right-hand side of Eq. (3.17) will equal zero, and we again come to correlation (3.23), from which the uniqueness follows. Thus, conditions (3.20), (3.28) ensure the unique solution to the equations system (3.12), (3.16) with specifying data on the "liquid" side surface.

However, it should be noted that from the differential equations theory, it is only possible to specify f_1 and f_2 in Eq. (3.26) as arbitrary at that part of the contour where the flux of particles is directed into the internal part of domain D. In the remainder of σ these functions are determined from the solution of the problem. Analogous results have been reported from Charney, Fjortofft and Neumann (1950).

3. Let us now consider another formulation of the problem, connected with the equations system (3.12). Let us assume that on the free sea surface, instead of condition

$$w = 0 \qquad \text{at } z = 0\,, \tag{3.32}$$

the kinematics condition is set up

$$g\rho_0 w = \frac{dP'}{dt} \qquad \text{at } z = 0\,. \tag{3.33}$$

Thus, we have the following boundary conditions:

$$g\rho_0 w = \frac{dP'}{dt} \qquad \text{at } z = 0$$

$$u_n = 0 \quad \text{at } \Sigma \qquad \left.\begin{aligned} u &= f_1 \\ v &= f_2 \\ T' &= f_3 \\ S' &= f_4 \end{aligned}\right\} \quad \text{at } \sigma + \sigma_0\,. \tag{3.34}$$

Here, Σ is the solid part of surface S', σ is the "liquid contour", selected from the domain of the World Ocean, and σ_0 is part of the plane $z = 0$, where conditions (3.33) are set up. Let us consider the functions f_i as continuous, together with their first derivatives at S. We will take

$$u = u_0\,, \quad v = v_0\,, \quad T' = T'_0\,, \quad S' = S'_0\,, \quad P' = P'_0 \quad \text{at } z = 0\,. \tag{3.35}$$

as the initial data.

As a result we will obtain the function:

$$\frac{d}{dt}\iiint\limits_D \pi dD = -\iint\limits_S (u_n^j \pi + u_n P')dS'\,,$$

where

$$\iint\limits_S (u_n^j \pi + u_n P')dS = \iint\limits_\Sigma (u_n^j \pi + u_n P')dS + \iint\limits_\sigma (u_n^j \pi + u_n P')dS$$
$$+ \iint\limits_{\sigma_0} (u_n^j \pi + u_n P')dS\,. \tag{3.36}$$

The first integral on the right-hand side of Eq. (3.36) equals zero due to the condition on the solid part of the contour:

$$u_n = 0\,, \quad u_n^j = 0 \quad \text{at } \Sigma\,. \tag{3.37}$$

The second integral changes to zero due to the condition:

$$u = 0\,, \quad v = 0\,, \quad T' = 0\,, \quad S' = 0 \quad \text{at } \sigma\,. \tag{3.38}$$

Since at the surface $\sigma_0(z=0)$

$$u=0\,,\quad v=0\,,\quad T'=0\,,\quad S'=0\,,$$

$$w=\frac{1}{g\rho_0}\frac{\partial P'}{\partial t}\,, \tag{3.39}$$

the last integral is transformed to the form:

$$\iint_{\sigma_0}(u_n^j\pi+u_nP')dS=\frac{1}{g\rho_0}\frac{d}{dt}\iint_{\sigma_0}\frac{P'^2}{2}dS\,. \tag{3.40}$$

Thus, we obtain:

$$\frac{d}{dt}\left\{\iiint_D \pi dD+\frac{1}{g\rho_0}\iint_{\sigma_0}\frac{P'^2}{2}dS\right\}=0\,. \tag{3.41}$$

As there are quadratic functions under the integral, the problem with homogeneous initial data will have only a trivial solution. So, the uniqueness of such a formulation of the problem is also proved.

It should be noted that the theorem of the uniqueness for the system of ocean dynamics equations, linearized relative to the quiescent state, was proved for the first time by L.V. Ovsyannikov (1966).

In conclusion, one important fact related to the solution of the discussed problems should be noted. Namely, if we do not linearize the initial system and choose conditions (3.34) as the boundary conditions, taking $f_i=0$, then the law of energy conservation exists in the form of Eq. (3.41) or

$$\iiint_D \pi dD+\frac{1}{g\rho_0}\iint_{\sigma_0}\frac{P'^2}{2}dS=\text{const}\,. \tag{3.42}$$

In this case the constant is determined by the initial data of the problem.

The theorems of the uniqueness can be proved likewise when the turbulent viscosity is present in the equations of dynamics.

3.2 The Operator Representation of the Problem and the Principal Algorithm of the Splitting

Let us consider one of the problems formulations on the basis of the equations system (3.12). On the whole boundary surface for the solution determination domain D, we take the normal component of the velocity vector equal to zero. At the initial moment of time the values u, v, T and S and considered to be specified.

Thus, we have the following problem in the final formulation

$$\frac{\partial u}{\partial t}+\operatorname{div}\mathbf{u}^j u-lv+\frac{1}{\rho_0}\frac{\partial P}{\partial x}=0\,,$$

$$\frac{\partial v}{\partial t}+\operatorname{div}\mathbf{u}^j v+lu+\frac{1}{\rho_0}\frac{\partial P}{\partial y}=0\,,$$

$$\frac{\partial P}{\partial z} = g(\alpha_T T + \alpha_s S) \qquad \text{in } D \times D_t\,,$$

$$\frac{\partial u}{\partial x} + \frac{\partial v}{\partial y} + \frac{\partial w}{\partial z} = 0\,,$$

$$\frac{\partial T}{\partial t} + \operatorname{div} \mathbf{u}^j T + \gamma_T w = 0\,,$$

$$\frac{\partial S}{\partial t} + \operatorname{div} \mathbf{u}^j S + \gamma_s w = 0\,. \tag{3.43}$$

Here, the primes at the functions P, S and T are omitted for simplicity and ρ is excluded from consideration. Let us assume that the velocity vector $\mathbf{u}^j$ is specified in this system. At the end of the present section when considering the organization of the computational process, this question will be discussed in detail.

Let us add to system (3.43) the boundary conditions:

$$\begin{aligned} w &= 0 \qquad \text{at } z = 0\,, \\ w &= 0 \qquad \text{at } z = H\,, \\ u_n &= 0 \qquad \text{at } \sigma\,, \end{aligned} \tag{3.44}$$

where σ is a coastal cylinder surface which consists of parts of the coordinate planes.

Assume

$$u = u^0\,, \quad v = v^0\,, \quad T = T^0\,, \quad S = S^0 \quad \text{at } t = 0 \tag{3.45}$$

as the initial data.

We will further assume that input data possess sufficient smoothness, which ensures the uniqueness of the solution.

Now we will introduce vectors φ, F and matrixes A and B:

$$\varphi = \begin{bmatrix} u \\ v \\ w \\ P \\ T \\ S \end{bmatrix}, \qquad A = \left\| \begin{matrix} \rho_0 \operatorname{div} \mathbf{u}^j & -l\rho_0 & 0 & \frac{\partial}{\partial x} & 0 & 0 \\ l\rho_0 & \rho_0 \operatorname{div} \mathbf{u}^j & 0 & \frac{\partial}{\partial y} & 0 & 0 \\ 0 & 0 & 0 & \frac{\partial}{\partial z} & -g\alpha_T & -g\alpha_S \\ \frac{\partial}{\partial x} & \frac{\partial}{\partial y} & \frac{\partial}{\partial z} & 0 & 0 & 0 \\ 0 & 0 & g\alpha_T & 0 & \frac{g\alpha_T}{\gamma_T} \operatorname{div} \mathbf{u}^j & 0 \\ 0 & 0 & g\alpha_S & 0 & 0 & \frac{g\alpha_S}{\gamma_S} \operatorname{div} \mathbf{u}^j \end{matrix} \right\|$$

$$F = \begin{bmatrix} u^0 \\ v^0 \\ 0 \\ 0 \\ T^0 \\ S^0 \end{bmatrix}, \qquad B = \begin{Vmatrix} \rho_0 & 0 & 0 & 0 & 0 & 0 \\ 0 & \bar{\rho} & 0 & 0 & 0 & 0 \\ 0 & 0 & 0 & 0 & 0 & 0 \\ 0 & 0 & 0 & 0 & 0 & 0 \\ 0 & 0 & 0 & 0 & \dfrac{g\alpha_T}{\gamma_T} & 0 \\ 0 & 0 & 0 & 0 & 0 & \dfrac{g\alpha_S}{\gamma_S} \end{Vmatrix}$$

Thus, the system of equations (2.1) can be formulated in the form:

$$B\frac{\partial \varphi}{\partial t} + A\varphi = 0 \tag{3.46}$$

where by

$$B\varphi = BF \qquad \text{at } t = 0 \tag{3.47}$$

are the initial data.

In order to formulate the system of equations (3.43) in the form of (3.46), we previously changed the form of Eqs. (3.43), multiplying them correspondingly by $\rho_0, \rho_0, 1, 1, \dfrac{g\alpha_T}{\gamma_T}$ and $\dfrac{g\alpha_S}{\gamma_S}$.

Now we will multiply Eq. (3.46) scalarly by φ. Then we will obtain:

$$\frac{1}{2}\frac{\partial}{\partial t}(B\varphi, \varphi) + (A\varphi, \varphi) = 0 \,. \tag{3.48}$$

It is easy to check that the following correlation:

$$(A\varphi, \varphi) = 0 \tag{3.49}$$

takes place, if we take into account conditions (3.44). In this correlation the scalar product is determined in the usual way:

$$(a, b) = \sum_{i=1}^{6} \int_D a^i b^i dD \,,$$

where a^i and b^i are the components of the vector functions a and b. Therefore,

$$(B\varphi, \varphi) = (BF, \varphi^0) = \text{const} \,, \tag{3.50}$$

where φ^0 is an initial vector (at $t = 0$).

Expressing Eq. (3.50) in the coordinate form, we obtain the correlation:

$$(B\varphi, \varphi) = 2 \iiint_D \pi dD = 2 \iiint_D \pi^0 dD \,, \tag{3.51}$$

where

$$\pi = \frac{1}{2}\left[\rho_0(u^2 + v^2) + \frac{g\alpha_T}{\gamma_T}T^2 + \frac{g\alpha_S}{\gamma_S}S^2\right]$$

is the function, determined earlier by formula (3.15).

Further, we will introduce the matrixes

$$A_1 = \left\| \begin{matrix} \rho_0 \operatorname{div} \mathbf{u}^j & 0 & 0 & 0 & 0 & 0 \\ 0 & \rho_0 \operatorname{div} \mathbf{u}^j & 0 & 0 & 0 & 0 \\ 0 & 0 & 0 & 0 & 0 & 0 \\ 0 & 0 & 0 & 0 & 0 & 0 \\ 0 & 0 & 0 & 0 & \dfrac{g\alpha_T}{\gamma_T} \operatorname{div} \mathbf{u}^j & 0 \\ 0 & 0 & 0 & 0 & 0 & \dfrac{g\alpha_S}{\gamma_S} \operatorname{div} \mathbf{u}^j \end{matrix} \right\|$$

$$A_2 = \left\| \begin{matrix} 0 & -l\rho_0 & 0 & \dfrac{\partial}{\partial x} & 0 & 0 \\ l\rho_0 & 0 & 0 & \dfrac{\partial}{\partial y} & 0 & 0 \\ 0 & 0 & 0 & \dfrac{\partial}{\partial z} & -g\alpha_T & -g\alpha_S \\ \dfrac{\partial}{\partial x} & \dfrac{\partial}{\partial y} & \dfrac{\partial}{\partial z} & 0 & 0 & 0 \\ 0 & 0 & g\alpha_T & 0 & 0 & 0 \\ 0 & 0 & g\alpha_S & 0 & 0 & 0 \end{matrix} \right\| .$$

Matrix A is related to A_1 and A_2 by the correlation

$$A = A_1 + A_2 .$$

It is not difficult to check, that if the components of the solution satisfy conditions (3.44), then the following correlations take place:

$$(A_1\varphi, \varphi) = 0 , \qquad (A_2\varphi, \varphi) = 0 . \tag{3.52}$$

Conditions (3.52) are necessary for constructing stable difference splitting schemes, possessing the second-order accuracy by τ.

In order to construct such schemes, one of the authors worked out the two-cyclic method of componentwise splitting (Marchuk 1974), which we will now present. Let us consider the evolutional equation:

$$\frac{\partial\varphi}{\partial t} + A\varphi = f ,$$

$$\varphi = g \quad \text{in } D \qquad \text{at } t = 0 , \tag{3.53}$$

where the operator $A \geqslant 0$ does not depend on time, and present in the form

$$A = A_1 + A_2 , \tag{3.54}$$

if

$$A_1 \geqslant 0 , \qquad A_2 \geqslant 0 . \tag{3.55}$$

Further let us assume that the solution of problem (3.53) possesses the necessary smoothness. Now we consider the componentwise splitting methods, supposing that problem (3.53) has already been reduced to the difference form, and so, the operators A, A_1 and A_2 are matrixes.

Let $A_1(t) \geqslant 0$, $A_2(t) \geqslant 0$. We will consider the approximations of these matrixes at the interval $t_j \leqslant t \leqslant t_{j+1}$ in the form

$$\Lambda_\alpha^j = A_\alpha(t_{j+1/2}) ,$$

assuming that the elements of these matrixes have satisfactory smoothness. Let us formulate the difference system of equations, suggested by N.N. Yanenko (1967), which consists of the successive solution of the simplest Crank-Nickolson schemes:[4]

$$\frac{\varphi^{j+1/2} - \varphi^j}{\tau} + \Lambda_1^j \frac{\varphi^{j+1/2} + \varphi^j}{2} = 0 ,$$

$$\frac{\varphi^{j+1} - \varphi^{j+1/2}}{\tau} + \Lambda_2^j \frac{\varphi^{j+1} + \varphi^{j+1/2}}{2} = 0 . \tag{3.56}$$

The system of difference equations (3.56), if we exclude the auxiliary function $\varphi^{j+1/2}$, can be reduced to one equation:

$$\varphi^{j+1} = T_j \varphi^j , \tag{3.57}$$

where

$$T_j = \left(E + \frac{\tau}{2}\Lambda_2^j\right)^{-1}\left(E - \frac{\tau}{2}\Lambda_2^j\right)\left(E + \frac{\tau}{2}\Lambda_1^j\right)^{-1}\left(E - \frac{\tau}{2}\Lambda_1^j\right) . \tag{3.58}$$

First, we will study the approximation problem for this purpose. Let us expand the operator T_j by degrees of τ, assuming that $\frac{\tau}{2}\|\Lambda_\alpha^j\| < 1$. As a result, after some simple transformations we will obtain:

$$T_j = E - \tau \Lambda^j + \frac{\tau^2}{2}[(\Lambda_1^j)^2 + 2\Lambda_2^j \Lambda_1^j + (\Lambda_2^j)^2] - \cdots \tag{3.59}$$

If the operators Λ_α^j commutate, i.e. $\Lambda_1^j \Lambda_2^j = \Lambda_2^j \Lambda_1^j$, then expansion (3.59) can be written in the form:

$$T_j = E - \tau \Lambda^j + \frac{\tau^2}{2}(\Lambda^j)^2 - \cdots \tag{3.60}$$

---

[4] The theoretical substantiation and modifications of the scheme were given by G.I. Marchuk in his presentation at the symposium on the numerical methods of the solution of the equations with partial derivatives, which took place in the USA in 1970 (SYNSPADE, 1970).

Thus, if $A_1(t) \geqslant 0$, $A_2(t) \geqslant 0$ and the elements of these matrixes and the solution φ of problems (3.53)–(3.55) are sufficiently smooth, then the difference scheme (3.56) is absolutely stable (it immediately follows from the inequality $\|T_j\| < 1$, which is valid according to the Kellog lemma) and approximates the initial equation (3.53) with the second order by τ when Λ_1^j and Λ_2^j are commutative, and with the first order when they are non-commutative.

Now we will approximate the operators $A_1(t)$ and $A_2(t)$, however, not at the interval $t_j \leqslant t \leqslant t_{j+1}$, as in Eq. (3.56), but at the interval $t_{j-1} \leqslant t \leqslant t_{j+1}$, assuming

$$\Lambda_\alpha^j = A_\alpha(t_j)\,.$$

Let us consider the following two systems of difference equations:

$$\frac{\varphi^{j-1/2} - \varphi^{j-1}}{\tau} + \Lambda_1^j \frac{\varphi^{j-1/2} + \varphi^{j-1}}{2} = 0\,,$$

$$\frac{\varphi^{j} - \varphi^{j-1/2}}{\tau} + \Lambda_2^j \frac{\varphi^{j} + \varphi^{j-1/2}}{2} = 0\,, \tag{3.61}$$

and then

$$\frac{\varphi^{j+1/2} - \varphi^{j}}{\tau} + \Lambda_2^j \frac{\varphi^{j+1/2} + \varphi^{j}}{2} = 0\,,$$

$$\frac{\varphi^{j+1} - \varphi^{j+1/2}}{\tau} + \Lambda_1^j \frac{\varphi^{j+1} + \varphi^{j+1/2}}{2} = 0\,. \tag{3.62}$$

The cycle of calculations consists precisely of the alternative application of difference schemes (3.61), (3.62). Similarly to the previous case we can show that at the full cycle of calculations with the help of Eqs. (3.61) and (3.62), we have

$$\varphi^{j+1} = T_j \varphi^{j-1}\,, \tag{3.63}$$

where

$$T_j = \left(E + \frac{\tau}{2}\Lambda_1^j\right)^{-1}\left(E - \frac{\tau}{2}\Lambda_1^j\right)\left(E + \frac{\tau}{2}\Lambda_2^j\right)^{-1}\left(E - \frac{\tau}{2}\Lambda_2^j\right)\left(E + \frac{\tau}{2}\Lambda_2^j\right)^{-1}$$
$$\times \left(E - \frac{\tau}{2}\Lambda_2^j\right)\left(E + \frac{\tau}{2}\Lambda_1^j\right)^{-1}\left(E - \frac{\tau}{2}\Lambda_1^j\right) = E - 2\tau\Lambda^j + \frac{(2\tau)^2}{2}(\Lambda^j)^2 - \cdots \tag{3.64}$$

If we compare the operator of the step T_j with the operator of the step of the following Crank-Nickolson scheme:

$$\frac{\varphi^{j+1} - \varphi^{j-1}}{2\tau} + \Lambda^j \frac{\varphi^{j+1} + \varphi^{j-1}}{2} = 0\,, \tag{3.65}$$

then we can conclude that with the accuracy to the value τ^2, the operator of the step T_j for the two-cycle splitting method coincides with the operator of the step for the Crank-Nickolson scheme, applied to the doubled interval by time, regardless of the commutativity of the operators A_α. Thus, this method eliminates the essential limitation of the operators' commutativity.

Now we will discuss the numerical stability of the method. For this purpose let us consider correlation (3.57) and estimate it in the energetic form:

$$\|\varphi^{j+1}\| \leqslant \|T_j\|\,\|\varphi^j\| \,.$$

As it was shown above, $\|T_j\| \leqslant 1$ at $A_\alpha \geqslant 0$, then we obtain the estimation:

$$\|\varphi^{j+1}\| \leqslant \|\varphi^j\| \tag{3.66}$$

whereby

$$\|\varphi^j\| \leqslant \|g\| \tag{3.67}$$

follows directly from it.

If the two-cyclic method is being considered, then the estimations of type (3.66) take place at every step of the cycle. This means that the two-cyclic method is also absolutely stable. It should be noted that an analogous method of symmetrization was offered independently by Strang[5] while considering the method of variable directions.

So, if the matrixes $A_1(t) \geqslant 0$ and $A_2(t) \geqslant 0$, and the solution φ of problems (3.53)–(3.55) and the elements of the matrixes $A_1(t)$ and $A_2(t)$ are sufficiently smooth, then the systems of difference equations (3.61) and (3.62) are absolutely stable, and scheme (3.63) approximates the initial equation (3.53) with the second-order accuracy by τ.

Now we will consider the inhomogeneous problem and find its solution with the help of the two-cyclic full splitting method. For this purpose we will consider the system of difference equations of type (3.61) and (3.62), written in a more convenient form:

$$\left(E + \frac{\tau}{2}\Lambda_1^j\right)\varphi^{j-1/2} = \left(E - \frac{\tau}{2}\Lambda_1^j\right)\varphi^{j-1} \,,$$

$$\left(E + \frac{\tau}{2}\Lambda_2^j\right)(\varphi^j - \tau f^j) = \left(E - \frac{\tau}{2}\Lambda_2^j\right)\varphi^{j-1/2} \,,$$

$$\left(E + \frac{\tau}{2}\Lambda_2^j\right)\varphi^{j+1/2} = \left(E - \frac{\tau}{2}\Lambda_2^j\right)(\varphi^j + \tau f^j) \,,$$

$$\left(E + \frac{\tau}{2}\Lambda_1^j\right)\varphi^{j+1} = \left(E - \frac{\tau}{2}\Lambda_1^j\right)\varphi^{j+1/2} \,, \tag{3.68}$$

where $f^j = f(t_j)$. By resolving these equations relative to φ^{j+1}, we will obtain:

$$\varphi^{j+1} = T^j\varphi^{j-1} + 2\tau T_1^j T_2^j f^j \,, \tag{3.69}$$

where $T^j = T_1^j T_2^j T_2^j T_1^j$ (3.70)

[5] Strang W.G. Difference methods for mixed boundary problems. Duke Math. J. 1960, v. 27, No 2, pp. 286–289.

and

$$T_\alpha^j = \left(E + \frac{\tau}{2}\Lambda_\alpha^j\right)^{-1}\left(E - \frac{\tau}{2}\Lambda_\alpha^j\right). \tag{3.71}$$

With the help of the expansion into the degrees of the small parameter τ, we can transform expression (3.69) into the form:

$$\varphi^{j+1} = \left[E - 2\tau\Lambda^j + \frac{(2\tau)^2}{2}(\Lambda^j)^2\right]\varphi^{j-1} + 2\tau(E - \tau\Lambda^j)f^j + 0(\tau^3). \tag{3.72}$$

The expression (3.72) can be transformed into the form:

$$\frac{\varphi^{j+1} - \varphi^{j-1}}{2\tau} + \Lambda^j(E - \tau\Lambda^j)\varphi^{j-1} = (E - \tau\Lambda^j)f^j + 0(\tau^2). \tag{3.73}$$

Furthermore, φ^{j-1} is excluded from correlation (3.73). For this purpose we will make use of the expansion of the solution in a Taylor series in the vicinity of the point t_{j-1}. To the accuracy of τ^2, we obtain:

$$\varphi^j = \varphi^{j-1} + \left(\frac{\partial\varphi}{\partial t}\right)^{j-1}\tau + 0(\tau^2). \tag{3.74}$$

The derivative $\frac{\partial\varphi}{\partial t}$ is excluded with the help of the equation:

$$\left(\frac{\partial\varphi}{\partial t}\right)^{j-1} = -\Lambda^j\varphi^{j-1} + f^j + 0(\tau). \tag{3.75}$$

Substituting (3.75) into (3.74), we obtain:

$$\varphi^j = (E - \tau\Lambda^j)\varphi^{j-1} + \tau f^j + 0(\tau^2),$$

from which

$$(E - \tau\Lambda^j)\varphi^{j-1} = \varphi^j - \tau f^j + 0(\tau^2) \tag{3.76}$$

follows.

Substituting correlation (3.76) into (3.73), we obtain:

$$\frac{\varphi^{j+1} - \varphi^{j-1}}{2\tau} + \Lambda^j\varphi^j = f^j + 0(\tau^2). \tag{3.77}$$

It is obvious that Eq. (3.77) approximates the initial equation (3.53) at the interval $t_{j-1} \leqslant t \leqslant t_{j+1}$, with the second order by τ. Thus, we have found the difference approximation of the inhomogeneous, evolutional, second-order equation with the help of the two-cyclic method.

The stability of the method is proved easily in the energetic norm. Thus, let us estimate Eq. (3.69) by the norm:

$$\|\varphi^{j+1}\| \leqslant \|T^j\| \cdot \|\varphi^{j-1}\| + 2\tau\|T_1\| \cdot \|T_2\| \cdot \|f^j\|. \tag{3.78}$$

It was determined above that $\|T_\alpha\| \leqslant 1$, so

$$\|T^j\| \leqslant \|T_1\| \cdot \|T_2\| \cdot \|T_2\| \cdot \|T_1\| \leqslant 1.$$

Therefore, we have

$$\|\varphi^{j+1}\| \leqslant \|\varphi^{j-1}\| + 2\tau\|f^j\| . \tag{3.79}$$

With the help of the recurrent correlation (3.79) we will obtain:

$$\|\varphi^j\| \leqslant \|g\| + \tau j\|f\| , \tag{3.80}$$

where

$$\|f\| = \max \|f^j\| .$$

From correlation (3.80) the numerical stability of the scheme at any finite time interval follows.

We can also formulate the system of equations (3.68) in the following, equivalent form:

$$\begin{aligned}
&\left(E + \frac{\tau}{2}\Lambda_1^j\right)\varphi^{j-2/3} = \left(E - \frac{\tau}{2}\Lambda_1^j\right)\varphi^{j-1} ,\\
&\left(E + \frac{\tau}{2}\Lambda_2^j\right)\varphi^{j-1/3} = \left(E - \frac{\tau}{2}\Lambda_2^j\right)\varphi^{j-2/3} ,\\
&\varphi^{j+1/2} = \varphi^{j-1/3} + 2\tau f^j ,\\
&\left(E + \frac{\tau}{2}\Lambda_2^j\right)\varphi^{j+2/3} = \left(E - \frac{\tau}{2}\Lambda_2^j\right)\varphi^{j+1/3} ,\\
&\left(E + \frac{\tau}{2}\Lambda_1^j\right)\varphi^{j+1} = \left(E - \frac{\tau}{2}\Lambda_1^j\right)\varphi^{j+2/3} .
\end{aligned} \tag{3.81}$$

Excluding the unknown values with fractional indexes, we will obtain the resolved equation with the form:

$$\varphi^{j+1} = T_1 T_2 T_2 T_1 \varphi^{j-1} + 2\tau T_1 T_2 f^j , \tag{3.82}$$

which coincides with Eq. (3.69). In some cases it is preferable to formulate the equations in the form of Eq. (3.81) rather than in the form of Eq. (3.68).

Thus, if the matrixes $A_1(t) \geqslant 0$, $A_2(t) \geqslant 0$ and the solution φ, the functions $f(t)$, and the elements of the matrixes $A_1(t)$, $A_2(t)$ are smooth enough, then the difference equations system (3.68) is absolutely stable at the interval $0 \leqslant t \leqslant T$ and approximates the initial equation with the second order by τ.

Now, on the basis of the above formulated method, we will construct the schemes, which possess second-order accuracy. For this purpose let us assume that the whole time interval $0 \leqslant t \leqslant T$ is divided into the equal intervals $t_{j-1} \leqslant t \leqslant t_j$ with the width $t_j - t_{j-1} = \tau$, and for the realization of the two-cyclic splitting method, let us consider the wider interval $t_{j-1} \leqslant t \leqslant t_{j+1}$, which consists of two intervals.

Further, it is important to note that in Eqs. (3.43) the vector function $\mathbf{u}^j$ is considered the same for the whole interval $t_{j-1} \leqslant t \leqslant t_{j+1}$. Only in this case does the two-cyclic method lead to the solution of the second-order accuracy by τ.

Then at the first interval $t_{j-1} \leqslant t \leqslant t_j$, the system of the split equations takes the form:

$$B\frac{\partial \varphi_1}{\partial t} + A_1\varphi_1 = 0\,, \qquad B\varphi_1^{j+1} = B\varphi^{j-1}\,,$$

$$B\frac{\partial \varphi_2}{\partial t} + A_2\varphi_2 = 0\,, \qquad B\varphi_2^{j-1} = B\varphi_1^{j}\,, \tag{3.83}$$

and then at the interval $t_j \leqslant t \leqslant t_{j+1}$

$$B\frac{\partial \varphi_3}{\partial t} + A_2\varphi_3 = 0\,, \qquad B\varphi_3^{j} = B\varphi_2^{j}\,,$$

$$B\frac{\partial \varphi_4}{\partial t} + A_1\varphi_4 = 0\,, \qquad B\varphi_4^{j} = B\varphi_3^{j+1}\,, \tag{3.84}$$

where φ^{j-1} represents the solution to problems (3.43)–(3.45) in the moment of time $t_{j-1} \leqslant t \leqslant t_{j+1}$. Furthermore, it should be noted that the second equation of system (3.83) and the first equation of system (3.84) coincide, therefore they can be solved in the united cycle at the interval $t_{j-1} \leqslant t \leqslant t_{j+1}$. Then the splitting scheme (3.84) can be presented as a scheme, consisting only of three equations:

$$B\frac{\partial \varphi_1}{\partial t} + A_1\varphi_1 = 0 \quad (t_{j-1} \leqslant t \leqslant t_j)\,,$$

$$B\varphi_1^{j-1} = B\varphi^{j-1}\,,$$

$$B\frac{\partial \varphi_2}{\partial t} + A_2\varphi_2 = 0 \quad (t_{j-1} \leqslant t \leqslant t_{j+1})\,,$$

$$B\varphi_2^{j-1} = B\varphi_1^{j}\,,$$

$$B\frac{\partial \varphi_3}{\partial t} + A_1\varphi_3 = 0 \quad (t_j \leqslant t \leqslant t_{j+1})\,,$$

$$B\varphi_3^{j} = B\varphi_2^{j+1}\,. \tag{3.85}$$

Problem (3.85) can be rewritten in the componentwise form at the first step $(t_{j-1} \leqslant t \leqslant t_j)$, thus

$$\frac{\partial u_1}{\partial t} + \operatorname{div} \mathbf{u}^j u_1 = 0\,,$$

$$\frac{\partial v_1}{\partial t} + \operatorname{div} \mathbf{u}^j v_1 = 0\,,$$

$$\frac{\partial T_1}{\partial t} + \operatorname{div} \mathbf{u}^j T_1 = 0\,,$$

$$\frac{\partial S_1}{\partial t} + \operatorname{div} \mathbf{u}^j S_1 = 0 \tag{3.86}$$

if

$$\operatorname{div} \mathbf{u}^j = 0 ,$$

$$u_n = 0 \qquad \text{at } \sigma \tag{3.87}$$

and the initial data are

$$u_1^{j-1} = u^{j-1} , \quad v_1^{j-1} = v^{j-1} , \quad T_1^{j-1} = T^{j-1} , \quad S_1^{j-1} = S^{j-1} ; \tag{3.88}$$

at the second step ($t_{j-1} \leqslant t \leqslant t_{j+1}$) in the form:

$$\frac{\partial u_2}{\partial t} - l v_2 + \frac{1}{\rho_0} \frac{\partial P_2}{\partial x} = 0 ,$$

$$\frac{\partial v_2}{\partial t} + l u_2 + \frac{1}{\rho_0} \frac{\partial P_2}{\partial y} = 0 ,$$

$$\frac{\partial T_2}{\partial t} + \gamma_T w_2 = 0 ,$$

$$\frac{\partial S_2}{\partial t} + \gamma_S w_2 = 0 , \tag{3.89}$$

where between the components of the solution there are connections in the form of the equations:

$$\frac{\partial P_2}{\partial z} = g(\alpha_T T_2 + \alpha_S S_2) ,$$

$$\frac{\partial u_2}{\partial x} + \frac{\partial v_2}{\partial y} + \frac{\partial w_2}{\partial z} = 0 , \tag{3.90}$$

if

$$w_2 = 0 \qquad \text{at } z = 0 ,$$

$$w_2 = 0 \qquad \text{at } z = H ,$$

$$(u_2)_h = 0 \qquad \text{at } \sigma \tag{3.91}$$

and the initial data are

$$u_2^{j-1} = u_1^j , \quad v_2^{j-1} = v_1^j , \quad T_2^{j-1} = T_1^j , \quad S_2^{j-1} = S_1^j . \tag{3.92}$$

Finally, at the last step of splitting ($t_j \leqslant t \leqslant t_{j+1}$):

$$\frac{\partial u_3}{\partial t} + \operatorname{div} \mathbf{u}^j u_3 = 0 ,$$

$$\frac{\partial v_3}{\partial t} + \operatorname{div} \mathbf{u}^j v_3 = 0 ,$$

$$\frac{\partial T_3}{\partial t} + \operatorname{div} \mathbf{u}^j T_3 = 0 , \tag{3.93}$$

if

$$\frac{\partial S_3}{\partial t} + \operatorname{div} \mathbf{u}^j S_3 = 0$$

$$\operatorname{div} \mathbf{u}^j = 0 ,$$

$$u_n^j = 0 \qquad \text{at } \sigma \tag{3.94}$$

and the initial data are

$$u_3^j = u_2^{j+1} , \quad v_3^j = v_2^{j+1} , \quad T_3^j = T_2^{j+1} , \quad S_3^j = S_2^{j+1} . \tag{3.95}$$

Note that the functions φ_α do not possess any definite physical meaning, i.e. they are auxiliary values, with the help of which we can find the approximate solution of the problem

$$\varphi_3^{j+1} = \varphi^{j+1} , \tag{3.96}$$

which has the physical meaning of the solution of Eqs. (3.43)–(3.45),

$$\varphi^{j+1} = \varphi(t_{j+1}) .$$

Now we will analyze the approximation of problem (3.43)–(3.45) with the help of split systems (3.83), (3.84). It should be noted that if the initial problems (3.43)–(3.45) were evolutional, the approximation by time according to the system of split equations could be solved trivially on the basis of the general theory presented above.

However, our problem, unfortunately, is not evolutional as it is not resolved relative to the first derivatives by time for all the components of the solution. Thus, we cannot determine directly from the system such derivatives as $\partial P/\partial t$ and $\partial w/\partial t$. So, in order to solve the problem of the approximation we will carry out some additional investigations.

3.3 The Evolutional Statement of the Problem

Let us show that initial problem (3.43)–(3.45) can be reduced to an evolutional one. For this purpose we will first consider problem (3.43) with more general boundary conditions. Namely, we will consider the system of equations:

$$\frac{\partial u}{\partial t} + \operatorname{div} \mathbf{u}^j u - lv + \frac{1}{\rho_0} \frac{\partial P}{\partial x} = 0 ,$$

$$\frac{\partial v}{\partial t} + \operatorname{div} \mathbf{u}^j v + lu + \frac{1}{\rho_0} \frac{\partial P}{\partial y} = 0 ,$$

$$\frac{\partial P}{\partial z} = g(\alpha_T T + \alpha_S S) ,$$

$$\frac{\partial u}{\partial x} + \frac{\partial v}{\partial y} + \frac{\partial w}{\partial z} = 0 ,$$

$$\frac{\partial T}{\partial t} + \operatorname{div} \mathbf{u}^j T + \gamma_T w = 0\,,$$

$$\frac{\partial S}{\partial t} + \operatorname{div} \mathbf{u}^j S + \gamma_S w = 0\,, \tag{3.97}$$

where the boundary conditions are:

$$\frac{\partial P}{\partial t} + g\rho_0 w = 0 \qquad \text{at } z = 0\,,$$

$$w = 0 \qquad \text{at } z = H\,,$$

$$u_n = 0 \qquad \text{at } \sigma\,, \tag{3.98}$$

and initial data are

$$u = u^0\,, \quad v = v^0\,, \quad T = T^0\,, \quad S = S^0 \qquad \text{at } t = 0\,,$$

$$P = P_0 \qquad \text{at } z = 0, t = 0\,. \tag{3.99}$$

Let us consider the first of the condition (3.98) in detail. We can show that this condition is the consequence of the fact that the SST is a free surface, the equation of which we will write in the form:

$$z = -\zeta(x, y, t)\,.$$

Further, we will differentiate this equation by t. Then we obtain:

$$w = -\left(\frac{\partial \zeta}{\partial t} + u\frac{\partial \zeta}{\partial x} + v\frac{\partial \zeta}{\partial y}\right) \qquad \text{at } z = -\zeta(x, y, t)\,,$$

where $u = \frac{dx}{dt}$, $v = \frac{dy}{dt}$, $w = \frac{dz}{dt}$ are the components of the velocity vector. Obviously, the pressure P is related to the free surface level ζ by the correlation

$$P = P_a + g\rho_0\zeta \qquad \text{at } z = 0\,,$$

where P_a is an atmospheric pressure. As the solution of the problem is given with the accuracy of $P = \text{const}$, let us take $P_a = 0$.

From the last equality the value ζ can be expressed through P. Then, taking perturbations ζ as small (of an order of 0.5 m) and neglecting values $u\frac{\partial \zeta}{\partial x} + v\frac{\partial \zeta}{\partial y}$ in comparison with $\frac{\partial \zeta}{\partial t}$, we obtain to the correlation:

$$\frac{\partial P}{\partial t} + g\rho_0 w_0 = 0 \qquad \text{at } z = 0\,.$$

Under the boundary condition let us take $z = 0$ and rewrite it in the form:

$$\frac{\partial P_0}{\partial t} + g\rho_0 w_0 = 0\,,$$

where the fact that values P and w are taken at $z = 0$, is marked by the subscript "zero".

From the statics equation we have:

$$P = P_0 + g \int_0^z (\alpha_T T + \alpha_S S) dz \,. \tag{3.100}$$

Further, let us integrate the continuity equation, taking

$$w = 0 \qquad \text{at } z = H \,,$$

then we obtain:

$$w = \int_z^H \left(\frac{\partial u}{\partial x} + \frac{\partial v}{\partial y} \right) dz \tag{3.101}$$

and, so,

$$w_0 = \int_0^H \left(\frac{\partial u}{\partial x} + \frac{\partial v}{\partial y} \right) dz \,. \tag{3.102}$$

With due regard to correlation (3.102) the first one from expression (3.98) will take the form:

$$\frac{\partial P_0}{\partial t} + g\rho_0 \int_0^H \left(\frac{\partial u}{\partial x} + \frac{\partial v}{\partial y} \right) dz = 0 \,. \tag{3.103}$$

Now the functions P and w are excluded from system (3.97) with the help of correlations (3.100) and (3.101). As a result, we obtain the evolution problem:

$$\frac{\partial u}{\partial t} + \operatorname{div} \mathbf{u}^j u - lv + \frac{1}{\rho_0} \frac{\partial P_0}{\partial x} + \frac{g}{\rho_0} \frac{\partial}{\partial x} \int_0^z (\alpha_T T + \alpha_S S) dz = 0 \,,$$

$$\frac{\partial v}{\partial t} + \operatorname{div} \mathbf{u}^j v + lu + \frac{1}{\rho_0} \frac{\partial P_0}{\partial y} + \frac{g}{\rho_0} \frac{\partial}{\partial y} \int_0^z (\alpha_T T + \alpha_S S) dz = 0 \,,$$

$$\frac{\partial T}{\partial t} + \operatorname{div} \mathbf{u}^j T + \gamma_T \int_z^H \left(\frac{\partial u}{\partial x} + \frac{\partial v}{\partial y} \right) dz = 0 \,,$$

$$\frac{\partial S}{\partial t} + \operatorname{div} \mathbf{u}^j S + \gamma_S \int_z^H \left(\frac{\partial u}{\partial x} + \frac{\partial v}{\partial y} \right) dz = 0 \,,$$

$$\frac{\partial P_0}{\partial t} + g\rho_0 \int_0^H \left(\frac{\partial u}{\partial x} + \frac{\partial v}{\partial y} \right) dz = 0 \,, \tag{3.104}$$

where the boundary condition is

$$u_n = 0 \qquad \text{at } \sigma \tag{3.105}$$

and the initial data are

$$u = u^0 \,, \quad v = v^0 \,, \quad T = T^0 \,, \quad S = S^0 \,, \quad \bar{P} = \bar{P}^0 \quad \text{at } t = 0 \,. \tag{3.106}$$

Problem (3.104)–(3.106) is equivalent to problem (3.97)–(3.99). As the splitting method is proved very easily for the evolution problem, let us carry out the two-cyclic splitting method for problem (3.104)–(3.106) at the interval $t_{j-1} \leqslant t \leqslant t_{j+1}$. Then at the first step ($t_{j-1} \leqslant t \leqslant t_j$) we will have the problem of the substances' transfer:

$$\frac{\partial u_1}{\partial t} + \operatorname{div} \mathbf{u}^j u_1 = 0\,,$$

$$\frac{\partial v_1}{\partial t} + \operatorname{div} \mathbf{u}^j v_1 = 0\,,$$

$$\frac{\partial T_1}{\partial t} + \operatorname{div} \mathbf{u}^j T_1 = 0\,,$$

$$\frac{\partial S_1}{\partial t} + \operatorname{div} \mathbf{u}^j S_1 = 0\,, \tag{3.107}$$

with the condition

$$u_n^j = 0 \qquad \text{at } \sigma \tag{3.108}$$

and the initial data

$$u_1^{j-1} = u^{j-1}\,, \quad v_1^{j-1} = v^{j-1}\,, \quad T_1^{j-1} = T^{j-1}\,, \quad S_1^{j-1} = S^{j-1}\,. \tag{3.109}$$

The adaptation of fields at the interval $t_{j-1} \leqslant t \leqslant t_{j+1}$ can be formulated as follows:

$$\frac{\partial u_2}{\partial t} - l v_2 + \frac{1}{\rho_0}\frac{\partial P_0}{\partial x} + \frac{g}{\rho_0}\frac{\partial}{\partial x}\int_0^z (\alpha_T T_2 + \alpha_S S_2)dz = 0\,,$$

$$\frac{\partial v_2}{\partial t} + l u_2 + \frac{1}{\rho_0}\frac{\partial P_0}{\partial y} + \frac{g}{\rho_0}\frac{\partial}{\partial y}\int_0^z (\alpha_T T_2 + \alpha_S S_2)dz = 0\,,$$

$$\frac{\partial T_2}{\partial t} + \gamma_T \int_z^H \left(\frac{\partial u_2}{\partial x} + \frac{\partial v_2}{\partial y}\right)dz = 0\,,$$

$$\frac{\partial S_2}{\partial t} + \gamma_S \int_z^H \left(\frac{\partial u_2}{\partial x} + \frac{\partial v_2}{\partial y}\right)dz = 0\,,$$

$$\frac{\partial P_0}{\partial t} + g\rho_0 \int_0^H \left(\frac{\partial u_2}{\partial x} + \frac{\partial v_2}{\partial y}\right)dz = 0\,, \tag{3.110}$$

with the boundary condition

$$u_n = 0 \qquad \text{at } \sigma \tag{3.111}$$

and the initial data

$$u_2^{j-1} = u_1^j\,; \quad v_2^{j-1} = v_1^j\,; \quad T_2^{j-1} = T_1^j\,; \quad S_2^{j-1} = S_1^j\,; \quad \bar{P}^{j-1} = \bar{P}_0^j\,. \tag{3.112}$$

Finally, at the time interval $t_j \leqslant t \leqslant t_{j+1}$ we will again consider the problem of the substances' transfer:

$$\frac{\partial u_3}{\partial t} + \operatorname{div} \mathbf{u}^j u_3 = 0\,,$$

$$\frac{\partial v_3}{\partial t} + \operatorname{div} \mathbf{u}^j v_3 = 0\,,$$

$$\frac{\partial T_3}{\partial t} + \operatorname{div} \mathbf{u}^j T_3 = 0 \,,$$

$$\frac{\partial S_3}{\partial t} + \operatorname{div} \mathbf{u}^j S_3 = 0 \,, \tag{3.113}$$

with the condition

$$u_n^j = 0 \qquad \text{at } \sigma \tag{3.114}$$

and the initial data

$$u_3^j = u_2^{j+1}\,, \quad v_3^j = v_2^{j+1}\,, \quad T_3^j = T_2^{j+1}\,, \quad S_3^j = S_2^{j+1}\,. \tag{3.115}$$

Now we will analyze the adaptation problem (3.110)–(3.112). First, let us take into consideration the fact that in the last of Eqs. (3.110) the term $\partial P_0/\partial t$ is smaller than each of the other terms by approximately two orders of magnitude, therefore it is possible to neglect this term. It is equivalent to the filtering of acoustic waves, which are not essential for non-stationary problems of ocean dynamics.

Second, let us introduce the average according to the values of the ocean's depth $\bar{P}$, $\bar{u}$ and $\bar{v}$, by the formulas:

$$\bar{P} = \frac{1}{H}\int_0^H P dz\,; \qquad \bar{u} = \frac{1}{H}\int_0^H u dz\,; \qquad \bar{v} = \frac{1}{H}\int_0^H v dz\,,$$

assuming that

$$u_2 = \bar{u} + u'\,,$$

$$v_2 = \bar{v} + v'\,,$$

$$P = \bar{P} + P'\,,$$

$$T_2 = T'\,,$$

$$S_2 = S'\,. \tag{3.116}$$

Then, by substituting values (3.116) into (3.110) we come to the problem of barotropic motion:

$$\frac{\partial \bar{u}}{\partial t} - l\bar{v} + \frac{1}{\rho_0}\frac{\partial \bar{P}}{\partial x} = 0\,,$$

$$\frac{\partial \bar{v}}{\partial t} + l\bar{u} + \frac{1}{\rho_0}\frac{\partial \bar{P}}{\partial y} = 0\,,$$

$$\frac{\partial \bar{u}}{\partial x} + \frac{\partial \bar{v}}{\partial y} = 0\,. \tag{3.117}$$

To this system of equations we add the boundary conditions:

$$(\bar{u})_n = 0 \qquad \text{at } \sigma \tag{3.118}$$

and the initial data

$$\bar{u}^{j-1} = \frac{1}{H}\int_0^H u_1^j dz\,, \qquad \bar{v}^{j-1} = \frac{1}{H}\int_0^H v_1^j dz\,. \tag{3.119}$$

Now we can formulate the system of the adaptation equations for the deviations:

$$\frac{\partial u'}{\partial t} - lv' + \frac{g}{\rho_0}\frac{\partial}{\partial x}\int_0^z (\alpha_T T' + \alpha_S S')dz = 0\,,$$

$$\frac{\partial v'}{\partial t} + lu' + \frac{g}{\rho_0}\frac{\partial}{\partial y}\int_0^z (\alpha_T T' + \alpha_S S')dz = 0\,,$$

$$\frac{\partial T'}{\partial t} + \gamma_T \int_z^H \left(\frac{\partial u'}{\partial x} + \frac{\partial v'}{\partial y}\right) dz = 0\,,$$

$$\frac{\partial S'}{\partial t} + \gamma_S \int_z^H \left(\frac{\partial u'}{\partial x} + \frac{\partial v'}{\partial y}\right) dz = 0\,. \tag{3.120}$$

As boundary conditions we assume:

$$u_n = 0 \qquad \text{at } \sigma \tag{3.121}$$

and initial data

$$u'^{j-1} = u^j - \bar{u}^{j-1}\,; \quad v'^{j-1} = v_1^j - \bar{v}^{j-1}\,; \quad T'^{j-1} = T_1^j\,, \quad S'^{j-1} = S_1^j\,. \tag{3.122}$$

It is easy to show that the evolution problem (3.117)–(3.122) is equivalent to problem (3.89)–(3.92).

Thus, the algorithm of the splitting of problem (3.43)–(3.45) in the form of (3.86)–(3.95) with respect to conditions (3.82) is theoretically grounded.

On the basis of the general results of Section 3.2 we can state that if each of the three problems (3.86)–(3.88), (3.89)–(3.92) and (3.93)–(3.95) is solved with the help of difference methods by time with the second-order accuracy, then the splitting method leads to the resultant scheme, which is also of second-order accuracy. The absolute stability of splitting schemes follows from the conditions of definiteness of the operators A_α.

The above formulated assumption on the filtering of the acoustic waves is not principal. All the subsequent algorithms in solving ocean dynamics problems are obviously generalized in this case also. However, the solution of this problem is left to our reader.

Now we will consider the numerical solution of the split "elementary" problem.

3.4 The Difference Schemes for the Equations of Motion

We will now introduce a grid net before discussing the construction of difference approximations for problems (3.86)–(3.95). As the boundary of the domain D, according to our assumption, consists of parts of the coordinate planes, then while crossing the coordinate planes $x = x_k$, $y = y_l$, $z = z_m$ we obtain the set of the main grid points. Let us assume that the indexes k, l and m vary in D_h in the limits:

$$k^0(l,m) \leqslant k \leqslant k^1(l,m)\,,$$

$$l^0(k,m) \leqslant l \leqslant l^1(k,m)\,,$$

$$m^0(k,l) \leqslant m \leqslant m^1(k,l)\,.$$

In this case we assume that the boundary points, the numbering of which is seen from the inequality for the indexes, coincide with the boundary ∂D of the domain D. We will denote the set of such points by ∂D_h.

Let us introduce the main points x_k, y_l, z_m and the auxiliary ones $x_{k+1/2}$, $y_{l+1/2}$, $z_{m+1/2}$, which are the centres of the basic intervals and denote

$$\Delta x_{k+1/2} = x_{k+1} - x_k\,, \quad \Delta y_{l+1/2} = y_{l+1} - y_l\,, \quad \Delta z_{m+1/2} = z_{m+1} - z_m$$

and

$$\Delta x_k = \tfrac{1}{2}(x_{k+1} - x_{k-1})\,, \quad \Delta y_l = \tfrac{1}{2}(y_{l+1} - y_{l-1})\,, \quad z_m = \tfrac{1}{2}(z_{m+1} - z_{m-1})\,.$$

Further, we will consider problem (3.86)–(3.88). [Problem (3.93)–(3.95) will also be considered.] As all the equations of these systems are alike, we will consider, for instance, the following:[6]

$$\frac{\partial \varphi}{\partial t} + \operatorname{div} \mathbf{u}^j \varphi = 0 \qquad \text{in } D\,, \tag{3.123}$$

and the conditions

$$\begin{aligned} &\operatorname{div} \mathbf{u}^j = 0 && \text{in } D\,, \\ &\mathbf{u}_n^j = 0 && \text{at } \sigma\,, \end{aligned} \tag{3.124}$$

where φ is any of functions $\{u, v, T, S\}$. Let us note that no demands are made on the solution φ except for the conditions of the necessary smoothness, but, as it is seen from Eq. (3.124), the fulfillment of some conditions for the coefficients of the equation, i.e. the components of the vector $\mathbf{u}^j$, is necessary.

We will approximate Eq. (3.123) by the spacial variables in the following form:

$$\frac{d\varphi}{dt} + \Lambda \varphi = 0\,, \tag{3.125}$$

where

$$\Lambda = \Lambda_1 + \Lambda_2 + \Lambda_3\,,$$

and φ is a vector function with the components $\{\varphi_{k,l,m}\}$, which are determined in D_h; Λ_α are such matrixes, whereby the values:

$$(\Lambda_1 \varphi)_{k,l,m} = \frac{u^j_{k+1/2}\varphi_{k+1} - u^j_{k-1/2}\varphi_{k-1}}{2\Delta x_k}\,,$$

$$(\Lambda_2 \varphi)_{k,l,m} = \frac{v^j_{l+1/2}\varphi_{l+1} - v^j_{l-1/2}\varphi_{l-1}}{2\Delta y_l}\,,$$

$$(\Lambda_3 \varphi)_{k,l,m} = \frac{w^j_{m+1/2}\varphi_{m+1} - w^j_{m-1/2}\varphi_{m-1}}{2\Delta z_m}\,,$$

are the components of the vectors $\Lambda_\alpha \varphi$.

---

[6] One should not confuse the function φ in Eqs. (3.123)–(3.124) and the solution of problem (3.43)–(3.45), which are denoted by the same symbol.

Let us assume that the coefficients u^j, v^j and w^j satisfy the continuity equation in the form:

$$\frac{u^j_{k+1} - u^j_{k-1}}{2\Delta x_k} + \frac{v^j_{l+1} - v^j_{l-1}}{2\Delta y_l} + \frac{w^j_{m+1} - w^j_{m-1}}{2\Delta z_m} = 0\,, \tag{3.126}$$

and the normal component of the vector $\mathbf{u}^j$ becomes zero in the centre of each interval, which is in the direct neighbourhood of the domain boundary. We denote the set of these points by $\partial D_h^{1/2}$. Then we have

$$u^j_n = 0 \qquad \text{at } \partial D_h^{1/2}\,.$$

Further, we introduce the scalar product

$$(a, b) = \sum_{i=1}^{6} \sum_{k,l,m} a^{(i)}_{klm} b^{(i)}_{klm} \Delta x_k \Delta y_l \Delta z_m\,, \tag{3.127}$$

where the summation can be made with respect to all internal points of the domain D_h. Then it is not difficult to show that

$$(\Lambda_\alpha \varphi, \varphi) = 0 \qquad (\alpha = 1, 2, 3)\,. \tag{3.128}$$

If Eq. (3.125) is multiplied by φ scalarly, then, taking into account Eq. (3.128), we obtain:

$$\frac{1}{2}\frac{d}{dt}(\varphi, \varphi) = 0\,.$$

If we now multiply the difference analogues of Eqs. (3.86) and (3.93), constructed with the help of the above formulated algorithm, by $\rho_0, \rho_0, \dfrac{\alpha_T g}{\gamma_T}$ and $\dfrac{\alpha_S g}{\gamma_S}$ respectively, and add the result and afterward sum it together with the elements of the volumes Δx_k, Δy_l, Δz_m with respect to all the internal points of the domain D_h, then, taking into account condition (3.126), we obtain:

$$\frac{1}{2}\frac{d}{dt}(B\varphi, \varphi) = 0 \tag{3.129}$$

or

$$(B\varphi^j, \varphi^j) = (B\varphi^{j-1}, \varphi^{j-1})\,. \tag{3.130}$$

We can show that problem (3.125), on the basis of the results of Section 3.2, has the second-order approximation on a regular grid, and due to condition (3.128) it is prepared for the splitting into elementary ones. As the matrixes Λ_α are noncommutative, we will use the two-cyclic method of componentwise splitting.

Then we have the first cycle:

$$\frac{\varphi^{j-5/6} - \varphi^{j-1}}{\tau/2} + \Lambda_1 \frac{\varphi^{j-5/6} + \varphi^{j-1}}{2} = 0\,,$$

$$\frac{\varphi^{j-4/6} - \varphi^{j-5/6}}{\tau/2} + \Lambda_2 \frac{\varphi^{j-4/6} + \varphi^{j-5/6}}{2} = 0\,,$$

$$\frac{\varphi^{j-3/6} - \varphi^{j-4/6}}{\tau/2} + \Lambda_3 \frac{\varphi^{j-3/6} + \varphi^{j-4/6}}{2} = 0\,, \tag{3.131}$$

and then the second one:

$$\frac{\varphi^{j-2/6} - \varphi^{j-3/6}}{\tau/2} + \Lambda_3 \frac{\varphi^{j-2/6} + \varphi^{j-3/6}}{2} = 0\,,$$

$$\frac{\varphi^{j-1/6} - \varphi^{j-2/6}}{\tau/2} + \Lambda_2 \frac{\varphi^{j-1/6} + \varphi^{j-2/6}}{2} = 0\,,$$

$$\frac{\varphi^{j} - \varphi^{j-1/6}}{\tau/2} + \Lambda_1 \frac{\varphi^{j} + \varphi^{j-1/6}}{2} = 0\,. \tag{3.132}$$

It should be noted that the calculations can be reduced to some extent if we unite the last of the equations (3.131) with the first of the equations (3.132) and write these two problems in the form of one, thus:

$$\frac{\varphi^{j-2/6} - \varphi^{j-4/6}}{\tau} + \Lambda_3 \frac{\varphi^{j-2/6} + \varphi^{j-4/6}}{2} = 0\,. \tag{3.133}$$

The consequent solution of Eqs. (3.131), (3.132) allows one to obtain the solution of problem (3.125) with the second-order accuracy by τ (see Sect. 3.2). As a result we come to the difference scheme, which approximates the equations of motion with the second-order accuracy by τ, which is absolutely stable at the interval $t_{j-1} \leqslant t \leqslant t_j$:

$$(B\varphi^j, \varphi^j) = (B\varphi^{j-1/6}, \varphi^{j-1/6}) = \cdots = (B\varphi^{j-5/6}, \varphi^{j-5/6}) = (B\varphi^{j-1}, \varphi^{j-1})\,.$$

If the spacial grid net is regular we also come to the scheme of the second-order accuracy by Δx, Δy, Δz.

Likewise, the equations of the substances' transfer can be solved at the concluding third cycle when $t_j \leqslant t \leqslant t_{j+1}$.

3.5 The Approximation of Adaptation Equations by Spacial Variables

Now we will consider the difference approximation of the adaptation of the physical fields, which are slightly non-adjusted after the solution of the equations of the substances' transfer along the trajectories at the first stage of splitting. We will solve this problem at the interval $t_{j-1} \leqslant t \leqslant t_{j+1}$. As it was noted above, by uniting the second stage in splitting scheme (3.83) and the first stage in splitting scheme (3.84), we obtain the splitting scheme in the form of Eq. (3.85).

The approximation by spacial variables is carried out in order to solve problem (3.89)–(3.92). For this purpose we will cover the domain D with the grid net described in Section 3.4 and introduce the difference operators of two types:

$$\nabla_P^+ \varphi = \frac{\varphi_{P+1} - \varphi_P}{\Delta x_{P+1/2}}\,, \qquad \nabla_P^- \varphi = \frac{\varphi_P - \varphi_{P-1}}{\Delta x_{P-1/2}}\,,$$

where x_P is any of the coordinates $\{x_k, y_l, z_m\}$. These operators are connected by the relation:

$$\nabla_P^- \varphi = \nabla_{P-1}^+ \varphi\,.$$

Now we return to problem (3.89)–(3.92), omit index "2" for the variables and consider the following approximation of this problem

$$\rho_0 \frac{du}{dt} - \rho_0 lv + \nabla_k^+ P = 0 ,$$

$$\rho_0 \frac{dv}{dt} + \rho_0 lu + \nabla_l^+ P = 0 ,$$

$$\nabla_m^+ P - g(\alpha_T T + \alpha_S S) = 0 ,$$

$$\nabla_k^- u + \nabla_l^- v + \nabla_m^- w = 0 ,$$

$$\frac{g\alpha_T}{\gamma_T}\frac{dT}{dt} + g\alpha_T w = 0 ,$$

$$\frac{g\alpha_S}{\gamma_S}\frac{dS}{dt} + g\alpha_S w = 0 , \tag{3.134}$$

if the components of the velocity vector, normal to the boundary ∂D, become zero at the points of ∂D_h, i.e.

$$u_n = 0 \qquad \text{in } \partial D_h . \tag{3.135}$$

The properties involved in the solving problem (3.134), will now be discussed. First, the equations of system (3.134) are multiplied by $u\delta D^+$, $v\delta D^+$, $w\delta D^+$, $P\delta D^-$, $T\delta D^+$, $S\delta D^+$ respectively, where

$$\delta D^+ = \Delta x_{k+1/2}\Delta y_{l+1/2}\Delta z_{m+1/2} ,$$

$$\delta D^- = \Delta x_{k-1/2}\Delta y_{l-1/2}\Delta z_{m-1/2} ,$$

then the result is summed with respect to all the indexes from D_h, taking into account conditions (3.135). Then it is not difficult to obtain:

$$\frac{1}{2}\frac{d}{dt}(B\varphi, \varphi) = 0 . \tag{3.136}$$

Here, the vector function φ is the solution of problem (3.134), (3.135) with the components $\{u, v, w, P, T, S\}$ at the points of D_h, and B is the matrix, determined in Eq. (3.46). Unfortunately, the scalar product in Eq. (3.136) is determined in a somewhat different manner than when studying the difference equations in Eq. (3.128), namely, in the present case we have

$$(a, b) = \sum_{i=1}^{6} \sum_{k,l,m} a_{k,l,m}^{(i)} b_{k,l,m}^{(i)} \Delta x_{k+1/2}\Delta y_{l+1/2}\Delta z_{m+1/2} \tag{3.137}$$

since in Eq. (3.128) we had

$$(a, b) = \sum_{i=1}^{6} \sum_{k,l,m} a_{k,l,m}^{(i)} b_{k,l,m}^{(i)} \Delta x_k \Delta y_l \Delta z_m .$$

If the domain D is covered with regular grid by Δx, Δy and Δz, then the definitions of scalar products (3.127) and (3.137) coincide with each other. And in this case we deal with the coordinated scalar products for all splitting stages of the problem. However, on irregular grids these scalar products remain slightly non-coordinated. But in this case the discrepancy between the values of the scalar products diminishes if the grid steps decreases. In this very sense of such an asymptotic approach we will understand the coordination of two such determinations.

Now we will consider the method of solving problem (3.134), (3.135) with the corresponding initial conditions.

First, we will separate the barotropic component of the solution from problem (3.134), (3.135). According to the above performed analysis we will assume:

$$\begin{aligned}
u &= \bar{u} + u' , \\
v &= \bar{v} + v' , \\
w &= w' , \\
P &= \bar{P} + P' , \\
T &= T' , \\
S &= S' ,
\end{aligned} \tag{3.138}$$

where $\bar{u}$, $\bar{v}$ and $\bar{P}$ depend only on (x, y, t). Then two problems become apparent, the barotropic one:

$$\begin{aligned}
&\frac{d\bar{u}}{dt} - l\bar{v} + \frac{1}{\rho_0} \nabla_k^+ \bar{P} = 0 , \\
&\frac{d\bar{v}}{dt} + l\bar{u} + \frac{1}{\rho_0} \nabla_l^+ \bar{P} = 0 , \\
&\nabla_k^- \bar{u} + \nabla_l^- \bar{v} = 0 ,
\end{aligned} \tag{3.139}$$

and

$$\bar{u}_n = 0 \qquad \text{at } \partial D_h , \tag{3.140}$$

and the baroclinic one:

$$\begin{aligned}
&\frac{du'}{dt} - lv' + \frac{1}{\rho_0} \nabla_k^+ P' = 0 , \\
&\frac{dv'}{dt} + lu' + \frac{1}{\rho_0} \nabla_l^+ P' = 0 , \\
&\nabla_m^+ P' - g(\alpha_T T' + \alpha_S S') = 0 , \\
&\nabla_k^- u' + \nabla_l^- v' + \nabla_m^- w' = 0 , \\
&\frac{dT'}{dt} + \gamma_T w' = 0, \quad \frac{dS'}{dt} + \gamma_S w' = 0 ,
\end{aligned} \tag{3.141}$$

and

$$u_n' = 0 \qquad \text{at } \partial D_h . \tag{3.142}$$

Such a representation of the solution of the system of equations of adaptation is the result of some physical considerations. It is known, for instance, that for the large-scale ocean processes, the barotropic components of the Coriolis force is balanced by the force of the baric gradient, i.e. it is approximately

$$-l\bar{v} + \frac{1}{\rho_0}\frac{\partial}{\partial x}\bar{P} = 0\,,$$

$$l\bar{u} + \frac{1}{\rho_0}\frac{\partial}{\partial y}\bar{P} = 0\,, \tag{3.143}$$

from the equator.

Further, in a sufficiently large domain of the ocean the Coriolis parameter can be taken as a constant. Then it is easy to see that the components $\bar{u}$ and $\bar{v}$ from Eq. (3.143) satisfy the continuity equation:

$$\frac{\partial}{\partial x}\bar{u} + \frac{\partial}{\partial y}\bar{v} = 0\,. \tag{3.144}$$

This means that the processes, describing the barotropic component of the field, are balanced so well, that it is impossible to split them. Thus, here have such an elementary problem, in which further splitting is unnecessary. Now we will consider the determinations of the dynamic baroclinic components, described by problem (3.141), (3.142). After the separation of the barotropic component, the connection between the individual components of the process is not as strong as in the above considered case. For instance, in the continuity equation the components $\nabla_k^- u$ and $\nabla_l^- v$ no longer play an exclusive role, i.e. each of the terms has the same order as $\nabla_m^- w$. This circumstance opens new possibilities to the further splitting of our problem.

But first we will show that with the accurate division of the problem into two, the energetic equation (3.136) is not violated. Thus, let us introduce the vectors

$$\bar{\varphi} = \begin{Vmatrix} \bar{u} \\ \bar{v} \\ 0 \\ \bar{P} \\ 0 \\ 0 \end{Vmatrix}, \qquad \varphi' = \begin{Vmatrix} u' \\ v' \\ w' \\ P' \\ T' \\ S' \end{Vmatrix},$$

where

$$\bar{P} = \frac{1}{H}\int_0^H P\,dz\,, \quad \bar{u} = \frac{1}{H}\int_0^H u\,dz\,, \quad \bar{v} = \frac{1}{H}\int_0^H v\,dz$$

and, so

$$\int_0^H u'\,dz = 0\,, \quad \int_0^H v'\,dz = 0\,, \quad \int_0^H P'\,dz = 0\,.$$

Further, we will formulate an expression for the function

$$(B\varphi, \varphi) = (B\bar{\varphi}, \bar{\varphi}) + (B\bar{\varphi}, \varphi') + (B\varphi', \bar{\varphi}) + (B\varphi', \varphi')\,;$$

since

$$(B\bar{\varphi}, \varphi) = 0\,, \qquad (B\varphi', \bar{\varphi}) = 0\,,$$

then

$$(B\varphi, \varphi) = (B\bar{\varphi}, \bar{\varphi}) + (B\varphi', \varphi')\,, \tag{3.145}$$

and this means that from the conditions:

$$(B\bar{\varphi}, \bar{\varphi}) = \text{const}\,, \qquad (B\varphi', \varphi') = \text{const}$$

at the interval $t_{j-1} \leqslant t \leqslant t_{j+1}$ follows

$$(B\varphi, \varphi) = \text{const}\,.$$

which is equivalent to Eq. (3.136).

For this purpose, following the notations of Eq. (3.85), we can formulate the initial problem (3.134)–(3.135) without the barotropic component at the interval $t_{j-1} \leqslant t \leqslant t_{j+1}$ in the operator form, omitting the primes at the unknown functions

$$B\frac{d\varphi}{dt} + A_2\varphi = 0\,,$$

$$B\varphi = B\varphi^{j-1} \qquad \text{when } t = t_{j-1}\,, \tag{3.146}$$

where

$$A_2 = A_{2,1} + A_{2,2} + A_{2,3}$$

and

$$A_{2,1} = \begin{Vmatrix} 0 & -l\rho_0 & 0 & 0 & 0 & 0 \\ l\rho_0 & 0 & 0 & 0 & 0 & 0 \\ 0 & 0 & 0 & 0 & 0 & 0 \\ 0 & 0 & 0 & 0 & 0 & 0 \\ 0 & 0 & 0 & 0 & 0 & 0 \\ 0 & 0 & 0 & 0 & 0 & 0 \end{Vmatrix}$$

$$A_{2,2} = \begin{Vmatrix} 0 & 0 & 0 & \nabla_k^+ & 0 & 0 \\ 0 & 0 & 0 & 0 & 0 & 0 \\ 0 & 0 & 0 & \frac{1}{2}\nabla_m^+ & -\frac{1}{2}g\alpha_T & -\frac{1}{2}g\alpha_S \\ \nabla_k^- & 0 & \frac{1}{2}\nabla_m^- & 0 & 0 & 0 \\ 0 & 0 & \frac{1}{2}g\alpha_T & 0 & 0 & 0 \\ 0 & 0 & \frac{1}{2}g\alpha_S & 0 & 0 & 0 \end{Vmatrix}$$

$$A_{2,3} = \begin{Vmatrix} 0 & 0 & 0 & 0 & 0 & 0 \\ 0 & 0 & 0 & \nabla_l^+ & 0 & 0 \\ 0 & 0 & 0 & \frac{1}{2}\nabla_m^+ & -\frac{1}{2}g\alpha_T & -\frac{1}{2}g\alpha_S \\ 0 & \nabla_l^- & \frac{1}{2}\nabla_m^- & 0 & 0 & 0 \\ 0 & 0 & \frac{1}{2}g\alpha_T & 0 & 0 & 0 \\ 0 & 0 & \frac{1}{2}g\alpha_S & 0 & 0 & 0 \end{Vmatrix}\,.$$

Then the above formulated problems, which are the components of the split adaptation problem, will be formally written in the following form.

Unfortunately, the operators $A_{2,1}$, $A_{2,2}$ and $A_{2,3}$ are non-commutative, therefore we will introduce the two-cyclic method of componentwise splitting to solve problem (3.146).

At the interval $t_{j-1} \leqslant t \leqslant t_j$

$$B\frac{d\varphi_1}{dt} + A_{2,1}\varphi_1 = 0\,, \qquad B\varphi_1^{j-1} = B\varphi_1^{j-1}\,,$$

$$B\frac{d\varphi_2}{dt} + A_{2,2}\varphi_2 = 0\,, \qquad B\varphi_2^{j-1} = B\varphi_1^{j}\,,$$

$$B\frac{d\varphi_3}{dt} + A_{2,3}\varphi_3 = 0\,, \qquad B\varphi_3^{j-1} = B\varphi_2^{j}\,. \tag{3.147}$$

At the interval $t_j \leqslant t \leqslant t_{j+1}$

$$B\frac{d\varphi_4}{dt} + A_{2,3}\varphi_4 = 0\,, \qquad B\varphi_4^{j} = B\varphi_3^{j}\,,$$

$$B\frac{d\varphi_3}{dt} + A_{2,2}\varphi_5 = 0\,, \qquad B\varphi_5^{j} = B\varphi_4^{j+1}\,,$$

$$B\frac{d\varphi_6}{dt} + A_{2,1}\varphi_6 = 0\,, \qquad B\varphi_6^{j} = B\varphi_5^{j+1}\,. \tag{3.148}$$

Analogously to the manner in which the analysis of problem (3.134)–(3.135) was performed, it is not difficult to show that the correlation

$$(A_{2,\alpha}\varphi, \varphi) = 0$$

$$(\alpha = 1, 2, 3) \tag{3.149}$$

takes place, and so, for each of problems (3.147)–(3.148) the correlation

$$\frac{1}{2}\frac{d}{dt}(B\varphi, \varphi) = 0 \tag{3.150}$$

or

$$(B\varphi, \varphi) = \text{const}$$

takes place, which characterizes the fulfillment of the law of conservation of the full energy of the system.

The theoretical explanation of the possibility of such a splitting of the adaptation equation follows from the evolutional problem (3.120)–(3.122), which can be split (according to the general theory) into such components, which generate problems (3.147)–(3.148).

Thus, the problems (3.147)–(3.148) can be written in the componentwise form. For this purpose problem (3.134) is split at the interval $t_{j-1} \leqslant t \leqslant t_j$ by the physical processes:

$$\rho_0 \frac{du_1}{dt} - \rho_0 l v_1 = 0\,,$$

$$\rho_0 \frac{dv_1}{dt} + \rho_0 l u_1 = 0\,,$$

$$\frac{g\alpha_T}{\gamma_T} \frac{dT_1}{dt} = 0\,,$$

$$\frac{g\alpha_S}{\gamma_S} \frac{dS_1}{dt} = 0\,, \tag{3.151}$$

with the conditions

$$u_1^{j-1} = u^{j-1}, \quad v_1^{j-1} = v^{j-1}\,, \quad T_1^{j-1} = T^{j-1}, \quad S_1^{j-1} = S^{j-1}\,. \tag{3.152}$$

Here and in the following for simplicity, the primes of the components of the baroclinic solution are omitted, and taking the adaptation problem as an independent object of the investigation, we will introduce the indexing which is not connected with the previously used notations in the general splitting scheme of problem (3.43)–(3.45).

Further, at the same interval two problems can be solved.

The first problem is

$$\rho_0 \frac{du_2}{dt} + \nabla_k^+ P_2 = 0\,,$$

$$\rho_0 \frac{dv_2}{dt} = 0\,,$$

$$\frac{1}{2}\nabla_m^+ P_2 - \frac{1}{2} g(\alpha_T T_2 + \alpha_S S_2) = 0\,,$$

$$\nabla_k^- u_2 + \frac{1}{2}\nabla_m^- w_2 = 0\,,$$

$$\frac{g\alpha_T}{\gamma_T} \frac{dT_2}{dt} + \frac{1}{2} g\alpha_T w_2 = 0\,,$$

$$\frac{d\alpha_S}{\gamma_S} \frac{dS_2}{dt} + \frac{1}{2} g\alpha_S w_2 = 0\,, \tag{3.153}$$

where the boundary conditions are:

$$(u_2)_n = 0 \qquad \text{at } \partial D_h \tag{3.154}$$

and the initial data are:

$$u_2^{j-1} = u_1^j\,, \quad v_2^{j-1} = v_1^j\,, \quad T_2^{j-1} = T_1^j\,, \quad S_2^{j-1} = S_1^j\,. \tag{3.155}$$

The second problem is

$$\rho_0 \frac{du_3}{dt} = 0 ,$$

$$\rho_0 \frac{dv_3}{dt} + \nabla_l^+ P_3 = 0 ,$$

$$\frac{1}{2}\nabla_m^+ P_3 - \frac{1}{2} g(\alpha_T T_3 + \alpha_S S_3) = 0 ,$$

$$\nabla_l^- v_3 + \frac{1}{2}\nabla_m^- w_3 = 0 ,$$

$$\frac{g\alpha_T}{\gamma_T}\frac{dT_3}{dt} + \frac{1}{2} g\alpha_T w_3 = 0 ,$$

$$\frac{g\alpha_S}{\gamma_S}\frac{dS_3}{dt} + \frac{1}{2} g\alpha_S w_3 = 0 , \tag{3.156}$$

where the boundary conditions are:

$$(u_3)_n = 0 \qquad \text{at } \partial D_h \tag{3.157}$$

and the initial data are:

$$u_3^{j-1} = u_2^j , \quad v_3^{j-1} = v_2^j , \quad T_3^{j-1} = T_2^j , \quad S_3^{j-1} = S_2^j . \tag{3.158}$$

The calculation cycle ends with the solution of the identical problems at the interval $t_j \leqslant t \leqslant t_{j+1}$, which are performed in reverse order, namely, first we solve the equation which is analogous to Eq. (3.156), with respect to the function φ_4 with initial conditions

$$B\varphi_4^j = B\varphi_3^j ,$$

then Eq. (3.153) with respect to the new, unknown function with initial conditions

$$B\varphi_5^j = B\varphi_4^j ,$$

and finally the calculation cycle ends by determining the functions φ_6, which correspond to Eq. (3.151) and the initial conditions

$$B\varphi_6^j = B\varphi_5^{j+1} .$$

We can reduce the algorithm by uniting the third and fourth splitting steps and solving one problem at the interval $t_{j-1} \leqslant t \leqslant t_{j+1}$.

As the operators of each of the above considered problems satisfy the conditions:

$$(A_{2,\alpha}\varphi, \varphi) = 0 \qquad (\alpha = 1, 2, 3) ,$$

we obtain the condition:

$$(B\varphi'^{j+1}, \varphi'^{j+1}) = (B\varphi'^{j-1}, \varphi'^{j-1})$$

as a result of the successive omission of the intermediate values.

Further, we will take into consideration that the identity

$$(B\bar{\varphi}^{j+1}, \bar{\varphi}^{j+1}) = (B\bar{\varphi}^{j-1}, \bar{\varphi}^{j-1})$$

takes place. Then on the basis of Eq. (3.145) we obtain:

$$(B\varphi^{j+1}, \varphi^{j+1}) = (B\varphi^{j-1}, \varphi^{j-1}) . \tag{3.159}$$

3.6 The Approximation of the Adaptation Equations by Time

In order to construct the approximation of the elementary problem by time, in connection with the solution of the adaptation equations, we will use the Crank-Nickolson schemes, which guarantee the second order of approximation by time.

First, we will consider the problem of evolution of the barotropic component of solution (3.139)–(3.140). Let us approximate this problem in the following way:

$$\frac{\bar{u}^{j+1} - \bar{u}^{j-1}}{2\tau} - l\frac{\bar{v}^{j+1} + \bar{v}^{j-1}}{2} + \frac{1}{\rho_0}\nabla_k^+ \bar{P}^j = 0 ,$$

$$\frac{\bar{v}^{j+1} - \bar{v}^{j-1}}{2\tau} + l\frac{\bar{u}^{j+1} + \bar{u}^{j-1}}{2} + \frac{1}{\rho_0}\nabla_l^+ \bar{P}^j = 0 ,$$

$$\nabla_k^- \left(\frac{\bar{u}^{j+1} + \bar{u}^{j-1}}{2}\right) + \nabla_l^- \left(\frac{\bar{v}^{j+1} + \bar{v}^{j-1}}{2}\right) = 0 , \tag{3.160}$$

with the conditions

$$\bar{u}_n^j = 0 \qquad \text{at } \partial D_h .$$

Here, $\bar{P}^j$ is also an unknown value. The third of Eqs. (3.160) allows one to introduce the difference analogue of the stream function ψ by the formulas:

$$\frac{\bar{u}^{j+1} + \bar{u}^{j-1}}{2} = -\nabla_l^- \psi^j , \qquad \frac{\bar{v}^{j+1} + \bar{v}^{j-1}}{2} = \nabla_k^- \psi^j . \tag{3.161}$$

Then the continuity equation in (3.160) will be fulfilled identically, and the two first equations can be reduced to the form:

$$\nabla_l^- \psi^j + l\tau\nabla_k^- \psi^j - \frac{\tau}{\rho_0}\nabla_k^+ \bar{P}^j = -\bar{u}^{j-1} ,$$

$$l\tau\nabla_l^- \psi^j - \nabla_k^- \psi^j - \frac{\tau}{\rho_0}\nabla_l^+ \bar{P}^j = -\bar{v}^{j-1} . \tag{3.162}$$

Now the operator ∇_l^+ is applied to the first equation of the system; and to the second, ∇_k^+; and then subtracted one from another. Thus, the equation for the stream function is obtained:

$$\nabla_l^+(\nabla_l^- \psi^j) + \nabla_k^+(\nabla_k^- \psi^j) + \tau[\nabla_l^+(l\nabla_k^- \psi)^j - \nabla_k^+(l\nabla_l^- \psi)^j] = \nabla_k^+ \bar{v}^{j-1} - \nabla_l^+ \bar{u}^{j-1} . \tag{3.163}$$

If $\Delta x \to 0$ and $\Delta y \to 0$ the expression in brackets can be approximated in the following way:

$$\nabla_l^+(l\nabla_k^- \psi^j) - \nabla_k^+(l\nabla_l^- \psi^j) = \frac{\partial l}{\partial y} \frac{\nabla_k^+ + \nabla_k^-}{2} \psi^j - \frac{\partial l}{\partial x} \frac{\nabla_l^+ + \nabla_l^-}{2} \psi^j .$$

As a result we obtain:

$$\nabla_l^+(\nabla_l^- \psi^j) + \nabla_k^+(\nabla_l^- \psi^j) + \tau \left[\frac{\partial l}{\partial y} \frac{\nabla_k^+ + \nabla_k^-}{2} \psi^j - \frac{\partial l}{\partial x} \frac{\nabla_l^+ + \nabla_l^-}{2} \psi^j \right]$$
$$= \nabla_k^+ \bar{v}^{j-1} - \nabla_l^+ \bar{u}^{j-1} . \tag{3.164}$$

The condition, which is equivalent to $\bar{u}_n^j = 0$ at ∂D_h, will obviously be the requirement that the boundary of the domain ∂D_h is a stream function, i.e.

$$\psi^j = 0 \qquad \text{at } \partial D_h . \tag{3.165}$$

Thus, the problem of the barotropic state of the ocean at each interval $t_{j-1} \leqslant t \leqslant t_{j+1}$ has been reduced to the Dirichlet problem for the equation, the main part of which is the difference analogue of the Laplace equation.

After problem (3.164)–(3.165) is solved with the help of correlations (3.161), the components of the velocity vector

$$\bar{u}^{j+1} \quad \text{and} \quad \bar{v}^{j+1}$$

are determined, after which the solution of the baroclinic problem should be added to the latter in order to find the components of the total velocity vector when solving the whole dynamics problem.

Now we will consider the statement of the adaptation problem for the baroclinic component of the solution, which is described by the systems of Eqs. (3.147) and (3.148). By using the Crank-Nickolson scheme for the first problem (3.147), we obtain the difference approximation of the form:

$$\frac{u_1^j - u_1^{j-1}}{\tau} - l \frac{v_1^j + v_1^{j-1}}{2} = 0 ,$$

$$\frac{v_1^j - v_1^{j-1}}{\tau} + l \frac{u_1^j + u_1^{j-1}}{2} = 0 ,$$

$$\frac{T_1^j - T_1^{j-1}}{\tau} = 0 ,$$

$$\frac{S_1^j - S_1^{j-1}}{\tau} = 0 . \tag{3.166}$$

The initial data (when $t = t_{j-1}$) for problem (3.166) will be the solution of problem (3.132), which we can denote:

$$u_1^{j-1} = u^{j-1} , \quad v_1^{j-1} = v^{j-1} , \quad T_1^{j-1} = T^{j-1} , \quad S_1^{j-1} = S^{j-1} .$$

The next elementary problem is reduced to the difference problem:

$$\frac{u_2^j - u_2^{j-1}}{\tau} + \frac{1}{\rho_0} \nabla_k^+ P_2^{j-1/2} = 0\,,$$

$$v_2^j - v_2^{j-1} = 0\,,$$

$$\nabla_m^+ P_2^{j-1/2} - g(\alpha_T T_2^{j-1/2} + \alpha_S S_2^{j-1/2}) = 0\,,$$

$$\nabla_k^- u_2^{j-1/2} + \frac{1}{2} \nabla_m^- w_2^{j-1/2} = 0\,,$$

$$\frac{T_2^j - T_2^{j-1}}{\tau} + \frac{1}{2} \gamma_T w_2^{j-1/2} = 0\,,$$

$$\frac{S_2^j - S_2^{j-1}}{\tau} + \frac{1}{2} \gamma_S w_2^{j-1/2} = 0\,, \tag{3.167}$$

where the boundary condition is

$$(u_2^j)_n = 0 \qquad \text{at } \partial D_h\,. \tag{3.168}$$

Here, the notation

$$\varphi^{j-1/2} = \tfrac{1}{2}(\varphi^j + \varphi^{j-1})$$

is used, where φ is any of the functions u, v, S and T.

The second problem of system (3.147) has the form:

$$\frac{u_3^j - u_3^{j-1}}{\tau} = 0\,,$$

$$\frac{v_3^j - v_3^{j-1}}{\tau} + \frac{1}{\rho_0} \nabla_l^+ P_3^{j-1/2} = 0\,,$$

$$\nabla_m^+ P_3^{j-1/2} - g(\alpha_T T_3^{j-1/2} + \alpha_S S_3^{j-1/2}) = 0\,,$$

$$\nabla_l^- v_3^{j-1/2} + \frac{1}{2} \nabla_m^- w_3^{j-1/2} = 0\,,$$

$$\frac{T_3^j - T_3^{j-1}}{\tau} + \frac{1}{2} \gamma_T w_3^{j-1/2} = 0\,,$$

$$\frac{S_3^j - S_3^{j-1}}{\tau} + \frac{1}{2} \gamma_S w_3^{j-1/2} = 0\,. \tag{3.169}$$

This system of equations will be solved with the boundary conditions:

$$(u_3^j)_n = 0 \qquad \text{at } \partial D_h \tag{3.170}$$

and with the initial data

$$u_3^{j-1} = u_2^j\,, \quad v_3^{j-1} = v_2^j\,, \quad T_3^{j-1} = T_2^j\,, \quad S_1^{j-1} = S_2^j\,.$$

Systems (3.166), (3.167) and (3.169) approximate the systems of Eqs. (3.147) at the interval $t_{j-1} \leqslant t \leqslant t_j$. Further, at the interval $t_j \leqslant t \leqslant t_{j+1}$, the same three problems can be solved again, but in reverse order.

Now the numerical solution of the formulated problems will be considered. The solution of the system of Eqs. (3.166) can be found trivially:

$$u_1^j = \frac{1}{1+\left(\frac{l\tau}{2}\right)^2}\left[u^{j-1} + \frac{l\tau}{2}v^{j-1} + \frac{l\tau}{2}\left(v^{j-1} - \frac{l\tau}{2}u^{j-1}\right)\right],$$

$$v_1^j = \frac{1}{1+\left(\frac{l\tau}{2}\right)^2}\left[v^{j-1} - \frac{l\tau}{2}u^{j-1} - \frac{l\tau}{2}\left(u^{j-1} + \frac{l\tau}{2}v^{j-1}\right)\right], \tag{3.171}$$

and

$$T_1^j = T^{j-1},$$

$$S_1^j = S^{j-1}. \tag{3.172}$$

Now we will consider problem (3.167)–(3.169). The fourth equation of system (3.167) allows one to introduce the stream function ψ_2 by the correlations:

$$u_2^{j-1/2} = -\tfrac{1}{2}\nabla_m^- \Psi_2^{j-1/2},$$

$$w_2^{j-1/2} = \nabla_u^- \Psi_2^{j-1/2}. \tag{3.173}$$

From the third equation of (3.167) we find:

$$\alpha_T T_2^{j-1/2} + \alpha_s S_2^{j-1/2} = \frac{1}{g}\nabla_m^+ P_2^{j-1/2}.$$

Using the last three correlations and excluding all the unknown values in Eq. (3.167), except for $\Psi_2^{j-1/2}$ and $P^{j-1/2}$, after simple calculations we obtain the system of equations:

$$\frac{\rho_0}{\tau}\nabla_m^- \Psi_2^{j-1/2} - \nabla_k^+ P_2^{j-1/2} = -\frac{2\rho_0}{\tau}u_2^{j-1},$$

$$\nabla_m^+ P_2^{j-1/2} + \frac{\tau}{4}\gamma g \nabla_k^- \Psi_2^{j-1/2} = g\rho^{j-1}. \tag{3.174}$$

We reduce this system of equations to the one equation for the stream function:

$$\nabla_m^+(\nabla_m^- \Psi_2^{j-1/2}) + \frac{\tau^2\gamma g}{4\bar{\rho}}\nabla_k^+(\nabla_k^- \Psi_2^{j-1/2}) = -2f_2^{j-1/2},$$

$$f_2^{j-1/2} = \nabla_m^+ u^{j-1} - \frac{\tau g}{2\bar{\rho}}\nabla_k^+(\alpha_T T_2^{j-1} + \alpha_s S_2^{j-1}). \tag{3.175}$$

The natural boundary condition for the stream function, corresponding to the free-slip condition $u_n = 0$, will be

$$\Psi_2^{j-1/2} = 0 \quad \text{at } \Sigma_l, \tag{3.176}$$

where Σ_l is the contour, obtained by means of the section of the domain D_h by the plane $y = y_l$.

After the solution of problem (3.175)–(3.176) is determined, we obtain the components of the solution u^j, v^j, T^j and S^j with the help of the correlations:

$$u_2^j = -\nabla_m^- \Psi_2^{j-1/2} - u_2^{j-1},$$

$$v_2^j = v_2^{j-1},$$

$$T_2^j = T_2^{j-1} - \frac{\tau\gamma_T}{2} \nabla_k^- \Psi_2^{j-1/2},$$

$$S_2^j = S_2^{j-1} - \frac{\tau\gamma_s}{2} \nabla_k^- \Psi_2^{j-1/2}. \tag{3.177}$$

While solving problem (3.169), (3.170) we will introduce a new stream function by the correlations:

$$v_3^{j-1/2} = -\tfrac{1}{2}\nabla_m^- \Psi_3^{j-1/2},$$

$$w_3^{j-1/2} = \nabla_l^- \Psi_3^{j-1/2}. \tag{3.178}$$

Then, as in the previous case, we come to the problem for the stream function $\Psi^{j-1/2}$:

$$\nabla_m^+(\nabla_m^- \Psi_3^{j-1/2}) + \frac{\tau^2\gamma g}{4\rho_0} \nabla_l^+(\nabla_l^- \Psi_3^{j-1/2}) = -2f_3^{j+1/2},$$

$$f_3^{j+1/2} = \nabla_m^+ v_3^{j-1} - \frac{\tau g}{2\rho_0} \nabla_l^+(\alpha_T T_3^{j-1} + \alpha_s S_3^{j-1}) \tag{3.179}$$

with the boundary condition

$$\Psi_3^{j-1/2} = 0 \qquad \text{at } \Sigma_k, \tag{3.180}$$

where Σ_k is the section of the domain D_n by the plane $x = x_k$. The functions u_3^j, v_3^j, T_3^j and S_3^j will be found by the formulas:

$$u_3^j = u_3^{j-1},$$

$$v_3^j = -\nabla_m^- \Psi_3^{j-1} - v_3^{j-1},$$

$$T_3^j = T_3^{j-1} - \frac{\tau\gamma_T}{2} \nabla_l^- \Psi_3^{j-1/2},$$

$$S_3^j = S_3^{j-1} - \frac{\tau\gamma_s}{2} \nabla_l^- \Psi_3^{j-1/2}. \tag{3.181}$$

So, the algorithm of the solution of the adaptation problem is formulated completely.

It is necessary to note that the algorithm of the solution of the ocean dynamics problem does not assume that $H = \text{const}$. For instance, we can take the function $H(x, y)$ as a partial constant. This circumstance allows us to construct and solve the problems of oceanic circulations, taking into account the bottom relief. However, one should remember that in this case the theorem of the uniqueness of the solution is not proved. If we assume the uniqueness of the solution, then the algorithm allows one to carry out the solution of such a problem. Moreover, in the problems pertaining to a non-homogeneous bottom depth, we can use the linear interpolation of the relief, setting up the condition of equality of the normal component of the velocity vector to zero (the free-slip condition) for the bottom points. With regard to the components u^j, v^j and w^j at the surface $H(x, y)$, which are necessary for the solution of the equations of the substances transfer, they can be determined in the same way as in the case of the partial constant approximation of the relief, i.e. through the stream function. The problem of setting up the boundary conditions is simplified, if we take into account the vertical and horizontal macro-turbulent exchange. In this case we can set up the "non-slip" condition ($u = 0$, $v = 0$) at the solid surface.

In conclusion, we will consider the case when the ocean's depth is assumed to be constant ($H = \text{const}$). Here, the barotropic component of the adaptation problem can be calculated on the basis of the above presented algorithm, and it is not difficult to calculate it. Let us concentrate our attention on the calculations of a more complex baroclinic component of the adaptation problem. For this purpose we will use the Fourier method.

We will consider the spectral problem:

$$
\begin{aligned}
&-\nabla_m^+(\nabla_m^- Z) = \lambda z\,, \\
&z = 0 \qquad \text{at } z = 0\,, \\
&z = 0 \qquad \text{at } z = H\,.
\end{aligned}
\tag{3.182}
$$

The eigenfunctions and eigenvalues of this problem (in the case of a regular grid net) will be:

$$
z_n = \alpha_n \sin\frac{nm\pi\Delta z}{H}\,, \qquad \lambda_n = \frac{n^2\pi^2}{H^2}\,,
\tag{3.183}
$$

where α_n is a normalized constant.

We will look for the solution of the problem in the form of the series:

$$
\begin{Bmatrix} u \\ v \\ p \end{Bmatrix} = \sum_n \begin{Bmatrix} u_n^* \\ v_n^* \\ p_n^* \end{Bmatrix} z_n
$$

and

$$
\begin{Bmatrix} w \\ T \\ S \end{Bmatrix} = \sum_n \begin{Bmatrix} w_n^* \\ T_n^* \\ S_n^* \end{Bmatrix} \nabla_m^+ z_n\,.
$$

As a result it is easy to formulate the problems for the Fourier coefficients, corresponding to the splitting schemes (3.166), (3.167)–(3.168), (3.169)–(3.170). Then problem (3.166) will be transformed into the following:

$$\frac{u_1^{*j} - u_1^{*j-1}}{\tau} - l\frac{v_1^{*j} + v_1^{*j-1}}{2} = 0\,,$$

$$\frac{v_1^{*j} - v_1^{*j-1}}{\tau} + l\frac{u_1^{*j} + u_1^{*j-1}}{2} = 0\,,$$

$$T_1^{*j} = T_1^{*j-1}\,,$$

$$S_1^{*j} = S_1^{*j-1}\,; \tag{3.184}$$

and problem (3.167)–(3.168) into:

$$\frac{u_2^{*j} - u_2^{*j-1}}{\tau} + \frac{1}{\rho_0}\nabla_k^+ P_2^{*j-1/2} = 0\,,$$

$$v_2^{*j} - v_2^{*j-1}\,,$$

$$\lambda P_2^{*j-1/2} + g(\alpha_T T_2^{*j-1} + \alpha_s S_2^{*j-1/2}) = 0\,,$$

$$\frac{1}{2}\lambda w_2^{*j-1/2} - \nabla_k^- u_2^{*j-1/2} = 0\,,$$

$$\frac{T_2^{*j} - T_2^{*j-1}}{\tau} + \frac{1}{2}\gamma_T w_2^{*j-1/2} = 0\,,$$

$$\frac{S_2^{*j} - S_2^{*j-1}}{\tau} + \frac{1}{2}\gamma_s w_2^{*j-1/2} = 0 \tag{3.185}$$

with the boundary conditions

$$(u_2^{*j})_n = 0 \qquad \text{at } \partial D_n^*\,, \tag{3.186}$$

where ∂D_n^* is the projection of the cylindric surface on the plane $z = 0$.

Problem (3.169)–(3.170) is transformed into:

$$u_3^{*j} = u_3^{*j-1}\,,$$

$$\frac{v_3^{*j} - v_3^{*j-1}}{\tau} + \frac{1}{\rho_0}\nabla_l^+ P_3^{*j-1/2} = 0\,,$$

$$\lambda P_3^{*j-1/2} + g(\alpha_T T_3^{*j-1/2} + \alpha_s S_3^{*j-1/2}) = 0\,,$$

$$\frac{1}{2}\lambda w_3^{*j-1/2} - \nabla_l^- v_3^{*j-1/2} = 0\,,$$

$$\frac{T_3^{*j} - T_3^{*j-1}}{\tau} + \frac{\gamma_T}{2} w_3^{*j-1/2} = 0\,,$$

$$\frac{S_3^{*j} - S_3^{*j-1}}{\tau} + \frac{\gamma_s}{2} w_3^{*j-1/2} = 0\,. \tag{3.187}$$

This system of equations can be solved with the conditions

$$(u_3^{*j+1})_n = 0 \qquad \text{at } \partial D_n^*\,. \tag{3.188}$$

The analogous problem for the Fourier coefficients, which can be solved in reverse succession, must be formulated for the interval $t_j \leqslant t \leqslant t_{j+1}$.

The solution of problem (3.184) can be found easily. As far as the systems of Eqs. (3.185)–(3.186), (3.187), (3.188) are concerned, they turn out to be "one-dimensional" for the Fourier coefficients, i.e. they depend only on one index, either k or l, where one of them exists only parametrically. By reducing these difference equations to the equations for the stream function, we obtain the problems, corresponding to (3.175) and (3.179):

$$-\lambda \Psi_2^{*j-1/2} + \frac{\tau^2 \gamma g}{4\rho_0} V_k^+ (V_k^- \Psi_2^{*j-1/2}) = -2f^{*j-1/2} ,$$

$$f_2^{*j-1/2} = u_2^{*j-1} - \frac{\tau g}{2\rho_0} V_k^+ (\alpha_T T_2^{*j-1} + \alpha_s S_2^{*j-1}) \tag{3.189}$$

with the condition

$$\Psi_2^{*j-1/2} = 0 \qquad \text{at } \Sigma_l \tag{3.190}$$

and

$$-\lambda \Psi_3^{*j-1/2} + \frac{\tau^2 \gamma g}{4\rho_0} V_l^+ (V_l^- \Psi_3^{*j-1/2}) = -2f_3^{*j-1/2} ,$$

$$f_3^{*j-1/2} = v_3^{*j-1} - \frac{\tau g}{2\rho_0} V_l^+ (\alpha_T T_3^{*j-1} + \alpha_s S_3^{*j-1}) \tag{3.191}$$

with the condition

$$\Psi_3^{*j-1} = 0 \qquad \text{at } \Sigma_k . \tag{3.192}$$

Problems and (3.192) can be solved by the factorization method.

Let us note that we have used the expansions:

$$\Psi^{j-1/2} = \sum_n \Psi_n^{*j-1/2} V_m^+ z_n . \tag{3.193}$$

After the values $\Psi^{j-1/2}$ are found, all the necessary components of the solution of the adaptation problem can be calculated.

If we want to carry out the solution with the larger time step τ (by order of a few hours or days), then the algorithm of the splitting of adaptation equations must be substituted by the direct solution of the adaptation problem on the basis of the iterative methods.

3.7 The Choice of the Parameters for Approximation in the Simplest Model

The parameters of approximation τ, Δx, Δy and Δz in the ocean dynamics problem (at least for a regular grid net) must be chosen so, that they will guarantee the needed accuracy in the solution, not only by the order of the error value, but also the absolute value. As was repeatedly noted above, the presented calculating algorithm allows one to obtain the solution with the accuracy of the second order by τ and

Δx, Δy, Δz. But as these parameters are dimensional, then it only follows from this statement, that when $\tau \to 0$, $\Delta x \to 0$, $\Delta y \to 0$ and $\Delta z \to 0$ the error will diminish with the second order. However, for purposes of practical calculations, it is necessary to have such an algorithm of choice with the indicated parameters, which ensures the necessary accuracy of the solution. It is known that the greatest accuracy of the solution is needed at the stage when considering adaptation equations, as these equations, describe the quick-running processes of the internal gravitational waves.

The problem of approximation in such an interpretation can be solved most easily on the basis of studying the model problems, one of which we will consider later.

Let us assume that the domain D is a parallelepiped ($0 \leqslant x \leqslant a$, $0 \leqslant y \leqslant b$, $0 \leqslant z \leqslant H$). We will cover this domain by a regular grid net with the steps Δx, Δy, Δz. Along with spectral problem (3.182) we will consider two new ones:

$$-\nabla_k^+(\nabla_k^- X) = \mu X\,,$$
$$X = 0 \qquad \text{at } \partial D_h^{(1)} \tag{3.194}$$

and

$$-\nabla_l^+(\nabla_l^- Y) = \gamma Y\,,$$
$$Y = 0 \qquad \text{at } \partial D_h^{(2)}\,, \tag{3.195}$$

where $\partial D_h^{(1)}$ is the set of the points while fixing $y = y_l$, and $\partial D_h^{(2)}$ at $x = x_k$.

It is not difficult to determine that

$$X_k = \beta_k \sin\frac{k\pi\Delta x}{a}\,,$$

$$Y_l = \gamma_l \sin\frac{l\pi\Delta y}{b}\,,$$

where β_k and v_l are normalized constants.

Then the solution of the adaptation equations system (3.134) and the split systems can be carried out in the form of the Fourier series of the following construction.

The Fourier components for the velocity vector are

$$u = \sum_n u_n^* X_n \cdot \nabla_l Y_n \cdot \nabla_m Z_n\,,$$

$$v = \sum_n v_n^* \nabla_k X_n \cdot Y_n \cdot \nabla_m Z_n\,,$$

$$w = \sum_n w_n^* \nabla_k X_n \cdot \nabla_l Y_n \cdot Z_n \tag{3.196}$$

and for the other components of the solution

$$P = \sum_n P_n^* \nabla_k X_n \cdot \nabla_l Y_n \cdot \nabla_m Z_m\,,$$

$$T = \sum_n T_n^* \nabla_k X_n \cdot \nabla_l Y_n \cdot \nabla_m Z_n\,,$$

$$S = \sum_n S_n^* \nabla_k X_n \cdot \nabla_l Y_n \cdot \nabla_m Z_m\,. \tag{3.197}$$

Here, n is a generalized index of the summation, which reduces, the summation by three indexes to one.

We will substitute correlations (3.196)–(3.197) into Eq. (3.134) and after obvious operations, we obtain the equations for the Fourier coefficients:

$$\frac{du^*}{dt} + \frac{\mu}{\rho_0} P^* = 0 ,$$

$$\frac{dv^*}{dt} + \frac{v}{\rho_0} P^* = 0 ,$$

$$\lambda P^* + g(\alpha_T T^* + \alpha_s S^*) = 0 ,$$

$$u^* + v^* + w^* = 0 ,$$

$$\frac{dT^*}{dt} + \gamma_T w^* = 0 ,$$

$$\frac{dS^*}{dt} + \gamma_s w^* = 0 . \tag{3.198}$$

Here, we neglected the index n as non-essential and assumed $l = 0$, since the motions caused by the Coriolis force have a far larger characteristic time scale than the time scale for the internal gravitational waves. This means that if the approximation (by time) of the baroclinic effects is good, the equations of motion with due regard to the Coriolis forces will be described even more accurately.

We will exclude w^*, P^* from the system of Eqs. (3.198) and substitute T^* and S^* by $P^* = \alpha_T T^* + \alpha_s S^*$. Then we obtain:

$$\frac{du^*}{dt} + \frac{\mu}{\rho_0} P^* = 0 ,$$

$$\frac{dv^*}{dt} + \frac{v}{\rho_0} P^* = 0 ,$$

$$\lambda \frac{dP^*}{dt} - g_j(u^* + v^*) = 0 . \tag{3.199}$$

We will reduce the system of Eqs. (3.199) to the one equation for P^*:

$$\frac{d^2 P^*}{dt^2} + \frac{\mu + v}{\lambda} \frac{\gamma g}{\rho_0} P^* = 0 . \tag{3.200}$$

Let us assume that

$$u^* = 0 , \quad v^* = 0 , \quad P^* = f \quad \text{at } t = t_{j-1} . \tag{3.201}$$

This means that we have for P^* the initial data:

$$P^* = f ,$$

$$\frac{dP^*}{dt} = 0 \qquad \text{when } t = t_{j-1} . \tag{3.202}$$

The solution of problem (3.199), when $t = t_j$, has the form:

$$P^*(t_j) = f \cos \sqrt{\frac{\mu + v}{\lambda} \frac{g\gamma}{\rho_0}} \cdot \tau . \tag{3.203}$$

According to the splitting schemes (3.167)–(3.168) and (3.169)–(3.170), two problems become apparent.

The first problem is

$$\frac{u_2^{*j} - u_2^{*j-1}}{\tau} + \frac{\mu}{\rho_0} \frac{P_2^{*j} + P_2^{*j-1}}{2} = 0 ,$$

$$\frac{v_2^{*j} - v_2^{*j-1}}{\tau} = 0 ,$$

$$\lambda \frac{P_2^{*j} - P_2^{*j-1}}{\tau} - g_j \frac{u_2^{*j} + u_2^{*j-1}}{2} = 0 . \tag{3.204}$$

The second one is

$$\frac{u_3^{*j} - u_3^{*j-1}}{\tau} = 0 ,$$

$$\frac{v_3^{*j} - v_3^{*j-1}}{\tau} + \frac{v}{\rho_0} \frac{P_3^{*j} + P_3^{*j-1}}{2} = 0 ,$$

$$\lambda \frac{P_3^{*j} - P_3^{*j-1}}{\tau} - g_j \frac{v_3^{*j} + v_3^{*j-1}}{2} = 0 . \tag{3.205}$$

It is necessary to solve these equations with condition (3.201). As a result we obtain:

$$P_3^{*j} \equiv P^{*j} = f \frac{1 - \frac{\gamma g}{\rho_0} \frac{\mu}{\lambda} \frac{\tau^2}{4}}{1 + \frac{\gamma g}{\rho_0} \frac{\mu}{\lambda} \frac{\tau^2}{4}} \cdot \frac{1 - \frac{\gamma g}{\rho_0} \frac{v}{\lambda} \frac{\tau^2}{4}}{1 + \frac{\gamma g}{\rho_0} \frac{v}{\lambda} \frac{\tau^2}{4}} . \tag{3.206}$$

Now we will consider modulus $|P^*(t_j) - P^{*j}|$, where $P^*(t_j)$ is determined by function (3.203). This very expression will characterize the measure of the relative error in the solution. Obviously, the closeness of the approximate solution (3.206), obtained with the help of the splitting method, to the exact solution (3.203) will depend on the smallness of the dimensionless parameters:

$$\frac{\mu}{\lambda} \frac{g\gamma}{\rho_0} \frac{\tau^2}{4} \quad \text{and} \quad \frac{v}{\lambda} \frac{g\gamma}{\rho_0} \frac{\tau^2}{4} .$$

Let us assume that these parameters are small enough. Then we will expand the expression on the right-hand side of Eq. (3.206) into a series by these parameters. As a result we obtain:

$$P^{*j-1} = f \left[1 - \frac{\mu + v}{\lambda} \frac{g\gamma}{\rho_0} \frac{\tau^2}{2} + O(\tau^4) \right] . \tag{3.207}$$

Now the exact solution (3.203) is expanded into the series:

$$P^*(t_j) = f\left[1 - \frac{\mu + \nu}{\lambda}\frac{g\gamma}{\rho_0}\frac{\tau^2}{2} + O(\tau^4)\right]. \tag{3.208}$$

The comparison of expressions (3.207) and (3.208) shows that, at least with the accuracy of the values of the second order inclusively, relative to the dimensionless parameter

$$\varepsilon = \tau\sqrt{\frac{\mu + \nu}{2\lambda}\frac{g\gamma}{\rho_0}},$$

these expressions coincide with each other. Let us note that

$$c = \sqrt{\frac{\mu + \nu}{\lambda}\frac{g\gamma}{\rho_0}}$$

is proportional to the velocity of the baroclinic waves propagation.

Thus, on the basis of the simplest model we can substantiate the splitting method and at the same time we can obtain the practical criterion for the choice of the time interval τ while solving the ocean dynamics problems. For instance,

$$\tau\sqrt{\frac{\mu + \nu}{2\lambda}\frac{g\gamma}{\rho_0}} \leqslant \frac{1}{5}. \tag{3.209}$$

For a priori estimations it is necessary to determine the characteristic scale of the phenomena by the coordinates x, y, z, which are described with the help of the mathematical model. It is necessary to compare the eigenvalues λ, μ, ν of the spectral problem and then with the help of inequality (3.209) (where the value 1/5 is certainly conventional) to determine the value τ, which will guarantee the required accuracy of the result. As far as the stability is concerned, it is guaranteed at any time step τ.

For example we will estimate τ while describing the evolution of the processes in the ocean, the characteristic scales of which are $\langle L\rangle = 5 \times 10^7$, $\langle H\rangle = 10^5$, $\frac{\gamma}{\rho_0} = \frac{1}{\rho_0}\frac{d\rho_0}{dz} = 10^{-8}$. Then we have $\mu = \nu = \left(\frac{2\pi}{\langle L\rangle}\right)^2$, $\lambda = \left(\frac{2\pi}{\langle H\rangle}\right)^2$ and, thus $\sqrt{\frac{\mu + \nu}{2\lambda}\frac{g\gamma}{\rho_0}} = \frac{\langle H\rangle}{\langle L\rangle}\sqrt{\frac{g\gamma}{\rho_0}} \approx 10^{-4}\ s^{-1}$. Therefrom and from correlation (3.209) the conclusion follows that it is reasonable to choose the parameter τ to be approximately equal to an hour. Although our conclusions refer to the simplified model, they can be used also for more complete numerical schemes. It is important to note that the time interval of the order of 1 h is acceptable for reaching the required accuracy and it corresponds to the real possibilities of modern computers. In conclusion, it should be noted that it was possible to choose such an interval only because the barotropic component, which demands a time interval two orders smaller than that allowed by equality (3.209), is excluded from our solution.

3.8 The Organization of the Numerical Algorithm

We have described and established theoretically the algorithm for solving the ocean dynamics equations at each time interval $t_{j-1} \leqslant t \leqslant t_{j+1}$, assuming that at the time moment t_{j-1} the components of the solution u^{j-1}, v^{j-1}, T^{j-1} and S^{j-1} are known and sufficiently smooth. A certain smoothness was required for the components of the solution u, v, w, P, T and S, which was also a priori assumed. But together with these quite natural requirements with regard to the formulation of the problem, one more very essential assumption was made, namely that the coefficients u^j, v^j and w^j in Eqs. (3.43) are known. Though, strictly speaking, they should be determined in the process of solving the problem under consideration. Besides, we have assumed that the indicated coefficients should satisfy the continuity equation:

$$\frac{\partial u^j}{\partial x} + \frac{\partial v^j}{\partial y} + \frac{\partial w^j}{\partial z} = 0 \,. \tag{3.210}$$

Thus, such an organization of the numerical algorithm is required which will not violate the considered construction of the numerical process and basic theoretical conclusions.

From this point of view it is important that the indicated coefficients should possess the approximation order by τ, i.e. not lower than the first-order approximation. Then the formulated algorithm leads to the solution with the second-order accuracy by time.

Having this in mind, and also taking into account the a priori requirement (3.210), we suggest the following organization of the algorithm.

First, at the interval $t_{j-1} \leqslant t \leqslant t_j$ we solve the substances' transfer equations (3.86)–(3.88), where under the divergence sign instead of an unknown vector $\mathbf{u}^j$, the value $\mathbf{u}^{j-2}$ is taken, which was used when solving the dynamics equations at the previous time interval $t_{j-3} \leqslant t \leqslant t_{j-1}$. As it is easy to see, such a substitution guarantees the first-order accuracy in solving the motion equations.

After the transfer equations are solved with the help of the above developed numerical algorithm, the components u_1^j, v_1^j, T_1^j, S_1^j can be found. These data must be now initial at $t = t_{j-1}$ for the solution of the adaption problem (3.89)–(3.92), taking into account the separation of the dynamic barotropic component. In this case one should first solve problem (3.160), then solve consequently problems (3.166), (3.167)–(3.168) and (3.169)–(3.170).

For u^j, v^j and w^j we will take the following approximations:

$$u^j = \frac{\overline{u}^{j+1} + \overline{u}^{j-1}}{2} + u_2^{j-1/2} + u_s^{j+1/2} \,,$$

$$v^j = \frac{\overline{v}^{j+1} + \overline{v}^{j-1}}{2} + v_3^{j-1/2} + v_n^{j+1/2} \,,$$

$$w^j = \tfrac{1}{2}(w_2^{j-1/2} + w_3^{j-1/2} + w_4^{j+1/2} + w_s^{j+1/2}) \,.$$

These expressions can be taken for u^j, v^j and w^j, since they end the calculating time step of all the dynamics equations in the interval $t_{j-1} \leqslant t \leqslant t_{j+1}$. By solving the adaptation equation at the wider interval $t_{j-1} \leqslant t \leqslant t_{j+1}$, but having taken the half-sum of the solutions in the time moments t_{j-1} and t_{j+1}, in fact, we have also reduced the solution of the adaptation equation to the initial interval $t_{j-1} \leqslant t \leqslant t_j$.

It is not difficult to verify that the obtained components of the vectors $\mathbf{u}^j$ satisfy the continuity equation, and this is a necessary condition for the organization of the numerical algorithm.

After the coefficients u^j, v^j, w^j are found we can consider a more developed solution of the dynamics problem again at the time interval $t_{j-1} \leqslant t \leqslant t_{j+1}$. Though such an algorithm requires an approximately 1.5-fold increase of the calculating work in comparison with the subsequent counting, it guarantees the necessary accuracy of the calculations.

In conclusion, it should be noted that the developed algorithm assumes almost obvious generalizations with respect to many oceanographic problems. In particular, the dynamics problems with due regard to the turbulent viscosity, the problems of tides and others can be solved in this way.

Chapter 4

The Stationary Problems of Ocean Dynamics

The initial data play a crucial role in forecasting ocean currents. For the World Ocean these data are usually unknown. Indeed, the simplest scale analysis shows that for forecasting the ocean currents one needs a fine resolution (100 km and less), whereas for the weather forecast a resolution of 300 km for the initial data is satisfactory. At the same time the number of hydrological stations is so small that direct measurements cannot provide information on the hydrological fields of the ocean.

The question arises whether, in principal, it is possible to define the initial data by hydrological methods. We answer this question positively. But this answer needs some explanation, which will be given below. With respect to initial data, first, it is necessary to define those oceanographic processes, which are to be forecast. It is natural to assume that the initial fields for the processes are influenced by weather conditions, mainly by wind and thermal regimes, in a preceding period of a month or so, since our main task is a forecast of weather and currents of the World Ocean for a period of 3–4 weeks. This can be proved in particular by the known relaxation time of the main thermocline disturbances. For the time preceding the forecast of weather and ocean currents, only information on meteorological elements is available. Therefore, it is natural to use this information in determining the formation of flows and density fields in the upper ocean layers, under the influence of the weather situation. These fields are responsible for the air-sea interaction. Generally, one has to solve two problems. First, it is necessary to determine the stationary (climatic) ocean condition. Then, the climatic condition is used each time as "initial" data, in addition to available meteorological data near the ocean surface, to solve the problem of ocean circulation disturbances under the influence of concrete meteorological situations in the atmosphere for a more or less long period of time. The fields of hydrophysical elements, constructed by using real disturbances of meteorological processes, may be used as initial data for forecasting weather and ocean circulation.

4.1 The Statement of the Linearized Problem of the Ocean Climatic Condition

First, we will formulate the problem concerning the ocean climatic condition, under atmospheric situations acting over long time periods, which we will consider as stationary. As it is known, the ocean climate is formed over several tens of years and, below several hundreds of metres, it is rarely influenced by the seasonal oscillations of atmospheric circulation.

We will consider here two linearized models to clarify the peculiarities of a baroclinic ocean circulation. The first model gives an impression of the stationary regime of ocean currents and density fields, under the influence of the free-surface density flux of the ocean, when the effect of turbulent friction is absent in Euler's equations. The second model is more detailed: it allows one to take into account the effect of wind stress, together with the density flux, in the formation of ocean currents.

4.2 The Simplest Model of the Stationary Ocean Currents

Let us consider the linearized formulation of the problem to determine the ocean circulation under the influence of stationary disturbances of the density fluxes at the ocean surface. Thus, let us assume that the main equations of baroclinic ocean dynamics are linearized in the following way:

$$w = w' ,$$

$$P = P_a + g\rho_0 z + \tfrac{1}{2} g\Gamma z^2 + P' ,$$

$$\rho = \rho_0 + \Gamma z + \rho' . \tag{4.1}$$

Here, P_a is the atmospheric pressure at the ocean's free surface, which, for simplicity, we take to be constant; ρ_0 is the average for the domain density of the whole ocean, Γ is the average ocean density gradient. The deviations of the quantities from their standard values are marked by primes. With the above mentioned assumptions (4.1) the linearized equations of ocean dynamics take the form:

$$-lv' = -\frac{1}{\rho_0}\frac{\partial P'}{\partial x} + \mu \Delta u' ,$$

$$l_u = -\frac{1}{\rho_0}\frac{\partial P'}{\partial y} + \mu \Delta v' ,$$

$$g\rho' = \frac{\partial P'}{\partial z} ,$$

$$\frac{\partial u'}{\partial x} + \frac{\partial v'}{\partial y} + \frac{\partial w'}{\partial z} = 0 ,$$

$$\Gamma w' = v_1 \frac{\partial^2 \rho'}{\partial z^2} + \mu_1 \Delta \rho' . \tag{4.2}$$

For simplicity and convenience with regard to the formulation of boundary conditions, we write the system of equations (4.2), using the notations:

$$P = \tfrac{1}{2} g\Gamma z^2 + P' , \qquad \rho = \Gamma z + \rho' . \tag{4.3}$$

Then, passing from P' and ρ' to the new functions P and ρ, according to formulas (4.3), we obtain the following problem[6]:

$$\mu \Delta u + lv = \frac{1}{\rho_0}\frac{\partial P}{\partial x},$$

$$\mu \Delta v - lu = \frac{1}{\rho_0}\frac{\partial P}{\partial y},$$

$$g\rho = \frac{\partial P}{\partial z},$$

$$\frac{\partial u}{\partial x} + \frac{\partial v}{\partial y} + \frac{\partial w}{\partial z} = 0,$$

$$\mu_1 \Delta \rho + \nu_1 \frac{\partial^2 \rho}{\partial z^2} = \Gamma w. \tag{4.4}$$

As boundary conditions for system (4.4), we take:

$$\nu_1 \frac{\partial \rho}{\partial z} = \gamma, \quad w = 0 \quad \text{at } z = 0,$$

$$\frac{\partial \rho}{\partial z} = 0, \quad w = 0 \quad \text{at } z = H,$$

$$u = 0, \quad v = 0, \quad \frac{\partial \rho}{\partial n} = 0 \quad \text{at } \sigma. \tag{4.5}$$

Here, $\gamma(x, y)$ is a density flux, specified at the ocean surface. We suppose also that the ocean has the constant depth H and σ is a cylindric boundary surface.

In Eq. (4.5), as one of the boundary conditions at $z = 0$, we took a specified density flux at the ocean surface. This condition has been chosen for simplicity and definiteness only. All the following results may be easily generalized for the case when either a simpler condition:

$$\rho = \rho^{(0)} \qquad \text{at } z = 0,$$

or a more complex one:

$$\frac{\partial \rho}{\partial z} = \alpha(\rho - \rho^{(0)}) \qquad \text{at } z = 0$$

is taken, where α and $\rho^{(0)}$ are specified functions of coordinates and time.

Let us now prove the uniqueness of problem (4.4), (4.5). For this aim, as usual, we suppose that at least two different solutions of the equations system (4.4) exist, which satisfy the same boundary conditions (4.5). We now construct a quadratic function. Thus, we multiply the equations of system (4.4) by u, v, $\frac{1}{\rho_0}w$, $\frac{1}{\rho_0}P$ and $\frac{g\rho}{\rho_0 \Gamma}$ respectively. Then we sum the result and integrate via the whole domain of the

[6] The primes over u, v and w are also omitted here.

solutions definition. Afterward we have

$$\iiint_D \left[\frac{1}{\rho_0}\operatorname{div}(P\mathbf{u}) + \mu(u\Delta u + v\Delta v) + \frac{g}{\rho_0 \Gamma}\left(\mu_1 \rho\Delta\rho + \nu_1 \rho\frac{\partial^2 \rho}{\partial z^2}\right)\right] dD = 0 . \tag{4.6}$$

According to the Gauss-Ostrogradsky formula and using the boundary conditions:

$$\nu_1 \frac{\partial \rho}{\partial z} = 0 , \quad w = 0 \quad \text{at } z = 0 ,$$

$$\frac{\partial \rho}{\partial z} = 0 , \quad w = 0 \quad \text{at } z = H$$

$$u = 0 , \quad v = 0 , \quad \frac{\partial \rho}{\partial n} = 0 \quad \text{at } \sigma \tag{4.7}$$

the integral correlation (4.6) may be transformed to the form

$$\iiint_D \left[\mu(\operatorname{grad}^2 u + \operatorname{grad}^2 v) + \frac{g}{\rho_0 \Gamma}\left(\mu_1 \operatorname{grad}^2 \rho + \nu_1 \left(\frac{\partial \rho}{\partial z}\right)^2\right)\right] dD = 0 . \tag{4.8}$$

Moreover, taking into account the boundary conditions, it follows that for the differences of the solution components, the following equalities exist:

$$u(x, y, z) \equiv 0 , \quad v(x, y, z) \equiv 0 , \quad \rho(x, y, z) \equiv \text{const} . \tag{4.9}$$

which are identical for the whole domain D.

From the continuity equation, and taking into account the boundary condition $w(x, y, H) = 0$, it follows that

$$w(x, y, z) \equiv 0 . \tag{4.10}$$

The pressure field, as well as the density field, is defined with the accuracy of the additive constant, i.e.

$$P(x, y, z) \equiv \text{const} . \tag{4.11}$$

So, the uniqueness of the solution of problem (4.4), (4.5) in the above mentioned system is proved.

We will now consider the formulation of the dynamics model with the explicit separation of the barotropic component. For this purpose we will find the solution of equations system (4.4) in the form

$$u = \bar{u}(x, y) + u' ,$$

$$v = \bar{v}(x, y) + v' ,$$

$$w = w' ,$$

$$P = P_0(x, y) + P' ,$$

$$\rho = \rho' . \tag{4.12}$$

Here,

$$\bar{u} = \frac{1}{H}\int_0^H u dz\,, \qquad \bar{v} = \frac{1}{H}\int_0^H v dz\,,$$

and, consequently

$$\int_0^H u' dz = 0\,, \qquad \int_0^H v' dz = 0\,.$$

$P_0(x, y)$ in Eqs. (4.12) denotes the pressure at level $z = 0$. It should be noted that with such an expression of the pressure field, we will have

$$P' = 0 \qquad \text{at } z = 0\,. \tag{4.13}$$

Let us substitute expressions (4.12) into (4.4) and integrate the result in the whole ocean depth, using boundary conditions (4.5). Then, for the barotropic component of the solution we will have the following system of equations:

$$\mu\Delta\bar{u} + l\bar{v} = \frac{1}{\rho_0}\frac{\partial P_0}{\partial x} + \frac{1}{\rho_0 H}\int_0^H \frac{\partial P'}{\partial x} dz\,,$$

$$\mu\Delta\bar{v} - l\bar{u} = \frac{1}{\rho_0}\frac{\partial P_0}{\partial y} + \frac{1}{\rho_0 H}\int_0^H \frac{\partial P'}{\partial y} dz\,,$$

$$\frac{\partial \bar{u}}{\partial x} + \frac{\partial \bar{v}}{\partial y} = 0 \tag{4.14}$$

and boundary conditions

$$\bar{u} = 0\,, \qquad \bar{v} = 0 \qquad \text{at } \sigma\,. \tag{4.15}$$

The system of Eqs. (4.14) is not closed, because it contains the additional quantity P'.

To deduce the system of equations for the baroclinic component, we substitute once more expressions (4.12) into system (4.14), and use Eqs. (4.14) to exclude the barotropic components of the current and P_0 as well. Then we obtain the following system of equations:

$$\mu\Delta u' + l v' = \frac{1}{\rho_0}\frac{\partial P'}{\partial x} - \frac{1}{\rho_0 H}\int_0^H \frac{\partial P'}{\partial x} dz\,,$$

$$\mu\Delta v' - l u' = \frac{1}{\rho_0}\frac{\partial P'}{\partial y} - \frac{1}{\rho_0 H}\int_0^H \frac{\partial P'}{\partial y} dz\,,$$

$$\frac{\partial P'}{\partial z} = g\rho'\,,$$

$$\frac{\partial u'}{\partial x} + \frac{\partial v'}{\partial y} + \frac{\partial w'}{\partial z} = 0\,,$$

$$\mu_1 \Delta\rho' + \nu_1 \frac{\partial^2 \rho'}{\partial z^2} = \Gamma w'\,. \tag{4.16}$$

As boundary conditions for system (4.16) we take the following:

$$\nu_1 \frac{\partial \rho'}{\partial z} = \gamma\,, \quad w' = 0\,, \quad (P' = 0) \quad \text{at } z = 0\,,$$

$$\frac{\partial P'}{\partial z} = 0\,, \quad w' = 0 \quad \text{at } z = H\,,$$

$$u' = 0\,, \quad v' = 0\,, \quad \frac{\partial \rho'}{\partial n} = 0 \quad \text{at } \sigma\,. \tag{4.17}$$

It may seem that there are more boundary conditions at $z = 0$ than necessary. But one has to keep in mind that the additional condition for P' is natural for this statement of the problem.

Let us give a more convenient form to equations system (4.16). For this purpose, using the equations of statics and continuity, we have:

$$P' = g \int_0^z \rho' dz\,, \qquad w' = \int_z^H \left(\frac{\partial u'}{\partial x} + \frac{\partial v'}{\partial y} \right) dz\,. \tag{4.18}$$

Using Eq. (4.18), we transform the equations system (4.16) to the form:

$$\mu \Delta u' + l v' = \frac{g}{\rho_0} \int_0^z \frac{\partial \rho'}{\partial x} dz - \frac{g}{\rho_0 H} \int_0^H dz \int_0^z \frac{\partial \rho'}{\partial x} dz\,,$$

$$\mu \Delta v' - l u' = \frac{g}{\rho_0} \int_0^z \frac{\partial \rho'}{\partial y} dz - \frac{g}{\rho_0 H} \int_0^H dz \int_0^z \frac{\partial \rho'}{\partial y} dz\,,$$

$$\mu_1 \Delta \rho' + \nu_1 \frac{\partial^2 \rho'}{\partial z^2} = \Gamma \int_z^H \left(\frac{\partial u'}{\partial x} + \frac{\partial v'}{\partial y} \right) dz \tag{4.19}$$

with boundary conditions

$$\nu_1 \frac{\partial \rho'}{\partial z} = \gamma \qquad \text{at } z = 0\,,$$

$$\frac{\partial \rho'}{\partial z} = 0 \qquad \text{at } z = H$$

$$u' = 0\,, \quad v' = 0\,, \quad \frac{\partial \rho'}{\partial n} = 0 \qquad \text{at } \sigma\,. \tag{4.20}$$

It should be mentioned here that from the special expression of P in form (4.20), it follows that:

$$P' = 0 \qquad \text{at } z = 0\,. \tag{4.21}$$

This result may be important for the other statements of the ocean dynamics problems.

Further, we will examine problem (4.14), (4.15). It has only a trivial solution with any values of P'. This can be shown for the case of a rather smooth solution and specified data. For this purpose we multiply the first equation of system (4.14) by $\bar{u}$ and the second, by $\bar{v}$. Then we integrate the result by the whole domain of the solution's definition, which we denote by S. Then we obtain the expression:

$$\mu \iint_S (\text{gzad}^2 \bar{u} + \text{grad}^2 \bar{v}) dS + \frac{1}{\rho_0} \iint_S \left(\bar{u} \frac{\partial \bar{P}'}{\partial x} + \bar{v} \frac{\partial \bar{P}'}{\partial y} \right) dS = 0 \tag{4.22}$$

where the notation

$$\bar{P}' = \frac{1}{H} \int_0^H P' dz$$

is used. We transform the second integral of Eq. (4.22), by writing it in the form:

$$\frac{1}{\rho_0} \iint_S \left(\bar{u} \frac{\partial \bar{P}'}{\partial x} + \bar{v} \frac{\partial \bar{P}'}{\partial y} \right) dS = \frac{1}{\rho_0} \iint_S \left(\frac{\partial}{\partial x} \bar{u} \bar{P}' + \frac{\partial}{\partial y} \bar{v} \bar{P}' \right) dS$$
$$- \frac{1}{\rho_0} \iint_S \bar{P}' \left(\frac{\partial \bar{u}}{\partial x} + \frac{\partial \bar{v}}{\partial y} \right) dS . \tag{4.23}$$

The first integral of the right-hand side of Eq. (4.23) is equal to zero because of the Grin formula:

$$\frac{1}{\rho_0} \iint_S \left(\frac{\partial}{\partial x} \bar{u} \bar{P}' + \frac{\partial}{\partial y} \bar{v} \bar{P}' \right) dS = \frac{1}{\rho_0} \int_\sigma \bar{u}_n P' d\sigma = 0 ,$$

and the second integral, because of the equation of continuity:

$$\frac{\partial \bar{u}}{\partial x} + \frac{\partial \bar{v}}{\partial y} = 0 . \tag{4.24}$$

Thus, we obtain the expression:

$$\mu \iint_S (\text{grad}^2 \bar{u} + \text{grad}^2 \bar{v}) dS = 0 . \tag{4.25}$$

It follows from expression (4.24) and also from the boundary conditions at σ that

$$\bar{u} \equiv 0 , \qquad \bar{v} \equiv 0 ,$$

and from Eq. (4.14) that

$$\bar{P} = P_0 + \bar{P}' \equiv \text{const} .$$

The constant is not essential, we can consider it to be equal to zero, because the quantity P' is under differentials in our equations.

So, we come to the conclusion that the barotropic component of ocean dynamics is absent in this model.

Let us prove now the uniqueness in solving problem (4.16), (4.17) or the equivalent problem (4.19), (4.20). For this purpose it is necessary to show, for example, that problem (4.16), (4.17) has only a trivial solution, if $\gamma = 0$. We multiply the equations of system (4.16) by $u', v', w', \frac{\rho}{\rho_0}, \frac{g}{\Gamma} \frac{\rho'}{\rho_0}$ respectively, sum the result and integrate by the whole domain of the solution's definition D. Afterward we transform the resulting integrals, using the Gauss-Ostrogradsky theorem and homogeneous boundary conditions. Then, we obtain:

$$\iiint\limits_D \left[\mu(\mathrm{grad}^2 u' + \mathrm{grad}^2 v') + \frac{\mu_1 g}{\rho_0 \Gamma} \mathrm{grad}^2 \rho' + \frac{\nu_1 g}{\rho_0 \Gamma} \left(\frac{\partial \rho'}{\partial z} \right)^2 \right] dD$$

$$+ \frac{1}{\rho_0} \iint\limits_D dS \int\limits_0^H \left[u' \frac{\partial P'}{\partial x} + v' \frac{\partial P'}{\partial y} + w' \frac{\partial P'}{\partial z} - \left(u' \frac{\partial \bar{P}'}{\partial x} + v' \frac{\partial \bar{P}'}{\partial y} \right) \right] dz \,. \tag{4.26}$$

Here, we again used the notation:

$$\bar{P} = \frac{1}{H} \int\limits_0^H P' dz \,. \tag{4.27}$$

Let us take into account the correlation

$$u' \frac{\partial P'}{\partial x} + v' \frac{\partial P'}{\partial y} + w' \frac{\partial P'}{\partial z} = \frac{\partial}{\partial x} u' P' + \frac{\partial}{\partial z} w' P' - P' \left(\frac{\partial u'}{\partial x} + \frac{\partial v'}{\partial y} + \frac{\partial w'}{\partial z} \right) \tag{4.28}$$

and use the fact of the fulfillment of the continuity equations by u', v' and w'. Now we substitute Eq. (4.28) into (4.26), taking into account the Gauss-Ostrogradsky formula, homogeneous boundary conditions at σ and also the identity:

$$\bar{u}' \frac{\partial \bar{P}'}{\partial x} + \bar{v}' \frac{\partial \bar{P}'}{\partial y} \equiv 0 \,. \tag{4.29}$$

As a result, we obtain:

$$\iiint\limits_D \left[\mu[(\mathrm{grad}^2 u' + \mathrm{grad}^2 v') + \frac{\mu_1 g}{\rho_0 \Gamma} \mathrm{grad}^2 \rho' + \frac{\nu_1 g}{\rho_0 \Gamma} \left(\frac{\partial \rho'}{\partial z} \right)^2 \right] dD = 0 \,, \tag{4.30}$$

and from Eq. (4.30) the uniqueness of the solution

$$u' \equiv 0 \,, \quad v' \equiv 0 \,, \quad \rho' \equiv \mathrm{const} \,. \tag{4.31}$$

follows.

4.3 The Ocean Dynamics Model, Taking into Account the Wind-Driven Currents

We will now examine a more detailed model of ocean dynamics, by taking into account the effects of turbulent friction in the equations of motion. In this case problem (4.4), (4.5) is transformed into:

$$\mu \Delta u + \nu \frac{\partial^2 u}{\partial z^2} + l v = \frac{1}{\rho_0} \frac{\partial P}{\partial x} \,,$$

$$\mu \Delta v + \nu \frac{\partial^2 v}{\partial z^2} - l u = \frac{1}{\rho_0} \frac{\partial P}{\partial y} \,,$$

$$g \rho = \frac{\partial P}{\partial z} \,,$$

$$\frac{\partial u}{\partial x} + \frac{\partial v}{\partial y} + \frac{\partial w}{\partial z} = 0 \,,$$

$$\mu_1 \Delta \rho + \nu_1 \frac{\partial^2 \rho}{\partial z^2} = \Gamma w \,. \tag{4.32}$$

As boundary conditions we choose the following:

$$\nu\frac{\partial u}{\partial z} = -\frac{\tau_x}{\rho_0}, \quad \nu\frac{\partial v}{\partial z} = -\frac{\tau_y}{\rho_0}, \quad \nu_1\frac{\partial \rho}{\partial z} = \gamma, \quad w = 0 \quad \text{at } z = 0,$$

$$u = 0, \quad v = 0, \quad w = 0, \quad \frac{\partial \rho}{\partial z} = 0 \quad \text{at } z = H,$$

$$u = 0, \quad v = 0, \quad \frac{\partial \rho}{\partial n} = 0 \quad \text{at } \sigma. \tag{4.33}$$

The existence of the barotropic component in the flow field represents the essential difference of this model compared to the simplest one, described in section 4.2. Indeed, let us solve system (4.32) in the form:

$$u = \bar{u} + u',$$

$$v = \bar{v} + v',$$

$$w = w',$$

$$P = P_0 + P',$$

$$\rho = \rho', \tag{4.34}$$

where $\bar{u}$, $\bar{v}$ are the functions of variables x and y only, describing the barotropic component of the ocean flows field, and the quantities with primes, i.e. the derivations from the barotropic circulation. Let us note that by definition, the correlations of the form:

$$\int_0^H u'dz = 0, \quad \int_0^H v'dz = 0,$$

$$P' = 0 \quad \text{at } z = 0 \tag{4.35}$$

take place.

Let us choose the first two and the next to last equations from system (4.32) to deduce the equations of barotropic currents:

$$\mu\Delta u + \nu\frac{\partial^2 u}{\partial z^2} + lv = \frac{1}{\rho_0}\frac{\partial P}{\partial x},$$

$$\mu\Delta v + \nu\frac{\partial^2 v}{\partial z^2} - lv = \frac{1}{\rho_0}\frac{\partial P}{\partial y},$$

$$\frac{\partial u}{\partial x} + \frac{\partial v}{\partial y} + \frac{\partial w}{\partial z} = 0 \tag{4.36}$$

with the boundary conditions

$$\nu\frac{\partial u}{\partial z} = -\frac{\tau_x}{\rho_0}, \quad \nu\frac{\partial v}{\partial z} = -\frac{\tau_y}{\rho_0}, \quad w = 0 \quad \text{at } z = 0,$$

$$u = 0, \quad v = 0, \quad w = 0 \quad \text{at } z = H. \tag{4.37}$$

We substitute expressions (4.34) into (4.36), then integrate the result within the limits $0 \leqslant z \leqslant H$, using conditions (4.35) and boundary correlations (4.37). As a result we have

$$\mu \Delta \bar{u} + l\bar{v} = \frac{1}{\rho_0}\frac{\partial \bar{P}}{\partial x} - \frac{\tau_x}{\rho_0 H} - \frac{v}{H}\frac{\partial u'}{\partial z}\Big|_{z=H},$$

$$\mu \Delta \bar{v} - l\bar{u} = \frac{1}{\rho_0}\frac{\partial \bar{P}}{\partial y} - \frac{\tau_y}{\rho_0 H} - \frac{v}{H}\frac{\partial v'}{\partial z}\Big|_{z=H},$$

$$\frac{\partial \bar{u}}{\partial x} + \frac{\partial \bar{v}}{\partial y} = 0. \tag{4.38}$$

Here, the components of the flow velocity vectors, belonging to the baroclinic part of the solution, are denoted by primes and $\bar{P}$ is related to P_0 and $\bar{P}_0'$ by the correlation:

$$\bar{P} = \bar{P}_0 + \bar{P}'. \tag{4.39}$$

We add the boundary conditions

$$\bar{u} = 0, \quad \bar{v} = 0 \quad \text{at } \sigma \tag{4.38'}$$

to system (4.38). Now, differing from the previous model, the problem will have non-trivial solutions.

For the baroclinic component we obtain the system:

$$\mu \Delta u' + v\frac{\partial^2 u'}{\partial z^2} + lv' = \frac{1}{\rho_0}\frac{\partial (P' - \bar{P}')}{\partial x} + \frac{\tau_x}{\rho_0 H} + \frac{v}{H}\frac{\partial u'}{\partial z}\Big|_{z=H},$$

$$\mu \Delta v' + v\frac{\partial^2 v'}{\partial z^2} - lu' = \frac{1}{\rho_0}\frac{\partial (P' - \bar{P}')}{\partial y} + \frac{\tau_y}{\rho_0 H} + \frac{v}{H}\frac{\partial v'}{\partial z}\Big|_{z=H},$$

$$\frac{\partial P'}{\partial z} = g\rho',$$

$$\frac{\partial u'}{\partial x} + \frac{\partial v'}{\partial y} + \frac{\partial w'}{\partial z} = 0,$$

$$\mu_1 \Delta \rho' + v_1\frac{\partial^2 \rho'}{\partial z^2} = \Gamma w'. \tag{4.40}$$

The boundary conditions for system (4.40) are the following:

$$v\frac{\partial u'}{\partial z} = -\frac{\tau_x}{\rho_0}, \quad v\frac{\partial v'}{\partial z} = -\frac{\tau_y}{\rho_0}, \quad v_1\frac{\partial \rho'}{\partial z} = \gamma, \quad w' = 0 \quad \text{at } z = 0,$$

$$u' = 0, \quad v' = 0, \quad w' = 0, \quad \frac{\partial \rho'}{\partial z} = 0 \quad \text{at } z = H.$$

$$u' = 0, \quad v' = 0, \quad \frac{\partial \rho'}{\partial n} = 0 \quad \text{at } \sigma. \tag{4.41}$$

Now, because the barotropic component is separated from the solution, we can exclude P' and w' from system (4.40), using the formulas:

$$P' = g \int_0^z \rho' dz \,, \quad w' = \int_z^H \left(\frac{\partial u'}{\partial x} + \frac{\partial v'}{\partial y} \right) dz \,. \tag{4.42}$$

Here, we considered the fact that the baroclinic component of pressure $P' = 0$ at the ocean free level and $w' = 0$ at the bottom. It has been shown that the conditions $w' = 0$ and $P' = 0$ are equivalent at the ocean surface ($z = 0$). Substituting expressions (4.42) into the system of Eqs. (4.40), (4.41) we obtain the final form of the linearized problem of baroclinic ocean dynamics:

$$\mu \Delta u' + \nu \frac{\partial^2 u'}{\partial z^2} + l v' = \frac{g}{\rho_0} \int_0^z \frac{\partial \rho'}{\partial x} dz - \frac{g}{\rho_0 H} \int_0^H dz \int_0^z \frac{\partial \rho'}{\partial x} dz + \frac{\tau_x}{\rho_0 H} + \frac{\nu}{H} \frac{\partial u'}{\partial z}\bigg|_{z=H} \,,$$

$$\mu \Delta v' + \nu \frac{\partial^2 v'}{\partial z^2} - l u' = \frac{g}{\rho_0} \int_0^z \frac{\partial \rho'}{\partial y} dz - \frac{g}{\rho_0 H} \int_0^H dz \int_0^z \frac{\partial \rho'}{\partial y} dz + \frac{\tau_y}{\rho_0 H} + \frac{\nu}{H} \frac{\partial v'}{\partial z}\bigg|_{z=H} \,,$$

$$\mu_1 \Delta \rho' + \nu_1 \frac{\partial^2 \rho'}{\partial z^2} = \Gamma \int_z^H \left(\frac{\partial u'}{\partial x} + \frac{\partial v'}{\partial y} \right) dz \,, \tag{4.43}$$

with boundary conditions

$$\nu \frac{\partial u'}{\partial z} = -\frac{\tau_x}{\rho_0} \,, \quad \nu \frac{\partial v'}{\partial z} = -\frac{\tau_y}{\rho_0} \,, \quad \nu_1 \frac{\partial \rho'}{\partial z} = \gamma \quad \text{at } z = 0 \,,$$

$$u' = 0 \,, \quad v' = 0 \,, \quad \frac{\partial \rho'}{\partial z} = 0 \quad \text{at } z = H \,,$$

$$u' = 0 \,, \quad v' = 0 \,, \quad \frac{\partial \rho'}{\partial n} = 0 \quad \text{at } \sigma \,. \tag{4.44}$$

The principal scheme in solving problems (4.38), (4.38′) and (4.40) is the following: first, we determine the baroclinic component of ocean dynamics (4.40), (4.41), then the barotropic component, based on Eqs. (4.38), (4.38′).

It is possible, according to the above described methods, to prove that the solution of problem (4.38), (4.38′), with the specified values of τ_x, τ_y at $z = 0$ and $\nu \frac{\partial v'}{\partial z}$, $\nu \frac{\partial u'}{\partial z}$ at $z = H$, is unique. In contrast to the model examined in the previous section, the barotropic component of ocean dynamics will not equal zero, because of the effect of wind stress at the sea surface, and because of the bottom stress. The uniqueness of the baroclinic component of ocean dynamics, defined by the equivalent problem (4.43), (4.44), is proved in the same way. It is necessary only to take into account that when proving the uniqueness of this problem by the above described methods we obtain the quadratic function:

$$\iiint_D \left\{ \mu(\operatorname{grad}^2 u' + \operatorname{grad}^2 v') + \nu \left[\left(\frac{\partial u'}{\partial z} \right)^2 + \left(\frac{\partial v'}{\partial z} \right)^2 \right] + \frac{\mu_1 g}{\rho_0 \Gamma} \operatorname{grad}^2 \rho' \right.$$
$$\left. + \frac{\nu_1 g}{\rho_0 \Gamma} \left(\frac{\partial \rho'}{\partial z} \right)^2 \right\} dD = \frac{\nu}{H} \iint_S \left[\left(\int_0^H u' dz \right) u'_z \bigg|_{z=H} + \left(\int_0^H v' dz \right) v'_z \bigg|_{z=H} \right] dS \,. \tag{4.45}$$

Because conditions (4.35) take place, the expression on the right-hand side of Eq. (4.45) becomes zero and we obtain the quadratic function, from which the uniqueness of the solution for the baroclinic part of the problem follows.

We note in conclusion that it is possible to consider the conditions

$$\nu u'_z = 0\,, \quad \nu v'_z = 0 \quad \text{at } z = H$$

instead of the conditions

$$u' = 0\,, \quad v' = 0 \quad \text{at } z = H\,.$$

In this case some of the terms in Eqs. (4.38) and (4.40) will be absent and the problem becomes simpler.

4.4 The Difference Operators of the Ocean Dynamics Problem and the Methods of Approximation

First, it is necessary to present the principles of reducing the main equations of ocean dynamics to the difference form and formulating convenient notations for several difference operators.

We assume that the whole domain of the solution's definition is covered by the grid set, which is regular for each of the independent variables with grid steps $\Delta x = \Delta y$ and $\Delta z = h$. As the boundary of the grid set domain, we take the grid points nearest to γ. The set of such points we denote by σ^h. This procedure relates to the approximation of the domain boundary by the planes parallel to the coordinates $x = \text{const}$ and $y = \text{const}$. We assume also that the ocean's free surface coincides with the coordinate plane $z = 0$ and the bottom, with the plane $z = H$. We denote the grid variable along the x-axis by k; along the y-axis, by l; and along the z-axis, by m. These variables vary within the limits:

$$k_0(l) \leqslant k(l) \leqslant k^*(l)\,,$$

$$l_0(k) \leqslant l(k) \leqslant l^*(k)\,,$$

$$0 \leqslant m \leqslant m^*\,.$$

Here, $k_0(l)$, $k^*(l)$, $l_0(k)$, $l^*(k)$, $m = 0$ and $m = m^*$ are the points coinciding with the domain boundary D^h of the grid set. For simplicity, we assume $k_0 = 0$, $l_0 = 0$.

Now we will define the difference operators. The most convenient way is to begin with the description of one-dimensional operators, because the difference operator is formed on the basis of the differential one, taking into account the boundary conditions for the class of the functions on which this particular operator acts. First, let us examine the approximation of the operators $\dfrac{\partial}{\partial x}, \dfrac{\partial}{\partial y}$ and $\dfrac{\partial}{\partial z}$ with different classes of functions. Let us assume that the function $\varphi \in \Phi$, on which the operator $\dfrac{\partial}{\partial x}$ acts, is continuous, differentiable and equals $\varphi_0 = a$ at the point x_0. Then the natural approximation of this derivative is the following:

$$\left.\frac{\partial \varphi}{\partial x}\right|_{x=x_k} \approx \left| \begin{array}{l} \nabla_{k+1}^- \varphi^h - \dfrac{a}{\Delta x}, \quad k = 0, \\ \nabla_k^- \varphi^h - \dfrac{a}{\Delta x}, \quad k = 1 \\ \nabla_k^- \varphi^h, \quad k = 1, 2, 3, \ldots, k^*, \end{array} \right.$$

where

$$\nabla_k^- \varphi^h = \begin{cases} \dfrac{\varphi_1^h}{\Delta x}, & k = 1, \\ \dfrac{\varphi_k^h - \varphi_{k-1}^h}{\Delta x}, & k = 2, 3, \ldots, k^*. \end{cases}$$

If the function φ is specified at the point $x = x_{k^*}$ and equals $\varphi_{k^*} = b$, then the natural approximation of its derivative is:

$$\left.\frac{\partial \varphi}{\partial x}\right|_{x=x_k} \approx \left| \begin{array}{l} \nabla_k^+ \varphi^h, \quad k = k^* - 2, k^* - 3, \ldots, 0, \\ \nabla_k^+ \varphi^h + \dfrac{b}{\Delta x}, \quad k = k^* - 1, \\ \nabla_{k-1}^+ \varphi^h + \dfrac{b}{\Delta x}, \quad k = k^*, \end{array} \right.$$

where

$$\nabla_k^+ \varphi^h = \begin{cases} \dfrac{\varphi_{k+1}^h - \varphi_k^h}{\Delta x}, & k = k^* - 2, k^* - 3, \ldots, 0, \\ -\dfrac{\varphi_{k^*-1}^h}{\Delta x}, & k = k^* - 1. \end{cases}$$

Now we will approximate the second-order derivatives: $\dfrac{\partial^2}{\partial x^2}, \dfrac{\partial^2}{\partial y^2}$ and $\dfrac{\partial^2}{\partial z^2}$. The different cases are possible here, depending on the a apriori information on the boundary conditions, which the functions $\varphi \in \Phi$ satisfy. We will dwell only on the boundary conditions of the Dirichlet and Neumann types.

Let us consider the case, when at the boundaries $x = x_0$ and $x = x_{k^*}$ the values of the function $\varphi(x_0) = a$, $\varphi(x_{k^*}) = b$ are specified. Then we define the approximating difference operator in the form:

$$\frac{\partial^2 \varphi}{\partial x^2} \approx \begin{cases} \Delta_k^{1,1} \varphi^h, & k = 0, \\ \Delta_k^{1,1} \varphi^h + \dfrac{a}{\Delta x^2}, & k = 1, \\ \Delta_k^{1,1} \varphi^h, & k = 2, 3, \ldots, k^* - 2, \\ \Delta_k^{1,1} \varphi^h + \dfrac{b}{\Delta x^2}, & k = k^* - 1, \\ \Delta_k^{1,1} \varphi^h, & k = k^*, \end{cases}$$

where

$$\Delta_k^{1,1}\varphi^h = \begin{cases} 0\,, \quad k=0\,, \\ \dfrac{-2\varphi_1^h + \varphi_2^h}{\Delta x}\,, \quad k=1\,, \\ \dfrac{\varphi_{k-1}^h - 2\varphi_k^h + \varphi_{k+1}^h}{\Delta x^2}\,, \quad k = 2, 3, \ldots, k^* - 2\,, \\ \dfrac{\varphi_{k^*-1}^h - 2\varphi_{k^*}^h}{\Delta x^2}\,, \quad k = k^* - 1\,, \\ 0\,, \quad k = k^*\,. \end{cases}$$

When the Dirichlet condition $\varphi(x_0) = a$ is specified at one end of the interval $x = x_0$ and the Neumann condition $\dfrac{\partial \varphi}{\partial x} = b$ at $x = x_{k^*}$ at the other end, the difference operator of the second derivative takes the form:

$$\frac{\partial^2 \varphi}{\partial x^2} \approx \begin{cases} \Delta_k^{1,2}\varphi^h\,, \quad k=0\,, \\ \Delta_k^{1,2}\varphi^h + \dfrac{a}{\Delta x^2}\,, \quad k=1\,, \\ \Delta_k^{1,2}\varphi^h\,, \quad k = 2, 3, \ldots, k^* - 1\,, \\ \Delta_k^{1,2}\varphi^h - \dfrac{2b}{\Delta x}\,, \quad k = k^*\,, \end{cases}$$

where

$$\Delta_k^{1,2}\varphi^h = \begin{cases} 0\,, \quad k=0\,, \\ \dfrac{-2\varphi_1^h + \varphi_2^h}{\Delta x^2}\,, \quad k=1\,, \\ \dfrac{\varphi_{k-1}^h - 2\varphi_k^h + \varphi_{k+1}^h}{\Delta x^2}\,, \quad k = 2, 3, \ldots, k^* - 1\,, \\ \dfrac{-2\varphi_{k^*}^h + 2\varphi_{k^*-1}^h}{\Delta x^2}\,, \quad k = k^*\,. \end{cases}$$

The approximations may be constructed in the same way, with other combinations of the boundary conditions. We only explain here some principles and notations, connected with such approximations. The subscript k indicates the approximation by coordinate x at the grid set $\{x_k\}$. The analogous approximations of derivatives by y and z will be denoted by indexes l and m respectively. The superscripts denote the boundary problem type in order of the increasing indexes. Namely, the Dirichlet boundary condition (the condition of the first order) is denoted by index 1; the Neumann condition (the condition of the second order) is denoted by index 2. Let us note that when the region of the Dirichlet solution consists only of the internal points $k = 1, 2, \ldots, k^* - 1$, then the Neumann solution is also indefinite at the corresponding points.

Finally, it is important to note that when passing from differential expressions to the finite difference ones, the inhomogeneous boundary conditions generate additional sources in corresponding difference equations. Instead of a detailed description, we will combine all such sources together with the other sources of the initial differential equation in a form of united grid-set functions for each equation. This will simplify our examination and allow us to concentrate our attention on the essence of the problem.

The finite-difference form of problem (4.4), (4.5), according to the above introduced definitions, takes the form:

$$\mu \Delta_{k,l}^{1,1} u + lv = \frac{1}{\rho_0} \nabla_k^+ P ,$$

$$\mu \Delta_{k,l}^{1,1} v - lu = \frac{1}{\rho_0} \nabla_l^+ P ,$$

$$g\rho = \nabla \bar{m} P ,$$

$$\nabla_k^- u + \nabla_l^- v + \nabla \bar{m} w = 0 ,$$

$$\mu_1 \Delta_{k,l}^{2,2} \rho + \nu_1 \Delta_m^{2,2} \rho = \Gamma w + G , \tag{4.46}$$

where G is a grid-set function, generated by the source γ in the boundary conditions. For simplicity, we have omitted indexes k, l and m of the variables. Here and below we will use the following definition:

$$\Delta_{k,l}^{1,1} = \Delta_k^{1,1} + \Delta_l^{1,1} , \qquad \Delta_{k,l}^{2,2} = \Delta_k^{2,2} + \Delta_l^{2,2} .$$

The boundary conditions for Eq. (4.46) will be

$$P_0 = 0 , \qquad w_{m^*} = 0 . \tag{4.47}$$

In the same way as with the analysis of problem (4.4), (4.5), it is possible to show that conditions (4.47) are equivalent to the conditions

$$w_0 = 0 , \qquad w_{m^*} = 0 . \tag{4.47}$$

It allows one to simplify problem (4.46), (4.47) by excluding the quantities P and w with the help of the correlations:

$$P_m = gh \sum_{m'=0}^{m-1} \rho m' ,$$

$$w_m = h \sum_{m'=m+1}^{m^*} (\nabla_k^- um' + \nabla_l^- v_{m'}) . \tag{4.48}$$

As a result we have:

$$\mu \Delta_{k,l}^{1,1} u_m + lv_m = \frac{gh}{\rho_0} \sum_{m'=0}^{m-1} \nabla_k^+ \rho_{m'} ,$$

$$\mu \Delta_{k,l}^{1,1} v_m - lu_m = \frac{gh}{\rho_0} \sum_{m'=0}^{m-1} \nabla_k^+ \rho_{m'} ,$$

$$\mu_1 \Delta_{k,l}^{2,2} \rho_m + \nu_1 \Delta_m^{2,2} \rho_m = \Gamma h \sum_{m'=m+1}^{m^*} (\nabla_k^- u_{m'} + \nabla_l^- v_{m'}) + G . \tag{4.49}$$

In Eqs. (4.48) and (4.49) the ocean level, for which the equations system is written, is denoted by the subscript m.

It is possible to show that problem (4.46), (4.47) as well as problem (4.49) has a unique solution. The method of proving the theorem of uniqueness does not differ principally from the one considered above in the differential statement.

Now let us consider the more detailed problem (4.32), (4.33). Then, for the barotropic component we will have the following difference analogue of problem (4.38), (4.38′).

$$\mu\Delta_{k,l}^{1,1}\bar{u} + l\bar{v} = \frac{1}{\rho_0}\nabla_k^+ \bar{P} + \bar{a},$$

$$\mu\Delta_{k,l}^{1,1}\bar{v} - l\bar{u} = \frac{1}{\rho_0}\nabla_l^+ \bar{P} + \bar{b},$$

$$\nabla_k^- \bar{u} + \nabla_l^- \bar{v} = 0, \tag{4.50}$$

where $\bar{a}$ and $\bar{b}$ are the grid-set functions, generated by the sources in equations system (4.38).

For the baroclinic component we obtain the analogue of problem (4.16), (4.17) in the form:

$$\mu\Delta_{k,l}^{1,1}u' + \nu\Delta_m^{2,1}u' + lv' = \frac{1}{\rho_0}\nabla_k^+ P' - \frac{1}{\rho_0 H}\sum_{m=1}^{m^*}\nabla_k^+ P_m' + a,$$

$$\mu\Delta_{k,l}^{1,1}v' + \nu\Delta_m^{2,1}v' - lu' = \frac{1}{\rho_0}\nabla_l^+ P' - \frac{1}{\rho_0 H}\sum_{m=1}^{m^*}\nabla_l^+ P_m' + b,$$

$$g\rho' = \nabla_m^- P',$$

$$\nabla_k^- u' + \nabla_l^- v' + \nabla_m^- w' = 0,$$

$$\mu_1\Delta_{k,l}^{2,2}\rho' + \nu_1\Delta_m^{2,2}\rho' = \Gamma w' + G. \tag{4.51}$$

Here, a, b and G are the sources generated by inhomogeneities in the boundary conditions and equations.

Taking into account expressions (4.48), we transform equations system (4.51) into the form:

$$\mu\Delta_{k,l}^{1,1}u_m' + \nu\Delta_m^{2,1}u_m' + lo_m' = \frac{gh}{\rho_0}\sum_{m'=0}^{m-1}\nabla_k^+\rho_{m'}' - \frac{gh^2}{\rho_0 H}\sum_{m=1}^{m^*}\sum_{m'=0}^{m-1}\nabla_k^+\rho_{m'}' + a_m,$$

$$\mu\Delta_{k,l}^{1,1}v_m' + \nu\Delta_m^{2,1}v_m' - lu_m' = \frac{gh}{\rho_0}\sum_{m'=0}^{m-1}\nabla_l^+\rho_{m'}' - \frac{gh^2}{\rho_0 H}\sum_{m=1}^{m^*}\sum_{m'=0}^{m-1}\nabla_l^+\rho_{m'}' + b_m,$$

$$\mu_1\Delta_{k,l}^{2,2}\rho_m' + \nu_1\Delta_m^{2,2}\rho_m' = \Gamma h\sum_{m'=m+1}^{m^*}(\nabla_k^- u_{m'}' + \nabla_l^- v_{m'}') + G_m. \tag{4.52}$$

The approximation in the form of Eqs. (4.51) or (4.52) does not violate the definiteness of the corresponding operators and guarantees the uniqueness of the solution, as in the case of the simplest model.

4.5 The Iterative Processes for Solving the Ocean Dynamics Difference Equations for the Barotropic Component

We now concentrate our attention on a more general problem (4.50), because problem (4.46) is a particular case of (4.50). First, we write problem (4.50) in a vectorial-matrix form. For this purpose we consider the matrix and vectors:

$$A = \begin{Vmatrix} -\mu(\Delta_k^{1,1} + \Delta_l^{1,1}) & -l & \dfrac{1}{\rho_0}\nabla_k^+ \\ l & -\mu(\Delta_k^{1,1} + \Delta_l^{1,1}) & \dfrac{1}{\rho_0}\nabla_l^+ \\ \dfrac{1}{\rho_0}\nabla_k^- & \dfrac{1}{\rho_0}\nabla_l^- & 0 \end{Vmatrix}$$

$$\varphi = \begin{vmatrix} \bar{u}_{k,l} \\ \bar{v}_{k,l} \\ \bar{P}_{k,l} \end{vmatrix}, \qquad f = \begin{vmatrix} -\bar{a}_{k,l} \\ -\bar{b}_{k,l} \\ 0 \end{vmatrix}.$$

Then we obtain the equations system:

$$A\varphi = f. \tag{4.53}$$

It is necessary to note that φ and f depend on k and l, and the operator equation (4.53) is actually a set of linear equation systems for all of k and l. Let us consider the space of elements φ with the scalar product:

$$(a, b) = \sum_{i=1}^{3} \sum_{k=1}^{k^*} \sum_{l=1}^{l^*} a_{k,l}^{(i)} b_{k,l}^{(i)},$$

where a, b are two elements belonging to the set $\{\varphi\}$. It is easy to verify that the operator A in this matrix is positive, i.e.

$$(A\varphi, \varphi) > 0. \tag{4.54}$$

This means that one can use different iterative methods to solve Eq. (4.53). When the Coriolis force exists, the spectrum of the operator becomes complex, so that neither Chebyshev's method of acceleration nor the method of overrelaxation is useful in solving Eq. (4.53). It is more natural to use the method of minimal residuals in the form:

$$\varphi^{j+1} = \varphi^j - \tau_j(A\varphi^j - f) \tag{4.55}$$

where

$$\tau_j = \frac{(A\xi^j, \xi^j)}{(A\xi^j, A\xi^j)}. \tag{4.56}$$

Here, ξ^j is a residual of the iterative process and is defined by the correlation:

$$\xi^j = A\varphi^j - f. \tag{4.57}$$

Krasnoselsky and Krein (1952) have shown that the iterative process (4.55) converges to the exact solution of Eq. (4.53), if the matrix A is positive.

But it is necessary to keep in mind that problem (4.53) has a unique solution with the accuracy of $\bar{P} = \text{const}$. This means that it is necessary to perform an orthogonalization of the solution with respect to P after every iterative made by scheme (4.55), where P does not depend on indexes k and l. For this purpose it is necessary to calculate the mean value of P by D^h and to subtract this value from every component $P_{k,l}$. According to this orthogonalization, not only the convergency of the components $u^j_{k,l}$, $v^j_{k,l}$ to $u_{k,l}$, $v_{k,l}$ will be guaranteed, but also $P^j_{k,l}$ to $P_{k,l}$.

Still we have to note that the convergency of iterative process (4.55) is slow, therefore we will consider another process, which is essentially more effective. In order to formulate this process it is necessary to transform the initial equation (4.50). For this purpose we introduce the difference analogues of the stream function $\Psi_{k,l}$ with the help of the third equation of system (4.50) by the formulas:

$$\bar{u} = -\nabla_l^- \Psi\,, \qquad \bar{v} = \nabla_k^- \Psi\,. \tag{4.58}$$

Then, we act on the first equation of system (4.53) by operator Δ_l^+; on the second one, by operator ∇_k^+; subtract one of the determined correlations from another and use correlation (4.58). As a result we obtain the equation for the difference stream function:

$$\mu(\nabla_k^+ \Delta_{k,l}^{1,1} \nabla_k^- + \nabla_l^+ \Delta_{k,l}^{1,1} \nabla_l^-)\Psi + (\nabla_l^+ l\nabla_k^- - \nabla_k^+ l\nabla_l^-)\Psi = F\,, \tag{4.59}$$

where

$$F = \nabla_l^+ a - \nabla_k^+ b$$

and the notation

$$\Delta_{k,l}^{1,1} = \Delta_k^{1,1} + \Delta_l^{1,1} \tag{4.60}$$

was inserted. When carrying out this reduction two more correlations have been used: first, the condition

$$\Psi = 0 \qquad \text{at } \sigma\,, \tag{4.61}$$

has been used, which results from the non-slip boundary condition; and second when writing Eq. (4.59), it was assumed that the expressions $\Delta_k u$ and $\Delta_l v$ are defined not only at all the internal points of domain D^h, but also at all boundary points. This appeared to be possible in the case of the odd continuation of functions u and v from the domain for one grid step.

Let us write Eq. (4.59) in the operator form:

$$A\Psi = F\,. \tag{4.62}$$

and introduce the scalar factor:

$$(a,b) = \sum_{k=1}^{k^*_{-1}} \sum_{l=1}^{l^*_{-1}} a_{k,l} \cdot b_{k,l}\,. \tag{4.63}$$

If elements a and b satisfy condition (4.61), then the correlations:

$$(\nabla_k^+ a, b) = -(a, \nabla_k^- b),$$

$$(\nabla_l^+ a, b) = -(a, \nabla_l^- b), \tag{4.64}$$

which can be checked easily, take place. By direct control it is possible to confirm that:

$$(\nabla_l^+ l\nabla_k^- \psi - \nabla_k^+ l\nabla_l^- \psi, \psi) = 0 \tag{4.65}$$

and

$$\begin{aligned}(\nabla_k^+ \Delta_{k,l}^{1,1} \nabla_k^- \psi + \nabla_k^+ \Delta_{k,l}^{1,1} \nabla_l^- \psi, \psi) &= -(\Delta_{k,l}^{1,1} \nabla_k^- \psi, \nabla_k^- \psi) - (\Delta_{k,l}^{1,1} \nabla_l^- \psi, \nabla_l^- \psi)\\ &= -(\Delta_{k,l}^{1,1} v, v) - (\Delta_{k,l}^{1,1} u, u).\end{aligned} \tag{4.66}$$

Let us remember now that

$$\Delta_{k,l}^{1,1} = \nabla_k^+ \nabla_k^- + \nabla_l^+ \nabla_l^-,$$

and components u and v satisfy condition (4.58). Then we have:

$$(\Delta_{k,l}^{1,1} v, v) = -(\nabla_k^- v, \nabla_k^- v) - (\nabla_l^- v, \nabla_l^- v),$$

$$(\Delta_{k,l}^{1,1} u, u) = -(\nabla_k^- u, \nabla_k^- u) - (\nabla_l^- u, \nabla_l^- u). \tag{4.67}$$

As a result we obtain the expression:

$$(A\psi, \psi) = \mu[(\nabla_k^- u, \nabla_k^- u) + (\nabla_l^- u, \nabla_l^- u) + (\nabla_k^- v, \nabla_k^- v) + (\nabla_l^- v, \nabla_l^- v)], \tag{4.68}$$

from which the positiveness of operator A follows directly, i.e.,

$$(A\psi, \psi) > 0.$$

Now, the iterative process of the method of minimal residuals for the solution of Eq. (4.62) is obtained:

$$\psi^{j+1} = \psi^j - \tau_j(A\psi^j - F) \tag{4.69}$$

with the condition

$$\tau_j = \frac{(A\xi^j, \xi^j)}{(A\xi^j, \xi^j)}, \tag{4.70}$$

where

$$\xi^j = A\psi^j - F$$

converges more rapidly than process (4.55), (4.56).

To accelerate the convergency of method (4.69) it is advisable to use the method of the universal algorithm in the following way. Let us examine the spectral problem:

$$-\Delta_{k,l}^{1,1} \omega = \lambda\omega. \tag{4.72}$$

According to Lusternic's iterative method we find the maximal $\beta(-\Delta_{k,l}^{1,1})$ and minimal $\alpha(-\Delta_{k,l}^{1,1})$ eigenvalues of problem (4.72). Afterward, we propose the following construction of the iterative process:

$$B\frac{\psi^{j+1} - \psi^j}{\tau_j} + A\psi^j = F\,, \tag{4.73}$$

where

$$B = (E - \sigma\Delta_k^{1,1})(E - \sigma\Delta_l^{1,1})\,,$$

$$\tau_j = \frac{(AB^{-1} - \xi^j, \xi^j)}{(AB^{-1}\xi^j, AB^{-1}\xi^j)}\,, \qquad \sigma = \frac{1}{\sqrt{\alpha\beta}}\,. \tag{4.74}$$

The performance scheme has the form[7]:

$$g^j = A\psi^j - F\,,$$

$$(E - \sigma\Delta_k^{1,1})\xi^{j+1/2} = -g^j\,,$$

$$(E - \sigma\Delta_l^{1,1})\xi^{j+1} = \xi^{j+1/2}\,,$$

$$\psi^{j+1} = \psi^j + \tau_j\xi^{j+1}\,. \tag{4.75}$$

It is easy to verify that if ξ^{j+1} is found, then

$$B^{-1}\xi^j = \xi^{j+1}\,.$$

Hence,

$$\tau_j = \frac{(A\xi^{j+1}, \xi^j)}{(A\xi^{j+1}, A\xi^{j+1})}\,.$$

With respect to σ, the following should be noted. Iterative scheme (4.73) is weakly sensitive to σ, therefore, in the spectral problem (4.72), instead of exact boundary values of the spectrum α and β, we examine the approximate ones, determined in the following way. We insert the solution's definition domain into a rectangle with sides a and b, and examine the spectral problem, instead of problem (4.72) with the real definition domain. Then, instead of α and β we will find the corresponding spectrum boundaries $\bar{\alpha}$ and $\bar{\beta}$ for the rectangle. It is simple to find them, according to:

$$\bar{\alpha} = \left(\frac{1}{a^2} + \frac{1}{b^2}\right)\pi^2\,,$$

$$\bar{\beta} = \frac{b}{h^2}\,,$$

and one can take these values for the approximate definition of

$$\sigma = \frac{h}{2\pi}\frac{ab}{\sqrt{2(a^2 + b^2)}}\,.$$

The problem is then prepared for the solution at each iterative step by Fourier's rapid transformation (Marchuk 1973).

---

[7] Here, parameter σ should not be classified with the domain boundary.

4.6 The Solution of the Difference Equations of the Ocean Dynamics Baroclinic Component

Now let us examine the baroclinic component, defined by the system of difference equations (4.52). For this purpose we take into consideration a difference operator of the matrix form and vectors:

$$A = \left\| \begin{matrix} -(\mu\Delta_{k,l}^{1,1} + \nu\Delta_m^{2,1}) & -l & -G_1 \\ l & -(\mu\Delta_{k,l}^{1,1} + \nu\Delta_m^{2,1}) & -G_2 \\ -\Gamma h \sum\limits_{m'=m+1}^{m^*} \nabla_k^- & -\Gamma h \sum\limits_{m'=m+1}^{m^*} \nabla_l^- & -(\mu_1\Delta_{k,l}^{2,2} + \nu_1\Delta_m^{2,2}) \end{matrix} \right\|$$

$$G_1 = \frac{gh}{\rho_0} \sum_{m'=0}^{m-1} \nabla_k^+ - \frac{gh^2}{\rho_0 H} \sum_{m=1}^{m^*} \sum_{m'=0}^{m-1} \nabla_k^+ ,$$

$$G_2 = \frac{gh}{\rho_0} \sum_{m'=0}^{m-1} \nabla_l^+ - \frac{gh^2}{\rho_0 H} \sum_{m=1}^{m^*} \sum_{m'=0}^{m-1} \nabla_l^+$$

$$\varphi = \left| \begin{matrix} u'_{k,l,m} \\ v'_{k,l,m} \\ \rho'_{k,l,m} \end{matrix} \right| , \qquad f = \left| \begin{matrix} -a_{k,l,m} \\ -b_{k,l,m} \\ -c_{k,l,m} \end{matrix} \right| .$$

Then, problem (4.52) will be written in the operator form:

$$A\varphi = f . \tag{4.76}$$

We specify the scalar product in the following way:

$$(a, b) = \sum_{i=1}^{3} \sum_{k=0}^{k^*} \sum_{l=0}^{l^*} \sum_{m=0}^{m^*} a_{k,l,m}^{(i)} b_{k,l,m}^{(i)} . \tag{4.77}$$

Then, in this matrix we will have:

$$(A\varphi, \varphi) > 0 . \tag{4.78}$$

Because of the positive definiteness of operator A, we will find the solution of problem (4.76) by the method of minimal residuals in the form of Eqs. (4.55), (4.56), the convergency of which is guaranteed. But it is possible to accelerate the method of the successive approximation by examining a more general iterative process. For this purpose let us first examine three spectral problems:

$$-\Delta_{k,l}^{1,1} \omega^{(1)} = \lambda^{(1)} \omega^{(1)} ,$$

$$-\Delta_{k,l}^{2,1} \omega^{(2)} = \lambda^{(2)} \omega^{(2)} ,$$

$$-\Delta_{k,l}^{2,2} \omega^{(3)} = \lambda^{(3)} \omega^{(3)} , \tag{4.79}$$

and find the spectrum boundaries for each. Suppose they are α_1, β_1, α_2, β_2 and α_3, β_3. Then we introduce the matrix in the following way:

$$B = \begin{Vmatrix} C & 0 & 0 \\ 0 & C & 0 \\ 0 & 0 & R \end{Vmatrix},$$

where

$$C = (E - \sigma_1 \Delta_k^{1,1})(E - \sigma_1 \Delta_l^{1,1})(E - \sigma_2 \Delta_m^{2,1}),$$
$$R = (E - \sigma_3 \Delta_k^{2,2})(E - \sigma_3 \Delta_l^{2,2})(E - \sigma_3 \Delta_m^{2,2}). \tag{4.80}$$

Let us examine the iterative process:

$$B\frac{\varphi^{j+1} - \varphi^j}{\tau_j} + A\varphi^j = f, \tag{4.81}$$

where

$$\tau_j = \frac{(AB^{-1}\xi^j, \xi^j)}{(AB^{-1}\xi^j, AB^{-1}\xi^j)}, \qquad \xi^j = A\varphi^j - f.$$

When performing this algorithm, the parameter σ may be found according to the simpler formula:

$$\sigma_i = \frac{1}{\sqrt{\bar{\alpha}_i \bar{\beta}_i}},$$

as in the case of Eq. (4.74), where $\bar{\alpha}_i$ and $\bar{\beta}_i$ are the spectrum boundaries of the corresponding spectral problems, the definition domain of which is a parallelepiped which involves domain D. Thus, if the minimal eigenvalue of the spectral problem equals zero, one has to take the smallest eigennumber as $\bar{\alpha}_i$. The essence of the above inserted operators C and R becomes understandable from the following considerations. In problem (4.76) the most essential considerations are the difference analogues of the Laplace operators in the class of the functions, which satisfy one or another boundary condition. So, when acting on the corresponding component of the solution by a smoothing operator (for instance, by the inverse Laplace operator), the residual of the iterative process will be suppressed in the whole domain simultaneously with due regard to the influence domain, natural to this or that problem.

4.7 The Modified Iterative Process

The intention to construct rapidly converging iterative methods for solving equations has stimulated the appearance of other effective methods for solving ocean dynamics difference equations. We shall consider one such method in this section.

First, let us consider the simplest problem of ocean dynamics in the form of Eq. (4.49). We shall attempt its solution according to the following method of successive iterations. First, we find the approximation for density:

$$\mu_1 \Delta_{k,l}^{2,2} \rho_m^n + \nu_1 \Delta_m^{2,2} \rho_m^n = \Gamma h \sum_{m'=m+1}^{m^*} (\nabla_k^- u_{m'}^{n-1} + \nabla_l^- v_{m'}^{n-1}) + G_m, \tag{4.82}$$

and then the components of velocity are found

$$\mu \Delta_{k,l}^{1,1} u_m^n + l v_m^n = \frac{gh}{\rho_0} \sum_{m'=0}^{m-1} \nabla_k^- \rho_{m'}^n ,$$

$$\mu \Delta_{k,l}^{1,1} v_m^n - l u_m^n = \frac{gh}{\rho_0} \sum_{m'=0}^{m-1} \nabla_l^- \rho_{m'}^n . \tag{4.83}$$

The convergency criteria of this method will be discussed later; here, we first consider the problems involved in the methods for algorithmic realization. For this purpose we first consider problem (4.82). Omitting indexes n and m we rewrite it in the form:

$$\mu_1 \Delta_{k,l}^{2,2} \rho + v_1 \Delta_m^{2,2} \rho = f , \tag{4.84}$$

where the value of f is known for every iterative step and is defined by the formula:

$$f = \Gamma h \sum_{m'=m+1}^{m^*} (\nabla_k^- u_{m'}^{n-1} + \nabla_l^- v_{m'}^{n-1}) + G_m . \tag{4.85}$$

First, let us solve the spectral problem:

$$-\Delta_m^{2,2} \omega = \lambda \omega \tag{4.86}$$

and find the orthogonal and normalized basis of the vectors $\{\omega_q\}$ and corresponding real and non-negative spectrum $\{\lambda_q\}$. Let us note that one of the eigennumbers is equal to zero. The eigenfunction, corresponding to $\lambda = 0$ is a vector, which does not depend on sea level and is not essential for calculations. Therefore, we omit the component, corresponding to $\lambda = 0$ and in this way orthogonalize the solution with respect to the constant. Algorithmically, this is quite simple. First, it is necessary to find the arithmetic mean value of ρ for every point (x_k, y_l) and for every level, then to subtract this constant from each component of the determined vector ρ_m.

Now let us consider the decomposition:

$$\rho_m = \sum_{q=1}^{m^*} \bar{\rho}_q \omega_q , \qquad f_m = \sum_{q=1}^{m^*} \bar{f}_q \omega_q . \tag{4.87}$$

Here, $\bar{\rho}_q$ and $\bar{f}_q$ are the Fourier coefficients of the expansion according to the orthogonalized system of the eigenfunctions:

$$\bar{\rho}_q = (\rho, \omega_q) , \qquad \bar{f}_q = (f, \omega_q) , \tag{4.88}$$

where ρ is a vector with components ρ_m, and the scalar product is defined in the form:

$$(a, b) = \sum_{m'=1}^{m^*} a_{m'} b_{m'} . \tag{4.89}$$

Problem (4.84) with the help of Eq. (4.87) is reduced to the set of problems for the Fourier coefficients:

$$\mu_1 \Delta_{k,l}^{2,2} \bar{\rho}_q - v_1 \lambda_q \bar{\rho}_q = \bar{f}_q ,$$

$$(q = 1, 2, \ldots, m^*) . \tag{4.90}$$

We will write the problems of type (4.90) in the operator form:

$$A_q \bar{\rho}_q = -\bar{f}_q \,, \tag{4.91}$$

where

$$A_q = \nu_1 \lambda_q E - \mu_1 \Delta_{k,l}^{2,2} \,. \tag{4.92}$$

We will determine the maximal and minimal eigenvalues of operator $-\Delta_{k,l}^{2,2}$, which we denote by $\beta(-\Delta_{k,l}^{2,2})$ and $\alpha(-\Delta_{k,l}^{2,2})$ respectively. Then the minimal and maximal eigenvalues of operator A_q will be found in the form:

$$\alpha(A_q) = \nu_1 \lambda_q + \mu_1 \alpha(-\Delta_{k,l}^{2,2}) \,,$$

$$\beta(A_q) = \nu_1 \lambda_q + \mu_1 \beta(-\Delta_{k,l}^{2,2}) \,. \tag{4.93}$$

Possessing now $\alpha(A_q)$ and $\beta(A_q)$, we organize the iterative process:

$$\bar{\rho}_q^{j+1} = \bar{\rho}_q^j - \tau_{q_j}(A_q \bar{\rho}_q^j + \bar{f}_q) \,, \tag{4.94}$$

where τ_{q_j} is chosen on the basis of the information on $\alpha(A_q)$ and $\beta(A_q)$ according to Chebyshev's acceleration.

Further, let us consider the solution method of equations system (4.83). We rewrite this problem in the form:

$$\mu \Delta_{k,l}^{1,1} u_m + l v_m = a_m \,,$$

$$\mu \Delta_{k,l}^{1,1} v_m - l u_m = b_m \,,$$

$$(m = 1, 2, \ldots, m^*) \,. \tag{4.95}$$

Here, a_m and b_m are the specified right-hand sides of Eq. (4.83). We take into consideration the matrix and the vectors of the form:

$$A = \begin{Vmatrix} \mu \Delta_{k,l}^{1,1} & -l \\ l & \mu \Delta_{k,l}^{1,1} \end{Vmatrix} \,, \qquad \varphi = \begin{vmatrix} u_m \\ v_m \end{vmatrix} \,, \qquad f = \begin{vmatrix} a_m \\ b_m \end{vmatrix} \,.$$

Then, equations system (4.95) will be written in the form:

$$A\varphi = -f \,, \tag{4.96}$$

where

$$(A\varphi, \varphi) > 0 \tag{4.97}$$

on the set of vectors $\{\varphi\}$, among which the solution of problem (4.96) is being sought. The scalar product in (4.97) is determined in the way:

$$(a, b) = \sum_{i=1}^{2} \sum_{k=1}^{k^*-1} \sum_{l=1}^{l^*-1} \sum_{m=1}^{m^*-1} a_{k,l,m}^{(i)} b_{k,l,m}^{(i)} \,.$$

It is important to note that the spectrum of the difference operator A is complex, although the operator itself is positive. Therefore, the optimization of iterative processes, based on Chebyshev's acceleration or on the successive overrelaxation, fall away. Then the method of minimal residuals is left

$$\varphi^{j+1} = \varphi^j - \tau_j(A\varphi^j + f), \tag{4.98}$$

where

$$\tau_j = \frac{(A\xi^j, \xi^j)}{(A\xi^j, A\xi^j)}. \tag{4.99}$$

Let us consider the combination of minimal residuals and splitting for accelerating the convergency of the iterative method. Let us assume that $\alpha(-\Delta_{k,l}^{1,1})$ and $\beta(-\Delta_{k,l}^{1,1})$ are the spectrum boundaries of operator $\Delta_{k,l}^{1,1}$. We consider the matrix:

$$B = \begin{Vmatrix} G & 0 \\ 0 & G \end{Vmatrix},$$

where

$$G = (E - \sigma\Delta_k^{1,1})(E - \sigma\Delta_l^{1,1}), \qquad \sigma = \frac{1}{\sqrt{\alpha\beta}},$$

and write the iterative scheme:

$$B\frac{\varphi^{j+1} - \varphi^j}{\tau_j} + A\varphi^j = -f. \tag{4.100}$$

Resolving this recurrent scheme with respect to φ^{j+1}, we obtain:

$$\varphi^{j+1} = \varphi^j - \tau_j B^{-1}(A\varphi^j + f),$$

$$\tau_j = \frac{(AB^{-1}\xi^j, \xi^j)}{(AB^{-1}\xi^j, AB^{-1}\xi^j)}, \qquad \xi^j = A\varphi^j + f. \tag{4.101}$$

When calculating σ_i, as in the case of Eq. (4.74), instead of α_i and β_i it is possible to take $\bar{\alpha}_i$ and $\bar{\beta}_i$, i.e. the spectrum boundaries of the respective spectral problems for the parallelepiped enclosing the real domain D. Here, as in Eq. (4.81), if the smallest eigenvalue of the problem equals zero, then for the optimization, the smallest non-zero value should be taken.

Now we will discuss the successive approximation method for the more detailed problem. Assuming that we deal with problem (4.52). We formulate the iterative process in analogy with the previous one. First, the equation for density:

$$\mu_1\Delta_{k,l}^{2,2}\rho_m^n + \nu_1\Delta_m^{2,2}\rho_m^n = \Gamma h \sum_{m'=m+1}^{m^*} (\nabla_k^- u_{m'}^{n-1} + \nabla_l^- v_{m'}^{n-1}) + G_m \tag{4.102}$$

is solved, and then the equations of motion:

$$\mu\Delta_{k,l}^{1,1}u_m^n + \nu\Delta_m^{2,1}u_m^n + lv_m^n = \frac{gh}{\rho_0}\sum_{m'=0}^{m-1}\nabla_k^+\rho_{m'}^n - \frac{gh^2}{\rho_0 H}\sum_{m=1}^{m^*}\sum_{m=0}^{m-1}\nabla_k^+\rho_{m'}^n + a_m,$$

$$\mu\Delta_{k,l}^{1,1}v_m^n + \nu\Delta_m^{2,1}v_m^n - lu_m^n = \frac{gh}{\rho_0}\sum_{m'=0}^{m-1}\nabla_l^+\rho_{m'}^n - \frac{gh^2}{\rho_0 H}\sum_{m=1}^{m^*}\sum_{m'=0}^{m-1}\nabla_l^+\rho_{m'}^n + b_m. \tag{4.103}$$

As far as Eq. (4.102) is concerned, its solution was discussed above. Some difference will be found in the algorithm of the solution of problem (4.103), i.e. the solution of these two equations should be sought in the form of the Fourier series by the eigenfunction of the problem:

$$-\Delta_m^{2,1}\Omega_q = \lambda_q \Omega_q \,. \tag{4.104}$$

Representing u_m and v_m in the form of the series:

$$u_m^k = \sum_{q=1}^{m^*} \bar{u}_q^n \Omega_q \,, \qquad v_m^n = \sum_{q=1}^{m^*} \bar{v}_q^n \Omega_q \,, \tag{4.105}$$

we obtain the system of equations for the Fourier coefficients $\bar{u}_q^n, \bar{v}_q^n$, which is solved on the basis of the method of minimal residuals in combination with the splitting method by a scheme similar to that described above.

Now we will consider the theoretical analysis of convergency of the iterative method (4.82), (4.83) and (4.102), (4.103). To carry out this analysis to the end, instead of denominated problems, we will consider others, similar to them in the differential form, the solution of which we assume to be periodic. So, first we consider the iterative process for the problem, corresponding to (4.82), (4.83), but written in the form where P and w are not excluded, thus

$$\mu_1 \Delta \rho^{n+1} + v_1 \frac{\partial^2}{\partial z^2} \rho^{n+1} = \Gamma w^{n+1/2} \,,$$

$$\frac{\partial w^{n+1/2}}{\partial z} = -\left(\frac{\partial u^{n+1/2}}{\partial x} + \frac{\partial v^{n+1/2}}{\partial y}\right),$$

$$\mu \Delta u^{n+1/2} + l v^{n+1/2} = \frac{1}{\rho_0}\frac{\partial P^n}{\partial x} \,,$$

$$\mu \Delta v^{n+1/2} - l u^{n+1/2} = \frac{1}{\rho_0}\frac{\partial P^n}{\partial y} \,,$$

$$\frac{\partial P^n}{\partial z} = g\rho^n \,. \tag{4.106}$$

We assume that any component of the solution has the form:

$$\varphi(x, y, z) = \bar{\varphi} e^{i\alpha x + i\beta y + i\gamma z} \,, \tag{4.107}$$

where $\bar{\varphi}$ is the Fourier coefficient, α, β and γ are the wave parameters

$$\alpha = \frac{2\pi k}{L} \,, \quad \beta = \frac{2\pi l}{L} \,, \quad \gamma = \frac{2\pi m}{H} \quad (k, l, m = 1, 2, 3, \ldots) \,.$$

We insert Eq. (4.107) into (4.106) and bring system (4.106) to the correlation for each of the disturbance harmonics of Eq. (4.107)

$$\bar{\varphi}^{n+1} = T\bar{\varphi}^n \,, \tag{4.108}$$

where T is the operator of the step, the value of which will be assessed on the basis of the characteristic scales of the disturbances.

Inserting Eq. (4.107) into (4.106) we obtain:

$$-(\mu_1 r^2 + \nu_1 \gamma^2)\bar{\rho}^{n+1} = \Gamma \bar{w}^{n+1/2},$$

$$\gamma \bar{w}^{n+1/2} = -(\alpha \bar{u}^{n+1/2} + \beta \bar{v}^{n+1/2}),$$

$$-\mu r^2 \bar{u}^{n+1/2} + l \bar{v}^{n+1/2} = \frac{i\alpha}{\bar{\rho}} \bar{P}^n,$$

$$-\mu r^2 \bar{v}^{n+1/2} - l \bar{u}^{n+1/2} = \frac{i\beta}{\bar{\rho}} \bar{P}^n,$$

$$i\gamma \bar{P}^n = g\bar{\rho}^n, \tag{4.109}$$

where

$$r^2 = \alpha^2 + \beta^2 .$$

Now we will exclude all the components of the solution, except $\bar{\rho}^{n+1}$ and $\bar{\rho}^n$. Because of the correlation

$$\bar{P}^n = \frac{ig}{\gamma} \bar{\rho}^n$$

we have the system for $\bar{u}^{n+1/2}$ and $\bar{v}^{n+1/2}$

$$\mu r^2 \bar{u}^{n+1/2} - l \bar{v}^{n+1/2} = \frac{g\alpha}{\gamma \rho_0} \bar{\rho}^n,$$

$$l \bar{u}^{n+1/2} + \mu r^2 \bar{v}^{n+1/2} = \frac{g\beta}{\gamma \rho_0} \bar{\rho}^n . \tag{4.110}$$

Solving the equations system (4.110) we obtain:

$$\bar{u}^{n+1/2} = \frac{g}{\gamma \rho_0} \frac{\alpha \mu r^2 + \beta l}{(\mu r^2)^2 + l^2} \bar{\rho}^n,$$

$$\bar{v}^{n+1/2} = \frac{g}{\gamma \rho_0} \frac{\beta \mu r^2 - \alpha l}{(\mu r^2)^2 + l^2} \bar{\rho}^n . \tag{4.111}$$

After inserting these quantities into the second equation of system (4.109) we have:

$$\bar{w}^{n+1/2} = -\frac{g}{\gamma^2 \rho_0} \frac{\mu r^4}{(\mu r^2)^2 + l^2} \bar{\rho}^n . \tag{4.112}$$

Now we insert the determined expression for $\bar{w}^{n+1/2}$ into the first equation of system (4.109) and obtain:

$$\bar{\rho}^{n+1} = \frac{g\Gamma}{\gamma^2 \rho_0} \frac{\mu r^4}{[(\mu r^2)^2 + l^2](\mu_1 r^2 + \nu_1 \gamma^2)} \bar{\rho}^n . \tag{4.113}$$

By comparing Eqs. (4.113) and (4.108) we obtain the expression for the operator of the step:

$$T = \frac{g\Gamma}{\gamma^2 \rho_0} \frac{\mu r^4}{(\mu_1 r^2 + \nu_1 \gamma^2)[(\mu r^2)^2 + l^2]} . \tag{4.114}$$

For the convergency of the iterative process, the condition

$$T < 1 \tag{4.115}$$

should be fulfilled.

When a more detailed system of equations, taking into account the turbulent friction in Euler's equation, is under consideration, we have a more detailed problem:

$$\mu_1 \Delta \rho^{n+i} + \nu_1 \rho_{zz}^{n+1} = \Gamma w^{n+1/2} ,$$

$$\frac{\partial w^{n+1/2}}{\partial z} = -\left(\frac{\partial u^{n+1/2}}{\partial x} + \frac{\partial v^{n+1/2}}{\partial y}\right),$$

$$\mu \Delta u^{n+1/2} + \nu u_{zz}^{n+1/2} + l v^{n+1/2} = \frac{1}{\rho_0} \frac{\partial P^n}{\partial x} ,$$

$$\mu \Delta v^{n+1/2} + \nu v_{zz}^{n+1/2} - l u^{n+1/2} = \frac{1}{\rho_0} \frac{\partial P^n}{\partial y} ,$$

$$P_z^h = g\rho^n . \tag{4.116}$$

Then in analogy with the foregoing we obtain the recurrent correlation (4.108), where operator T has the form:

$$T = \frac{g\Gamma r^2}{\rho_0 \gamma^2} \frac{(\mu r^2 + \nu\gamma^2)}{(\mu_1 r^2 + \nu_1 \gamma^2)[(\mu r^2 + \nu\gamma^2)^2 + l^2]} \tag{4.117}$$

and the same criterion of (4.115) will serve again for the convergency of the process

$$T < 1 . \tag{4.118}$$

Let us interpret the determined criterion of the iterative process convergency. For this purpose we consider a more detailed case. First, we note that for the periodic problem in the difference form we again come to formula (4.117), where r^2 and m^2 are related to h, H, $\Delta x = \Delta y$ and L (L is the maximal wave-length in the horizontal plane), by formulas:

$$r^2 = \alpha^2 + \beta^2 ,$$

$$\alpha^2 = \frac{4}{\Delta x^2} \sin^2 \frac{k\pi \Delta x}{L} , \qquad \beta^2 = \frac{4}{\Delta y^2} \sin^2 \frac{l\pi \Delta y}{L} ,$$

$$\gamma^2 = \frac{4}{h^2} \sin^2 \frac{m\pi h}{H} . \tag{4.119}$$

It is necessary now to calculate the limits of variation of the parameters α, β and γ and to check the fulfillment of condition (4.115) with the help of formulas (4.118). The choice of L remains conditional, of course, as well as the assumption of the problem's periodicity, but the iterative mechanism, related to the quantity T, depending on α, β, γ and the conclusion on convergency, as well as the efficiency of the method's application, becomes more constructive and understandable.

4.8 The Simplest Model of Ocean Dynamics, Taking into Account the Non-Linear Turbulent Exchange

The next generalization concerning the climate theory of the ocean is the introduction of the non-linear turbulence. As it is well known, in a stably-stratified liquid $\partial\rho/\partial z > 0$, the turbulent exchange may be described with sufficient accuracy by the constant coefficient of vertical turbulence ν. But such a description appears to be satisfactory only in this case. When the stratification of the liquid is unstable, irregular convective movements are immediately, generated in the liquid, under the influence of buoyancy forces and these movements agitate a large ocean volume. This means that the coefficient of the turbulent exchange is a non-linear function by nature. One can approximate it in the following, simple way:

$$\nu = \begin{cases} \bar{\nu} & \text{if } \dfrac{\partial\rho}{\partial z} > 0\,, \\[2ex] \dfrac{\bar{\nu}}{\varepsilon} & \text{if } \dfrac{\partial\rho}{\partial z} \leqslant 0\,, \end{cases}$$

where ε is a specified, non-dimensional constant of a phenomenological character, which may be chosen in the order of 10^{-2} or 10^{-3}. We shall assume that if $\partial\rho/\partial z \leqslant 0$ occurs at some ocean depth, then the whole ocean layer $0 \leqslant z \leqslant H$ is subjected to the intensive turbulence. This model of turbulence is similar to the one considered by Bryan (1969a, b).

The main equations of dynamics are not subjected to any essential, explicit variations in this model. Thus, let us take Eq. (4.49) as the initial problem and define the method of successive approximations in the form similar to Eqs. (4.82), (4.83):

$$\mu_1 \Delta_{k,l}^{2,2} \rho_m^n + \nu_1^{n-1} \Delta_m^{2,2} \rho_m^n = \Gamma h \sum_{m'=m+1}^{m^*} (\nabla_k^- u_{m'}^{n-1} + \nabla_l^- v_{m'}^{n-1}) + c_m\,,$$

$$\mu \Delta_{k,l}^{1,1} u_m^n + l v_m^n = \frac{gh}{\rho_0} \sum_{m'=0}^{m-1} \nabla_k^+ \rho_m^n\,,$$

$$\mu \Delta_{k,l}^{1,1} v_m^n - l u_m^n = \frac{gh}{\rho_0} \sum_{m'=0}^{m-1} \nabla_l^+ \rho_m^n\,. \tag{4.120}$$

In comparison with the foregoing, here the assumption is made that $\nu^{n-1} = \nu^{n-1}(x_k, y_l)$ is defined by scheme (4.120) and by assumptions of the agitation of the whole thickness of the unstably stratified liquid. Such a model of the turbulent exchange in principle does not violate the foregoing scheme, but demands its modification in that part, which is related to the solution of the density equations at the specified iterative step n. Indeed, unlike the foregoing problem (4.82), (4.83), in Eq. (4.84) the coefficient depends on the grid point (x_k, y_l). Therefore, we have to rewrite expression (4.91) in the form:

$$\mu \Delta_{k,l}^{2,2} \bar{\rho}_q - \nu_{k,l} \lambda_q \bar{\rho}_q = \bar{f}_q\,, \tag{4.121}$$

where $\bar{\rho}_q$, $\bar{f}_q$ and $\nu_{k,l}$ are grid functions of k and l. The solution of this problem may be found according to the same method as for problem (4.91). It is important to note that the assumption on the mechanism of turbulent exchange does not violate the possibility of applying the method of variable separation by indexes k, l and m in these calculations.

A more detailed model of dynamics (4.52) may be treated in the same way.

4.9 The Statement of Several Non-Linear Problems

Let us examine a more detailed statement of the ocean dynamics problem in a quasi-static approximation without linearization. We shall have the system of equations:

$$\mu \Delta u + lv = \frac{1}{\rho_0}\frac{\partial P}{\partial x} + R(u)\,,$$

$$\mu \Delta v - lu = \frac{1}{\rho_0}\frac{\partial P}{\partial y} + R(v)\,,$$

$$\frac{\partial P}{\partial z} = g\rho\,,$$

$$\frac{\partial u}{\partial x} + \frac{\partial v}{\partial y} + \frac{\partial w}{\partial z} = 0\,,$$

$$\mu_1 \Delta \rho + \nu_1 \frac{\partial^2 \rho}{\partial z^2} = R(\rho)\,, \tag{4.122}$$

where R is an operator of the form:

$$R(\varphi) = u\frac{\partial \varphi}{\partial x} + v\frac{\partial \varphi}{\partial y} + \frac{\partial \varphi}{\partial z}\int_z^H \left(\frac{\partial u}{\partial x} + \frac{\partial v}{\partial y}\right) d\xi\,,$$

and φ is any of functions u, v, ρ.

We take the following boundary conditions for system (4.122):

$$\nu_1 \frac{\partial \rho}{\partial z} = \gamma\,, \quad w = 0 \quad \text{at } z = 0\,,$$

$$\frac{\partial \rho}{\partial z} = 0\,, \quad w = 0 \quad \text{at } z = H\,,$$

$$u = 0\,, \quad v = 0\,, \quad \frac{\partial \rho}{\partial n} = 0 \quad \text{at } \sigma\,. \tag{4.123}$$

We shall look for the solution in the form:

$$u = \bar{u} + u'\,,$$

$$v = \bar{v} + v'\,,$$

$$w = w'\,,$$

$$P = \bar{P} + P'\,,$$

$$\rho = \rho_0 + \bar{\rho} + \rho'\,. \tag{4.124}$$

Here, the quantities with the bar over the symbols depend only on x and y, and φ' is the deviation from these values which satisfies the relation:

$$\int_0^H \varphi' dz = 0 . \tag{4.125}$$

We insert expressions (4.124) into the equations system (4.122), integrate them by z within the limits $0 \leqslant z \leqslant H$ and use correlation (4.125) and boundary conditions (4.123). Then we come to the equations for the barotropic component:

$$\mu \Delta \bar{u} + l\bar{v} = \frac{1}{\rho_0} \frac{\partial \bar{P}}{\partial x} + \overline{R(u)} ,$$

$$\mu \Delta \bar{v} - l\bar{u} = \frac{1}{\rho_0} \frac{\partial P}{\partial y} + \overline{R(v)} ,$$

$$\frac{\partial \bar{u}}{\partial x} + \frac{\partial \bar{v}}{\partial y} = 0 ,$$

$$\mu_1 \Delta \bar{\rho} = \overline{R(\rho)} + \gamma \tag{4.126}$$

with the boundary conditions

$$\bar{u} = 0 , \quad \bar{v} = 0 , \quad \frac{\partial \bar{\rho}}{\partial n} = 0 \quad \text{at } \sigma . \tag{4.127}$$

For the baroclinic component we shall have the system of equations of the type:

$$\mu \Delta u' - lv' = \frac{g}{\rho_0} \int_0^z \frac{\partial \rho'}{\partial x} dz + R(u) - \overline{R(u)} ,$$

$$\mu \Delta v' + lu' = \frac{g}{\rho_0} \int_0^z \frac{\partial \rho'}{\partial y} dz + R(v) - \overline{R(v)} ,$$

$$\mu_1 \Delta \rho' + \nu_1 \frac{\partial^2 \rho'}{\partial z^2} = R(\rho) - \overline{R(\rho)} - \gamma . \tag{4.128}$$

To system (4.128) we add the boundary conditions:

$$\nu_1 \frac{\partial \rho'}{\partial z} = \gamma \quad \text{at } z = 0 ,$$

$$\frac{\partial \rho'}{\partial z} = 0 \quad \text{at } z = H ,$$

$$u' = 0 , \quad v' = 0 , \quad \frac{\partial \rho'}{\partial n} = 0 \quad \text{at } \sigma . \tag{4.129}$$

Assuming that the initial approximations of the values in problems (4.126), (4.127) and (4.128), (4.129) are found by means of solving the linear problems discussed in the previous sections, one may formulate the following method of successive approximations. For problem (4.126):

$$\mu \Delta \bar{u}^n + l\bar{v}^n = \frac{1}{\rho_0} \frac{\partial \bar{P}^n}{\partial x} + \overline{R^{n-1}(u)},$$

$$\mu \Delta \bar{v}^n - l\bar{u}^n = \frac{1}{\rho_0} \frac{\partial \bar{P}^n}{\partial y} + \overline{R^{n-1}(v)},$$

$$\frac{\partial \bar{u}^n}{\partial x} + \frac{\partial \bar{v}^n}{\partial y} = 0,$$

$$\mu_1 \Delta \bar{\rho}^n = \overline{R^{n-1}(\rho) + \gamma} \tag{4.130}$$

with conditions (4.127).

In the same way, the system of equations (4.128) may be brought to the form (the primes are omitted):

$$\mu \Delta u^n + l v^n = \frac{g}{\rho_0} \int_0^z \frac{\partial \rho^n}{\partial x} dz + R^{n-1}(u) - \overline{R^{n-1}(u)},$$

$$\mu \Delta v^n - l u^n = \frac{g}{\rho_0} \int_0^z \frac{\partial \rho^n}{\partial y} dz + R^{n-1}(v) - \overline{R^{n-1}(v)},$$

$$\mu_1 \Delta \rho^n + \nu_1 \frac{\partial^2 \rho^n}{\partial z^2} = R^{n-1}(\rho) - \overline{R^{n-1}(\rho)} - \gamma \tag{4.131}$$

with condition (4.129).

It can be seen that with this definition of the successive approximation method, both problems (4.130) and (4.131) coincide with the ones discussed above at the given step of iteration, and, consequently, is solved by the foregoing algorithm. It is only necessary to make a notation on the difference approximation of expressions $R(\rho)$. This approximation should be made in accordance with the methods presented in Section 1.3.

Finally, we will mention that the baroclinic component of the field is the most interesting for applications. The inverse relationship from the barotropic component to the baroclinic one is so weak, that it is possible for the approximate solution of the problem to find first u', v' and ρ' from problem (4.128), (4.129), assuming that $\bar{u} = \bar{v} = 0$, $\bar{P} = 0$ and then calculate $\bar{u}$, $\bar{v}$ and $\bar{P}$ on the basis of the solution of problem (4.126), (4.127).

The presented way of solving the problem does not suffer any violation when taking into account the vertical turbulent exchange in equations of motion.

4.10 The Problem of Non-Stationary Adjustment of Flow Fields to Atmospheric Disturbances

When solving the problem with due regard to air-sea interaction, and with the aim of weather forecasting, the definition of initial conditions in the ocean, which are formed under continuous, seasonal, non-stationary atmospheric action, is very important. For this purpose we will consider the following linear, but now non-stationary, ocean problem:

$$\frac{\partial u}{\partial t} - lv = -\frac{1}{\rho_0}\frac{\partial P}{\partial x} + \mu \Delta u + v\frac{\partial^2 u}{\partial z^2},$$

$$\frac{\partial v}{\partial t} + lu = -\frac{1}{\rho_0}\frac{\partial P}{\partial y} + \mu \Delta v + v\frac{\partial^2 v}{\partial z^2},$$

$$\frac{\partial P}{\partial z} = g\rho,$$

$$\frac{\partial u}{\partial x} + \frac{\partial v}{\partial y} + \frac{\partial w}{\partial z} = 0,$$

$$\frac{\partial \rho}{\partial t} + \Gamma w = \mu_1 \Delta \rho + v_1 \frac{\partial^2 \rho}{\partial z^2}. \tag{4.132}$$

We add to system (4.132) the boundary conditions:

$$v\frac{\partial u}{\partial z} = -\frac{\tau_x}{\rho_0}, \quad v\frac{\partial v}{\partial z} = -\frac{\tau_y}{\rho_0}, \quad v_1\frac{\partial \rho}{\partial z} = \gamma, \quad w = 0 \quad \text{at } z = 0,$$

$$u = 0, \quad v = 0, \quad \frac{\partial \rho}{\partial z} = 0, \quad w = 0 \quad \text{at } z = H$$

$$u = 0, \quad v = 0, \quad \frac{\partial \rho}{\partial n} = 0 \quad \text{at } \sigma. \tag{4.133}$$

Let us assume that quantities τ_x, τ_y and γ are functions (x, y, t) and are determined from the diagnosis of the atmospheric conditions near the ocean's free surface.

To solve problem (4.132), (4.133) the ocean climatic condition is chosen as the initial condition, described by one of the theoretical models discussed in previous sections of this chapter. Our task now is to solve problem (4.132), (4.133) with the chosen, initial (climatic) ocean condition. Assuming that $\tau_{x,z}$, $\tau_{y,z}$ and γ are specified for a sufficiently long period (an order of season) we shall solve the problem of ocean currents and the density field formation to some priori given time.

To solve problem (4.132), (4.133) we approximate it by time. For this purpose we use the scheme of the natural filter (Marchuk 1967a):

$$\frac{u^{j+1} - u^j}{\tau} - lv^{j+1} = -\frac{1}{\rho_0}\frac{\partial P^{j+1}}{\partial x} + \mu \Delta u^{j+1} + v\frac{\partial^2 u^{j+1}}{\partial z^2},$$

$$\frac{v^{j+1} - v^j}{\tau} + lu^{j+1} = -\frac{1}{\rho_0}\frac{\partial P^{j+1}}{\partial y} + \mu \Delta v^{j+1} + v\frac{\partial^2 u^{j+1}}{\partial z^2},$$

$$\frac{\partial P^{j+1}}{\partial z} = g\rho^{j+1},$$

$$\frac{\partial u^{j+1}}{\partial x} + \frac{\partial v^{j+1}}{\partial y} + \frac{\partial w^{j+1}}{\partial z} = 0,$$

$$\frac{\rho^{j+1} - \rho^j}{\tau} + \Gamma w^{j+1} = \mu_1 \Delta \rho^{j+1} + v_1 \frac{\partial^2 \rho^{j+1}}{\partial z^2}. \tag{4.134}$$

The system of equations (4.134) is solved with boundary conditions (4.133), taken at the corresponding moments of time. It is important to note that the implicit scheme of approximation in Eqs. (4.134) enables one to obtain an absolutely stable scheme which automatically filters out the high-frequency processes, which are non-essential for the dynamics of large-scale processes. The solution of problem (4.132), (4.133) serves as initial data for the more exact problem of non-stationary currents, which we have considered in Chapter 3. This problem should be solved for a period of 2 to 4 weeks. Now, the resulting solution may serve as initial ocean data for the joint solution of weather forecast and ocean currents dynamics.

4.11 The Formation of the Thermocline in the Ocean

The thermocline theory is a central problem in oceanography. The essence of the problem is a description of the vertical structure of temperature and density in a baroclinic ocean. There are different points of view on the problem, which has not been solved until now.

As it is known, the density gradient decreases exponentially in ocean areas, with the exception of the northern latitudes and below the friction layer. The density is constant in the layer $0 < z < h$ from the sea surface up to the upper boundary of the thermocline, because of the intense, turbulent mixing. At deeper layers the variation of density with depth is small and the currents become barotropic. Furthermore, there is no thermocline at the northern latitudes. The ocean zone with an explicitly expressed thermocline coincides more or less with the domain of density flux from the atmosphere to the ocean, and the zone where the thermocline is absent, with the domain of density flux from the ocean to the atmosphere. The effective thickness of the thermocline varies from several tens of metres in one ocean area to several hundreds of metres in another. These are the main factors, which should be explained by the thermocline theory. In this section the simplest mathematical model of the thermocline is presented, which, in our opinion, gives a qualitative explanation of the enumerated factors.

Let us suppose that the ocean dynamics equations are averaged for a period corresponding to the time of the thermocline formation, i.e. several tens of years. If φ is one of the hydrophysical substances and u_α is a component of the flow velocity, then, usually, for presenting the averaged terms of hydrodynamics equations, the following correlations are used:

$$\overline{\frac{\partial u_\alpha \varphi}{\partial x_\alpha}} = \frac{\partial \bar{u}_\alpha \bar{\varphi}}{\partial x_\alpha} + \frac{\partial \overline{u'_\alpha \varphi'}}{\partial x_\alpha}, \qquad \bar{\varphi}' = 0, \quad \bar{u}' = 0, \quad (\alpha = 1, 2, 3). \tag{4.135}$$

The lines above the functions denote the time average. On the basis of the semi-empirical turbulence theory, we have:

$$\overline{u'_\alpha \varphi'} = -k_\alpha \frac{\partial \bar{\varphi}}{\partial x_\alpha}, \tag{4.136}$$

where k_α is the coefficient of the turbulent exchange.

Usually the multi-year, averaged fields of substances and the averaged fluctuations are in the following correlations:

$$\left|\frac{\partial \bar{u}_\alpha \bar{\varphi}}{\partial x_\alpha}\right| \leqslant \left|\frac{\partial \overline{u'_\alpha \varphi'}}{\partial x_\alpha}\right| .$$

Taking this into account, we write correlation (4.135) in the following approximate form:

$$\frac{\partial \overline{u_\alpha \varphi}}{\partial x_\alpha} = -\frac{\partial}{\partial x_\alpha} k_\alpha \frac{\partial \bar{\varphi}}{\partial x_\alpha} . \tag{4.137}$$

One may note that the turbulent exchange coefficient is strongly inhomogeneous. We assume $k_x = k_y = \mu$ and $k_z = v$. Coefficient μ may be taken to be constant. As for the coefficient of the vertical turbulent exchange, it essentially depends on the vertical density stratification and is a function of the vertical density gradient, i.e.

$$v = v(\rho_z) .$$

After neglecting the advective and convective components of the averaged fields, we will write the equations of heat and salt transfer in the form:

$$\frac{\partial}{\partial z}\left[v(\rho_z)\frac{\partial T}{\partial z}\right] + \mu \Delta T = 0 ,$$

$$\frac{\partial}{\partial z}\left[v(\rho_z)\frac{\partial S}{\partial z}\right] + \mu \Delta S = 0 . \tag{4.138}$$

Here, for simplicity we have omitted the averaging lines over T, S and ρ. To system (4.138) we add the equation of state in the form:

$$\rho = f(T, S) . \tag{4.139}$$

Function f is known from the experimental data. It is necessary to find function $v(\rho_z)$ in order to enclose the system of equations (4.138), (4.139). We assume:

$$v = \begin{cases} v_0 & \text{when } \rho_z < 0 , \\ v_\infty & \text{when } \rho_z \geqslant 0 . \end{cases} \tag{4.140}$$

The coefficient of the vertical turbulence for the case of unstable or neutral stratification v_∞ has a much larger value that of the corresponding coefficient v_0 for the case of a stable stratification. For the special case one might take $v_\infty = \infty$.

As boundary conditions for the system (4.138)–(4.140) we take the following:

$$\frac{\partial T}{\partial n} = 0 , \quad \frac{\partial S}{\partial n} = 0 \quad \text{at } \sigma ,$$

$$v\frac{\partial T}{\partial z} = -\Gamma , \quad v\frac{\partial S}{\partial z} = -\gamma \quad \text{at } \sigma_0 , \tag{4.141}$$

where σ is a solid surface with the external normal n, σ_0 is the ocean's free surface, which is slightly disturbed from the equilibrium state; Γ and γ are the specificed, turbulent heat and salt fluxes at the sea surface (they are the functions of x and y).

Let us note that $\mu > 0$, $\nu > 0$. Besides, we have mentioned already that in the cases of neutral ($\rho_z = 0$) or unstable stratification ($\rho_z > 0$), the coefficient ν becomes extremely large. This fact corresponds to the active development of irregular convective motions in an unstably or neutrally stratified ocean, which may be described statistically by the vertical turbulent exchange.

The uniqueness of the solution of problem (4.138)–(4.141) takes place with the accuracy of an unessential arbitrary constant if the conditions are:

$$\iint_{\sigma_0} \Gamma dxdy = 0\,, \quad \iint_{\sigma} \gamma dxdy = 0\,,$$

and quite natural assumptions on the smoothness of the surface, which limits the solution's definition domain, are fulfilled. Here, obviously, the assumption on the positiveness of values μ and ν is used.

Now we will consider the qualitative analysis of the solution of problems (4.138)–(4.141). For simplicity we will ignore salinity and consider the temperature as the main informative function. Then we obtain a simpler problem:

$$\frac{\partial}{\partial z}\left[\nu(T_z)\frac{\partial T}{\partial z}\right] + \mu \Delta T = 0\,,$$

$$\frac{\partial T}{\partial n} = 0 \quad \text{at } \sigma\,,$$

$$\nu\frac{\partial T}{\partial z} = -\Gamma \quad \text{at } \sigma_0\,. \tag{4.142}$$

We will consider only the domain of the stably stratified liquid, where $\Gamma > 0$. As for the domains with an unstable stratification, we assume that there is an intensive, vertical, turbulent exchange, which transfers the stored heat from the ocean to the atmosphere. This means that at the boundary of such a domain one can make an approximate assumption on the existence of the heat outflow. We can solve problem (4.142) with the accuracy of an arbitrary constant, which we assume to be equal to zero. Then problem (4.142) is formulated for deviations.

We will assume that the ocean has an infinite depth, σ is the side boundary cylindric surface and Σ is a cylindric liquid surface, which separates the domain of the stable stratification from the unstable one. Then we have the problem:

$$\frac{\partial}{\partial z}\nu_0\frac{\partial T}{\partial z} + \mu \Delta T = 0\,,$$

$$\frac{\partial T}{\partial n} = 0 \quad \text{at } \sigma\,,$$

$$T = 0 \quad \text{at } \Sigma\,,$$

$$\nu_0\frac{\partial T}{\partial z} = -\Gamma(x, y) \quad \text{at } z = 0\,,$$

$$\nu_0\frac{\partial T}{\partial z} = 0 \quad \text{when } z = \infty\,. \tag{4.143}$$

Let us consider the spectral problem:

$$-\Delta\Psi = \lambda\Psi\,,$$

$$\frac{\partial\Psi}{\partial n} = 0 \quad \text{at } \sigma\,,$$

$$\Psi = 0 \quad \text{at } \Sigma\,. \tag{4.143'}$$

Problem (4.143) has the basis of eigenfunctions $\{\Psi_n\}$ and the corresponding set of positive eigenvalues in a quite general case of the smoothness of surfaces σ and Σ. We will solve problem (4.143) in the form of the Fourier series by the eigenfunctions of problem (4.143'):

$$T(x, y, z) = \sum_n T_n(z)\Psi_n(x, y)\,. \tag{4.144}$$

We will present function $\Gamma(x, y)$ also in the form of the series:

$$\Gamma(x, y) = \sum_n \Gamma_n \Psi_n(x, y)\,. \tag{4.145}$$

Then we come to the problem:

$$\frac{d}{dz}\nu_0\frac{dT_n}{dz} - \lambda_n\mu T_n = 0\,,$$

$$\nu_0\frac{dT_n}{dz} = -\Gamma_n\,, \quad z = 0\,,$$

$$\nu_0\frac{dT_n}{dz} = 0\,, \quad z = \infty \tag{4.146}$$

for the Fourier coefficients.

In a stably stratified liquid the coefficient of turbulent exchange ν_0 is a function of depth. In a friction layer (10–50 m) the value of the coefficient ν_0 varies in the limits: 10–1000 $\text{cm}^2\,\text{s}^{-1}$, depending on the cause of the turbulent exchange in the surface layer (convective or wave mixing). Below the friction layer the coefficient ν_0 equals approximately 1 $\text{cm}^2\,\text{s}^{-1}$ when the stratification is stable.

So, let us assume:

$$\nu_0 = \begin{cases} \nu_1\,, & z < h\,, \\ \nu_2\,, & z > h\,. \end{cases} \tag{4.147}$$

Then, it is not difficult to obtain the solution of problem (4.146) in the following form:

$$T_n(z) = \begin{cases} \dfrac{\Gamma_n}{\nu_2\sqrt{\dfrac{\mu\lambda_n}{\nu_2}}} + \dfrac{\Gamma_n}{\nu_1}(h - z)\,, & 0 \leqslant z < h\,, \\[3ex] \dfrac{\Gamma_n}{\nu_2\sqrt{\dfrac{\mu\lambda_n}{\nu_2}}}\, e^{\sqrt{\frac{\mu\lambda_n}{\nu_2}}(h - z)}\,, & h < z < \infty\,. \end{cases} \tag{4.148}$$

Here, the assumption is made that a thin friction layer with a large coefficient of mixing, the solution is almost a linear function of depth. This means that we can limit ourselves to the main part of the solution, ignoring the terms with a higher order of magnitude by parameter $\varepsilon_n = \sqrt{\frac{\mu\lambda_n}{\nu_2}}h$. It is not difficult to verify that solution (4.148), as well as the flux, is continuous at the friction layer boundary $z = h$.

We can find the Fourier coefficients for the temperature gradient distribution with the help of solution (4.148):

$$-\frac{dT_n}{dz} = \begin{cases} \frac{\Gamma_n}{\nu_1}, & 0 \leqslant z < h, \\ \frac{\Gamma_n}{\nu_2} e^{\sqrt{\frac{\mu\lambda_n}{\nu_2}}(h - z)}, & h < z < \infty. \end{cases} \quad (4.149)$$

As a result we can conclude that below the friction layer, the Fourier coefficients decay exponentially with depth, within such a mathematical formulation of the problem. With $n = 1$ we obtain the main thermocline, which exists at the middle and southern latitudes of the Atlantic and at the southern and northern latitudes of the Pacific Ocean.

Let us make some evaluations. We assume that $\mu = 10^8$ cm^2 s^{-1}, $\nu_1 = 10$ cm^2 s^{-1}, $\nu_2 = 1$ cm^2 s^{-1}, $L = 10^9$ cm. Here, L is a characteristic size of the ocean, which we assume to be rectangular for the evaluation. Then, the thermocline depth we calculate from the correlation:

$$\sqrt{\frac{\mu\lambda_n}{\nu_2}} H_n = \pi,$$

i.e. the depth at which the intensity of the temperature gradient decays n times, which we take as the conventional thermocline thickness. Then we have:

$$H_n = \frac{L}{n}\sqrt{\frac{\nu_2}{\mu}} = \frac{10^5}{n} \quad (n = 1, 2, \ldots).$$

This means that the main thermocline ($n = 1$) is approximately 1 km thick, and the other modes of depth are n times smaller.

Considering now the characteristic time of thermocline formation, one can easily find the relaxation time from the scale analysis of the non-stationary equation of the heat transport:

$$t_n^{(R)} = (\mu\lambda_n)^{-1}.$$

Assuming, as in the previous case, that

$$\lambda_n = \frac{n^2\pi^2}{L^2},$$

we obtain approximately

$$t_1^{(R)} = 30 \text{ years}, \qquad t_2^{(R)} = 8 \text{ years}, \qquad t_3^{(R)} = 3 \text{ years}.$$

These are, of course, only evaluations of the order of magnitude, however, we can conclude that the order of magnitude of the thermocline spin-up time can be described by a relatively simple model.

A few comments on this problem will be presented. As it has been already concluded, the solution of the thermocline may be reduced to the supersposition of the solutions for the Fourier coefficients. This means that the vertical distribution of the temperature in every point of the ocean can be described by some set of exponents. In the points where the first eigenfunction with specified Γ_1 dominates, one describes the main thermocline. In the points where the input of the first harmonics is small, the thermocline of smaller depth is expressed brightly, etc. So, it is possible to explain qualitatively one more experimental fact, i.e. the variability of the thermocline depth in different parts of the ocean. This fact was established experimentally long ago. Further, when approaching the areas with an unstable stratification, all the eigenvalues $\Psi_n(z)$ tend to zero and the thermocline diminishes.

On the basis of the presented, simple thermocline model, it is possible to introduce the whole physical process of the thermocline formation in a stratified liquid.

The ocean's thermocline is not only a consequence of local processes, but also the result of a large-scale horizontal diffusion interaction of a global character with a vertical turbulent exchange. And in this model the thermocline thickness depends on the spectrum of the horizontal diffusion operator for the whole ocean. We believe that this very fact is a key to the solution of the thermocline formation problem.

The physical picture of the temperature thermocline formation can be introduced now in the following form. In the ocean domains with a stable stratification, where, as a result of the vertical turbulent mixing, an inflow of heat takes place, the heat penetrates into internal deep layers. As a result of the horizontal turbulent diffusion, the heat is transported to the ocean domain, where the stratification is unstable and the heat is transferred from the ocean to the atmosphere. Such domains are usually located in polar regions.

Let us now formulate several problems, the solution of which will help us to develop the thermocline theory of the ocean. The first concerns the solution of a non-linear problem for the whole World Ocean. The second is the evaluation of the effect of advective and convective components in the equations of heat and salt transport. The theory of the thermocline will be improved with regard to of these components, mainly in the regions of regular mean velocities and jetlike currents. By the present time the essential results have been achieved by numerical modelling investigations on baroclinic ocean circulation. Finally, it is necessary to study the evolution of thermocline formation, its periodicity and the evaluation of the upper mixed layer, which is directly responsible for the atmosphere-ocean interaction.

Chapter 5

The Analysis of the Results of Calculations

The previous chapters dealt with many World Ocean circulation models. Such a variety of models is in agreement with the process of development in the sea current theory observed during the last decades. Still, this great number of models can be divided into two groups. (1) Prognostic models: The first group includes models used to calculate thermohydrodynamic ocean characteristics based on the known boundary conditions. The given fields of the elements inside the ocean itself (in the case of non-stationary problems, these are the initial values of the fields) are of no significant value and can be set in any idealized form. Many results have been obtained in this direction, and they are of great interest/Broshe and Sündermann 1985; Friedrich 1970; Haney 1980, 1985; Latif et al. 1985; Manabe and Bryan 1985; Marchuk and Kochergin 1968; Marchuk 1970; Marchuk 1967b; Sarkisyan et al. 1972; Seidov 1977/. Though for several reasons (difficulties with parameterization of the heat and salt turbulent diffusion processes, limited computer capabilities, etc.) they are of a quantitative interest, and provide for the present, rather approximate values of mean climatic characteristics of real oceanic basins. (2) Diagnostic models and models of adaptation: Both boundary conditions and values of one or two internal characteristics are specified in these models. For example, in diagnostic calculations, temperature and salinity fields are specified from observed data (naturally after being averaged and filtered). In the case of the adaptation models some adjustment of the observed data with the model equations, boundary conditions and basin geometry takes place. In this case, unlike prognostic calculations, the observed data (or initial data) play a main part. This chapter deals with a brief analysis of the results of new diagnostic and adaption calculations. A detailed description and analysis of calculations are given in a recently published book (Sarkisyan et al. 1986).

5.1 On the Results of Diagnostic Calculations of the Currents in Different Oceanic Basins

At present, the available (in atlases) large-scale maps of seasonal currents in the World Ocean are based partially on ship-drift data, and partially on the results of calculations with a dynamic method. It is noteworthy that atlases do not provide objective information on currents. They are a schematic representation, a result of subjective synthesis, visual interpolation and extrapolation of sparse and inaccurate data obtained from the two mentioned sources. A number of principal drawbacks characteristic of the dynamic method were already mentioned in Chapter 1. The calculations performed are based on very large grids, as a rule, with a 5° mesh. This results in inhibition of narrow, jetlike currents, the main elements of the World Ocean water circulation. On the other hand, in the near future there will be no maps of deep-sea currents compiled from the measurement data. The measurement data cannot therefore provide reliable information on such significant (e.g. short-range climate variations) characteristics, as the heat content anomalies and meridional heat transfer. So, there is no doubt, that development of methods for diagnostic calculations of currents is rather an important problem. The information on the density (temperature and salinity) field needed for diagnostic calculations represents the most reliable, abundant and ever increasing information on the ocean climate.

As it has been mentioned the sea current models can be divided into two types, prognostic and diagnostic. This type of subdivision is rather conditional, although widespread. The term "prognostic" refers to models in which temperature and salinity (or density) fields are determined alongside the current field. These models cannot yet predict the hydrological situation in the accepted sense of the term, i.e. provide its forecast for practical purposes, e.g. navigation, fishing industry, weather forecast, etc.

The dynamic method, the most primitive geostrophic model of the baroclinic ocean circulation, is up until now the most widespread in the practice of sea current calculations.

Lately, ever-increasing attention is placed on the development of modified methods of practical calculations of sea currents as an alternative to the dynamic method. The diagnostic calculations are the next step following the dynamic method. Their main advantages compared with the dynamic method lie in the ability to solve hydrodynamic problems for the closed basin with corresponding boundary conditions, and in the observance of the law of mass conservation. Therefore, it is possible to take the sea surface wind stress and basin geometry into account and to calculate both horizontal and vertical current velocity components.

Among models designed for a closed baroclinic basin with an arbitrary boundary shape and an arbitrary bottom relief, the models described Charney et al. (1950), Bryan (1969b), Marchuk (1976), Marchuk et al. (1977), Marchuk and Sarkisyan (1980) and Sarkisyan (1977b) are best known. The Bryan model alongside the Coriolis force and vertical viscosity (as many quasi-geostrophic models) considers non-linear effects and horizontal exchange, thus it is universal and applicable for current calculations, both at mid-latitudes and in the equatorial zone. The model is based on the equation for an integral stream function.

It is noteworthy that in recent years work is in progress to modify the dynamic method. The Stommel-Schott method (1977) presents the best-known modification. The advantage of this approach is that it allows one to minimize the effect of density (temperature and salinity) field measurement errors and calculation errors. However, principal errors common to the classic dynamic method also remain in this modification.

Diagnostic calculations of currents are divided into two groups:

1. calculations performed with the help of a sea surface topography;
2. calculations performed with the help of a total mass transport stream function.

Without dwelling upon review work, some preliminary results of diagnostic calculations of gradient currents with the help of a sea surface topography can be summarized.

1. The contradictoriness of the two approaches for the ocean current calculations was mentioned in Chapter 1: homogeneous ocean models consider only the direct effect of the sea surface wind stress when determining ζ, while the dynamic method considers only the density field ρ (Kozlov 1975; Sarkisyan 1977a, b). The diagnostic calculations, in which both factors are taken into account, showed that homogeneous ocean models cannot be used as a basis for some estimations, the more so for current calculations in actual seas and oceans. The main indicator is the ρ field, which reflects results of heat and salt exchange processes as well as the indirect force of the wind field (Sarkisyan 1977a, b).

2. While calculating large-scale circulation on a $1°$ grid or larger, the non-linear terms and lateral exchange effect can be neglected. While calculating the surface topography of the baroclinic ocean, it is of importance to consider the bottom relief. It was Ekman who studied how the ocean bottom topography effects the gradient currents. But in the Ekman model H changes only the picture of wind currents. An other fact is of greater importance: the joint effect of baroclinicity and bottom relief (JEBAR) on the right-hand side of the equation for ζ is an order of magnitude higher than the direct wind impact. The variable bottom topography in the upper baroclinic ocean layer has a major effect on the current in the upper ocean layer. H change in the deep sea has less effect on the ζ field. It is important to note that the assumption $H = \text{const.}$ results in a significant underestimation of the role of the sea water baroclinicity.

3. In addition to the β-effect, the importance of which has been shown by Stommel (1948), the equation for ζ incorporates a still more important factor, a baroclinic β-effect. The β-effect can only redistribute wind-driven ocean currents, while the baroclinic β-effect on the right-hand side of Eq. (2.1), conserves the summand exceeding the force of direct wind by an order of magnitude (Sarkisyan 1977a, b).

4. Beginning with the known work of Stommel (1948), the β-effect is supposed to be a dominating factor in intensifying the currents near the western ocean coast. This conclusion was made because of the ignored sea water baroclinicity and due to a number of other simplifications. Actually the situation is much more complicated: continental slope, heat and salt advection by the drift currents, etc. promote the intensification of currents near the ocean coasts (intensive currents are observed near the eastern coasts as well).

5. The vertical current velocity component is small in a sea-surface layer and increases with depth by one or two orders. It is first stipulated by the sea water baroclinicity and bottom topography. The wind stress vortex plays a secondary role in forming the w field in the deep sea. The vertical flow velocity component is generally sensitive to the density field error, to shortcomings of the model and

numerical method. In many cases the current velocity of the horizontal gradient in the upper ocean layers can be indifferent to some errors, while the vertical velocity component undergoes both quantitative and qualitative changes.

The diagnostic calculations of currents with the help of a total mass transport stream function with allowance for the joint effect of baroclinicity and bottom relief began after the work of Sarkisyan (1969) was published.

The solution of Eq. (1.116) with allowance for all its summands was first given in Sarkisyan and Ivanov (1971). It was thus possible to estimate the relative significance of the baroclinicity and H in the formation of the total mass transport stream function field. It was found that (1) a total mass transport stream function in principal is generated by the deep-layer density field, and mainly due to JEBAR; (2) τ is of less importance in forming the ψ field. The current field of the surface gradient differs from the integral circulation, while abyssal water circulation is in rather good qualitative agreement with the latter. It is essential that the integral circulation does not correlate with the sea-surface currents.

Intensive currents are observed in both the western and eastern coastal zones. In the latter they are first and foremost stipulated by the baroclinicity and cannot be obtained within the classic Sverdrup-Stommel-Munk theory. It is worth noting that despite a great difference between the two versions of calculations of ψ, the sea-surface currents obtained in both cases are similar to each other and quite realistic. They indicate two known calculation gyres with the western intensification. Only abyssal water currents differ widely. So, anomaly errors of the deep-layer density lead to significant errors in the deep-sea current field, considerably transforming the ψ field, but affecting insignificantly the sea-surface current field.

A review and analysis of the results of diagnostic investigations are given in more detail in Sarkisyan (1977a, b). Here, we shall dwell only upon major conclusions.

1. The density field is a major indicator of the large-scale sea currents. The diagnostic calculations are not satisfactory, if the density field is too smooth or if it contains significant errors. Consequently, an important oceanographic problem is the accumulation of hydrological data for the development of more reliable atlases of density and current field in the future.

2. Diagnostic models, as mentioned above, are generalized dynamic methods of calculating currents and are used as its alternative. The simplest version of diagnostic calculations of currents is as follows: the determination of ζ_d from the elementary formula (1.103), i.e. with the help of the dynamic method, where the bottom topography serves as a reference level; the determination of corrections for ζ_d according to the equation for ζ' with boundary conditions $\zeta' = 0$; the calculation of u, v, w according to common formulas of the quasi-geostrophic model.

3. In the case of diagnostic calculations of currents even a small-sized area can be considered, whereby its boundary can be totally liquid. In general, the solid boundary in all calculations is substituted by liquid, since the vertical wall depth is, as a rule, more than 100 m. The inaccuracy of the boundary values set by the dynamic method is observed only in the points closest to the boundary. The main factor in each internal point of the area, which forms ζ and the current fields, is the ρ field in the given point and those nearest to it.

4. The vortex of the surface wind stress plays a secondary role in generating gradient currents in the upper ocean layer. The major wind contribution here is a pure drift current.

5. The temperature and salinity observations in deep layers are poor. Horizontal changes in density are close to errors of measurements, averagings, etc. Besides, the ρ fields are in poor agreement with the H field, so deep-sea current fields are calculated with a high error. The errors are extremely high when calculating ψ, thus the ψ fields obtained by different authors, differ significantly from each other. However, the current fields of the surface gradient calculated from so different ψ are rather close to each other, if ρ fields in the upper 1–1.5 km layer are close. The current velocity in deep layers (the bulk of the World Ocean) and the total mass transport stream function are very sensitive to both the deep-layer density field errors and to drawbacks of the model and calculation method. This is due to the horizontal current velocity in deep ocean layers, which is approximately five to ten times less compared with that in surface layers, i.e. a small difference of large values and ψ is principally formed as an integral by height from the horizontal velocity component.

6. The historically based idea that the westward intensification of the ψ isolines reflects known intensive currents (Gulf Stream and Kuroshio) needs a revision. In many areas of the World Ocean the integral transfer does not coincide with mass transport of the surface water layers and is even opposite to it. To compare mean mass transport with currents in the surface layer, at least vertically mean currents of the upper 0.5–1 km layer and the lower layer, as shown by Sarkisyan (1977a, b), should be found. It is even better to determine the transports in several layers, as reported by Cox (1975). The currents averaged over the whole depth reflect in principal the mean deep-layer circulation, and thus incorporate the calculation error of currents in these layers integrated over the height.

7. The models of the deep barotropic ocean do not show such high velocities of the gradient current, i.e. 5–10 cm s^{-1}, which are characteristic of the real ocean. Intensive and stable gradient currents are observed only in places where there are available significant horizontal density gradients extending over the substantial ocean thickness. If there is no western intensification in the isopycnal field (regardless of whether it was based on observational data or obtained through prognostic calculations), it will be absent from the calculated current fields as well.

8. The diagnostic calculations already provide a partial answer to the main question of sea current dynamics, i.e. what are the principal factors responsible for the large-scale World Ocean circulation? These factors are obviously of a thermohydrodynamic character rather than of a net mechanic character. The surface wind stress produces a pure drift transport which plays a secondary role in the general water mass transport. The wind gradient component is small as well, at least in the upper, 0.5–1 km ocean layer. The processes of heat and possibly salt exchange are much more important. Thus, the question as to the role any factor (including rot τ) plays in the dynamics of the upper layer currents is also the question of the role it plays in the density field redistribution.

9. The diagnostic calculations are of practical interest. The atlases of sea and ocean density fields can be compiled after the adequate processing of observed data.

The horizontal and vertical circulation atlases can be compiled by diagnostic calculations according to the observed and processed density fields. There will be less necessity for diagnostic calculations only when more reliable and accurate density maps, in comparison with those presently constructed of observed data, can be compiled, based on the prognostic calculations.

5.2 The World Ocean Surface Topography and the Surface Gradient Currents

5.2.1 Density and Wind Fields for Diagnostic Calculations. Filtration of the Fields

Temperature and salinity observations, used as initial data for diagnostic calculations, can be transformed into a density field in a variety of ways. The density is calculated from temperature and salinity according to an equation of the state in any form. Several empirical relations are known, which more or less accurately describe the relation between density, temperature and salinity, e.g. Knudsen, Ekkart, Mamayev relations and the Bryan and Cox et al. formulas given in Chapter 1.

For the calculation of the World Ocean currents we used the formula recommended by UNESCO (1973) (it has been somewhat modified for more convenient use by computers):

$$\begin{aligned}\sigma_t = {} & 800.969062 \times 10^{-4} + 588.194023 \times 10^{-4} T + 797.018644 \times 10^{-3} S \\ & - 811.465413 \times 10^{-5} T^2 S - 325.310441 \times 10^{-5} TS + 131.710842 \\ & \times 10^{-6} S^2 + 476.600414 \times 10^{-7} ST^3 + 389.187483 \times 10^{-7} ST^2 \\ & + 287.971530 \times 10^{-8} S^2 T - 611.831499 \times 10^{-10} S^3 . \end{aligned} \quad (5.1)$$

Although the formula is more complicated compared with the compact relations, e.g. the Mamayev relation (1963), this does not create any significant limitation in calculations, particularly using algorithmic languages. At the same time numerical experiments indicate that formula (5.1) provides results similar to those obtained by the Mamayev formula, but Eq. (5.1) is more accurate.

The surface wind stress field serves as another specified field. Observed data are indicative either of the wind direction and its velocity modulus or of the atmospheric pressure at sea level. In the first case the surface wind stress components τ_θ and τ_λ are calculated with the following empirical formulas:

$$\tau_\theta = \rho_\alpha c_p w_\theta |w_\theta| \qquad \tau_\lambda = \rho_\alpha c_p w_\lambda |w_\lambda| . \quad (5.2)$$

In the most common case, which provides good results for moderate winds at mid-latitudes, the following parameter values are used:

$$\rho_\alpha = 1.25 \times 10^{-3}\,\mathrm{g\,cm^{-3}} ; \qquad c_p = 1.3 \times 10^{-3} .$$

More accurate calculations of the seasonal and annual mean values of τ_θ and τ_λ were performed by Hellerman (1967) in a 5° grid set. The following parameters of formula (2.2) were used:

$$\rho_\alpha = \begin{cases} (0.0022\varphi + 1.136) \times 10^{-3} & \text{for } \varphi > 20^\circ \\ 1.18 \times 10^{-3} & \text{for } -20^\circ \leqslant \varphi \leqslant 20^\circ \\ (-0.00284\varphi + 1.124) \times 10^{-3} & \text{for } \varphi < -20^\circ \end{cases} \tag{5.3}$$

$$c_p = \begin{cases} 0.0088 & \text{at } |w^b| \leqslant 6.6\,\text{c}\,\text{c}^{-1} \\ 0.0026 & \text{at } |w^b| > 6.6\,\text{m}\,\text{c}^{-1} \end{cases}, \tag{5.4}$$

($\varphi > 0$, northern latitude, $\varphi < 0$ southern latitude).

If the sea level pressure is known, then τ_θ and τ_λ can be determined with the help of the Akerblom model, i.e. through formulas (1.150). The most practically used value is $v' = 10^4\,\text{cm}^2\,\text{s}^{-1}$; the upper sign in formulas (1.150) corresponds to the Northern Hemisphere, the lower to the Southern Hemisphere. The intermediate calculations of fields τ_θ and τ_λ can be omitted if expressions (1.150) are substituted into model equations and boundary conditions for the sea surface topography after the necessary mathematic transformations.

From the experience gained in calculation we know that flow fields, calculated from the atmospheric pressure [directly or indirectly through such relations as τ_θ, $\tau_\lambda = f_1, f_2(Pa)$], are "smoother" compared with those obtained from data on wind directions and velocities. This is indicative of either a substantial spacial and temporal variability of wind fields or inadequate representativeness of the ship-wind observations, thus placing high demands upon statistical provision and processing of such observations.

While preparing data for diagnostic calculations, based particularly on the hydrological surveys in the majority of cases, the initial observation fields should be filtered. If the task is to compile generalized circulation maps or the area under investigation is poor in statistical observation data, the data must be filtered (or smoothed). A rather simple and efficient filtration of oceanographic fields can be performed with a two-dimensional isotropic cosine filter successfully used in practice. For a nine point pattern on a regular grid, it is as follows:

$$\begin{aligned} \bar{f}_{i,j} = {} & 0.28 f_{i,j} + 0.13(f_{i,j+1} + f_{i,j-1} + f_{i+1,j} + f_{i-1,j}) \\ & + 0.05(f_{j+1,i+1} + f_{j-1,i+1} + f_{j-1,i-1} + f_{j+1,i-1}) \end{aligned} \tag{5.5}$$

where f and $\bar{f}$ are the initial and smoothed functions, i and j are the numbers of grid points at the x- and y-axis.

The formula is true for the internal grid points. Weight factors change at the boundary grid points with the peculiarities of the boundary contour of the basin. Some points of the nine-point pattern are "lost". For example, for points located over the rectilinear western boundary of the basin, formula (5.5) will be as follows:

$$\begin{aligned} \bar{f}_{i,j} = {} & 0.28 f_{i,j} + \frac{0.52}{3}(f_{i,j+1} + f_{i,j-1} + f_{i+1,j}) \\ & + \frac{0.2}{2}(f_{i+1,j+1} + f_{i+1,j-1}). \end{aligned} \tag{5.6}$$

The corresponding alternatives of the general formula can be easily obtained for any other type of boundary. However, when using computers it is much more convenient to use the generalized calculation procedure rather than specific relations for each type of boundary. The computer itself reduces the general formula in accordance with a particular, definite case "looking through" the neighbourhood of each specific grid point, where the specified function is smoothed. The following procedure is convenient.

Let us rewrite Eq. (5.7) in a generalized form:

$$\bar{f}_{i,j} = 0.28 f_{i,j} + \frac{0.52}{M^{(1)}}(m_1^{(1)} f_{i,j+1} + m_2^{(1)} f_{i+1,j} + m_3^{(1)} f_{i-1,j}$$
$$+ m_1^{(1)} f_{i,j-1}) + \frac{0.2}{M^{(2)}}(m_1^{(2)} f_{i+1,j+1}$$
$$+ m_2^{(2)} f_{i+1,j-1} + m_3^{(2)} f_{i-1,j+1} + m_1^{(2)} f_{i-1,j-1}), \tag{5.7}$$

where

$$m_k^{(1)}, m_k^{(2)} = \begin{cases} 1, & \text{if the } k\text{th point is located at the boundary or within the area;} \\ 0, & \text{if the } k\text{th point is located beyond the area.} \end{cases}$$

The calculation program is designed in such a manner that the computer "looks through" all eight points in the neighbourhood of the basin point under calculation and assigns variables $m_k^{(1)}$, $m_k^{(2)}$ values equal to 0 or 1, depending on whether the corresponding point belongs to the grid space under study or not. Simultaneously, the computer storage cells for variables $M^{(1)}$, $M^{(2)}$ store the sums of these values. The algorithm described is rather efficient and simple for programming. Time and again it was used in specific calculations for basins with a complicated boundary shape.

5.2.2 Specified Data and Peculiarities of Calculating World Ocean Currents

The World Ocean current fields given in this chapter are obtained with the help of a quasi-geostrophic model. The multi-year mean temperature and salinity fields are used for the four seasons of the year. They are set in 1° grid points at 31 standard levels. These specified fields, together with ocean depth data, were prepared at Princeton University (USA) and represent one of the modified fields under analysis in Levitus and Oort (1977). The density was calculated by formula (5.1). The seasonal variations concern only the upper 200-m layer, and below, the annual average fields. The calculation of the surface wind stress is based on the multi-year mean seasonal atmospheric pressure fields obtained in the USSR Hydrometcentre with the Ackerblom formula (1.150). The input parameters were selected in such a manner that mean values τ_θ, τ_λ were similar to those given in the known tables of Hellerman (1967). The limitation characteristic of the Hellerman data, used in the first calculations (1967), is that they do not cover the pre-arctic zone, and in particular, the coastal regions (because of the 5° averaging scale). This results in the need for extrapolation, which is certainly very unreliable and distorts drift currents. Thus, a number of obvious drawbacks is characteristic of the map of sea surface currents, based on Hellerman's data. The atmospheric pressure fields used here are designed on the same averaging scale, and because they cover the whole globe, there is no need for extrapolation. The components fields of the surface wind stress obtained in points of a 5° trapezium were linearly interpolated to a 1° grid point set. The fields reflect major planetary particularities in the atmospheric circulation: subtropical anti-cyclones and mid-latitude cyclones. Maximum changes in τ take place in the convergence and divergence zones of air flows. As a whole the τ field is rather smooth, because more detailed seasonal climatic fields are presently unknown.

The specified bottom topography field was smoothed before calculations with the cosine filter (5.5). The resultant bottom relief field reflects all principal large-scale elements of the real topography.

The confidence of the obtained circulations is significantly related to the initial temperature and salinity fields. Statistical provision of measurement data differs certainly in various regions. The least observations are carried out in remote areas of the Pacific and Indian Oceans in the Southern Hemisphere and in deep and near-bottom layers.

Let us consider briefly the vertical structure of the temperature and salinity fields at levels of 0, 100, 250, 500, 1000, 4000 m for the summer season in the Northern Hemisphere. The selection of the levels is based on the generally accepted division of the ocean water column into surface water, subsurface intermediate (up to 200–250 m), intermediate (up to 1000–1500 m), deep sea (up to 4000 m) and near-bottom water. For filtering synoptic "noise" the fields given below were obtained after the initial fields were smoothed according to the cosine filter.

Temperature and salinity distributions at the sea surface (Figs. 1 and 2) generally conform to existing ideas, although due to the smaller grid size they also reflect fine peculiarities. The isotherms in the Southern Hemisphere and in the Northern Pacific extend in one zonal direction and temperature decreases from the tropics polewards from 28°–29° to $-1°$, in accordance to the decreased heat flux. The gradients are insignificant at tropical and subtropical latitudes. In moderate regions where major oceanographic fronts are located, the temperature drop approaches 20° per 500 km. Such drops are observed in places where waters of warm subtropical currents converge with cold mid-latitude water. In the North Atlantic it is the Gulf Stream. Cold Labrador water penetrates the far south-west along the American coast and converges with warm water forming the front. In the Pacific Ocean it is formed by the convergence of warm water of Kuroshio and North Pacific currents with cold water of the northern ocean. In the region of the Antarctic circumpolar current (ACC), the front is formed when cold Antarctic water converges with warm subtropical water carried by the western boundary currents. In the eastern Atlantic and Pacific Oceans the penetration of colder water from high latitudes to low latitudes is observed. It is much pronounced in the Southern Hemisphere, where the Benguela and Peruvian current water, leaving the ACC, spreads far north and turns westward near the equator which is manifested by the isoline "tongues". Cold water spreading in the Northern Hemisphere is less intense.

In the Indian Ocean, north of Madagascar island the isotherms near the African coast extend in a meridional direction displaying the monsoon circulation and spreading of colder water northward.

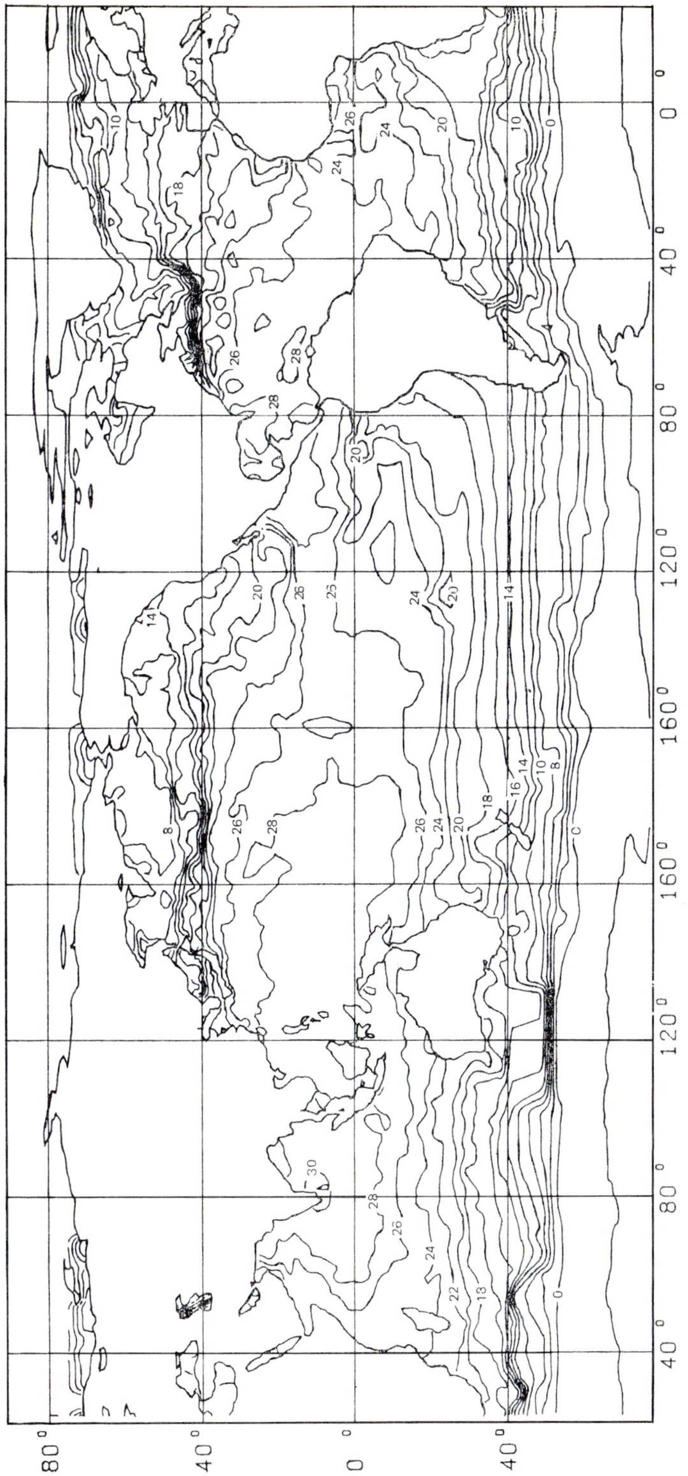

Fig. 1. The temperature of the World Ocean surface in the summer season (C°)

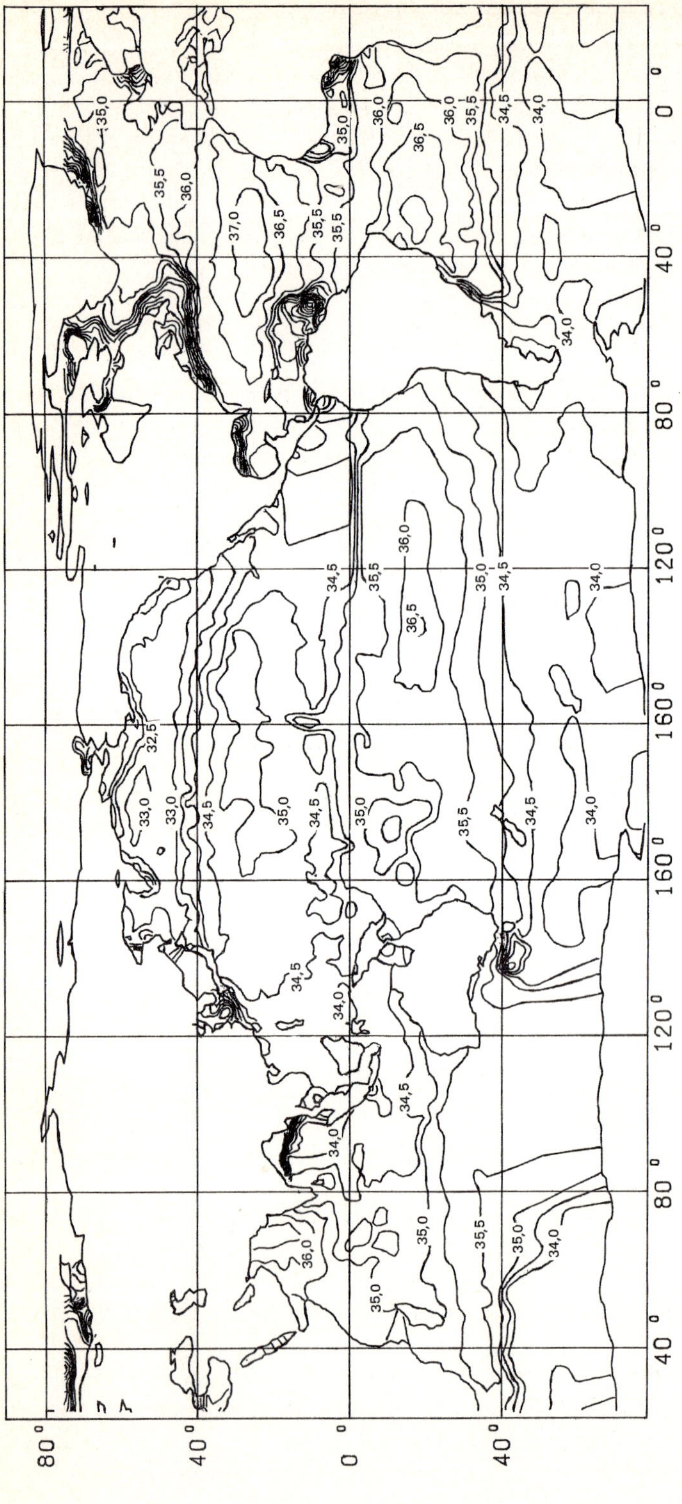

Fig. 2. The salinity of the World Ocean surface in the summer season (‰)

The main difference in temperature distribution compared with that given by Burkov (1980) and Stepanov (1974) is that major coastal currents are expressed in the considered isotherm field. The meridional currents in the case of a 5° scale are represented rather weakly in Burkov (1980) and Stepanov (1974).

In the salinity distribution we observe at the surface: a subtropical maxima (up to 37‰ in the Atlantic) in zones where evaporation exceeds the precipitation amount; the equatorial zone of lowered salinity, where precipitation exceeds evaporation; and polar regions of low salinity due to ice melting. The sharp salinity front is characteristic of the Gulf Stream and the North Atlantic current where warm saline water of the subtropics and cold fresh water of the polar latitudes are observed. In some coastal regions, such as the Gulf of Mexico, the Bay of Bengal and Guinea, there are zones of sharp lowering of salinity due to freshening caused by large rivers. The freshening impact of the Amazon River is observed north of its mouth, which indicates that the influx of water is transported in a north-west direction by the Guinea current along the coast. The salinity distribution in polar regions of the Northern Hemisphere is very complicated due to difficulties in measurements.

Subsurface intermediate waters are presented by levels of 100 and 250 m (Figs. 3–6). In the Atlantic and Pacific Oceans colder and fresher waters from high latitudes penetrate into the tropical region in the eastern parts of the oceans, producing cold anomalous areas. The water temperature in the belt 10°N–10°S drops to 14° and to the south and north, in subtropical regions, it rises to 24°. The cluster of isolines extends in the Atlantic from the southern African edge towards the north-west up to the Brazilian coasts, and then turns to the north and north-east, to the African coasts. The temperature drops east of this area.

Only in a narrow band in the region of 10°N is a warmer water crest observed, which might be produced by the eastern transport of warm and saline water from zones of subtropical convergence. In the Pacific, in the tropical region, cold anomalous waters reach approximately 140°W generating strong gradients with warm subtropical water. In the Indian Ocean the temperature distribution in the subsurface water is more complicated, and the isotherm situation is related to individual closed formations, but as a whole the cold intermediate water tends northward.

Planetary peculiarities of the salinity distribution are observed in the isohaline situation, like at the surface, though the field is smoother. The freshening induced by rivers is not yet pronounced at a depth of 50 m. A significant salinity gradient is preserved in the Gulf Stream region.

Deep-sea intermediate waters, represented at the 500- and 1000-m levels (Figs. 7–10) are characterized by subtropical areas of maximum temperature and salinity values and by tropical areas of lower values. The subtropical maxima are less pronounced in the Southern Hemisphere, and a larger space is covered by the tropical area with colder water. The temperature at the 1000-m level varies within 0°–8°. In the Atlantic it is related to the warm Mediterranean water spreading over the whole ocean.

The subtropical region of the temperature maximum is allocated near the North American coast. This is indicative of an intensive and deeply spread process of water downwelling in subtropical gyres. When leaving the Strait of Gibraltar the water

Fig. 3. The temperature at 100-m depth in the summer season

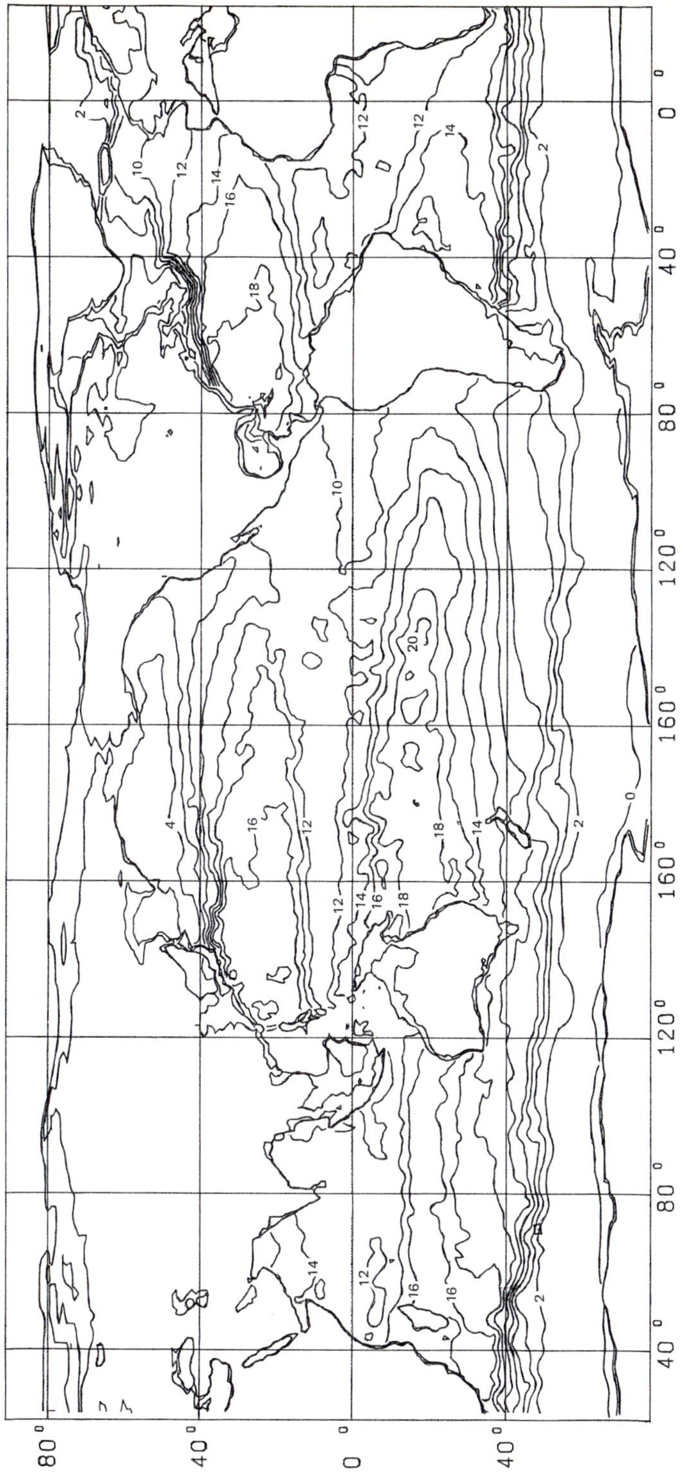

Fig. 4. The temperature at 250-m depth

Fig. 5. The salinity at 100-m depth in the summer season

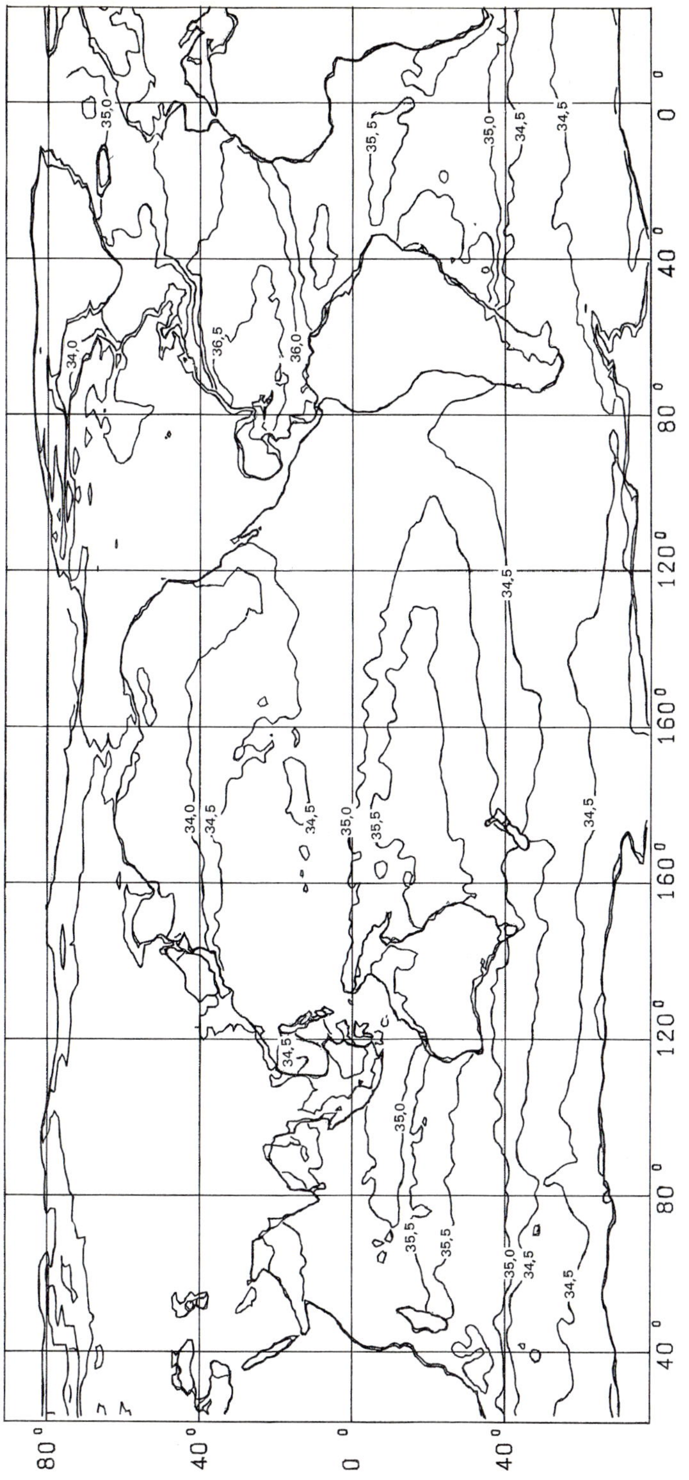

Fig. 6. The salinity at 250-m depth

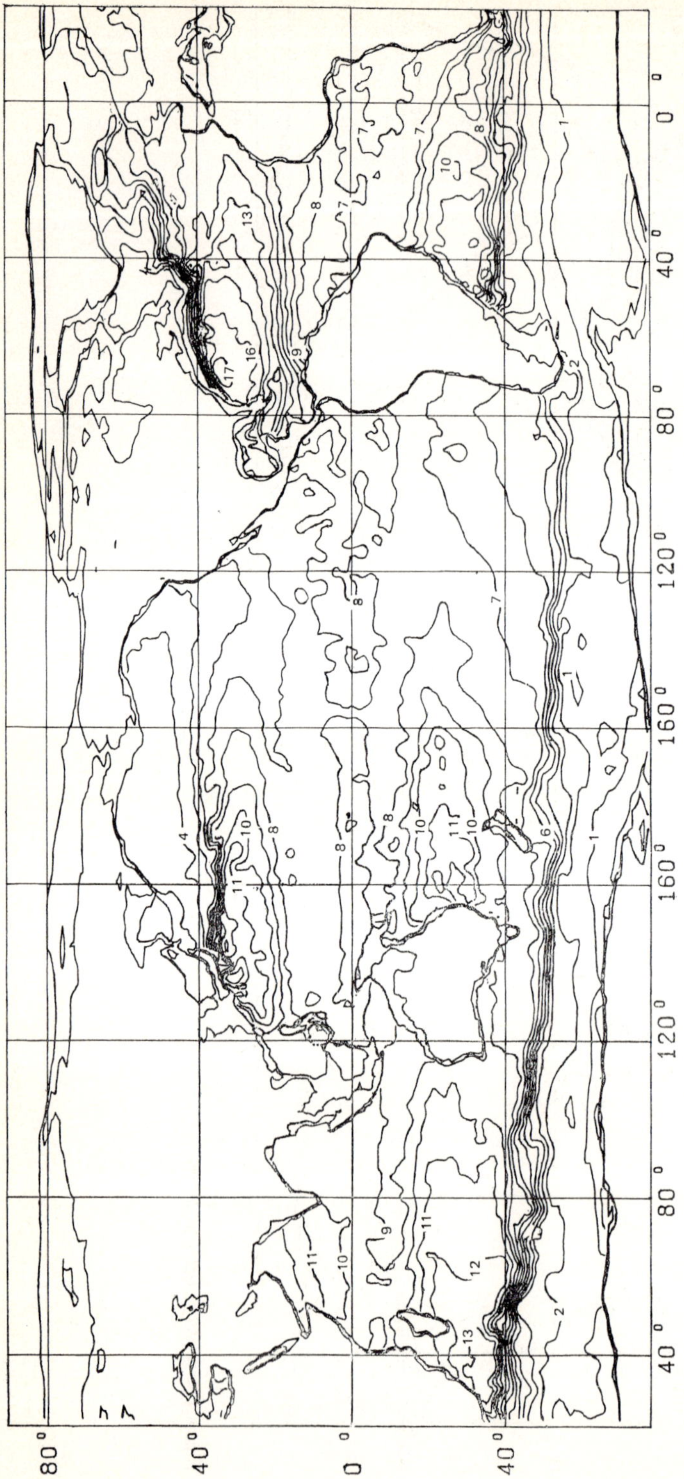

Fig. 7. The temperature at 500-m depth

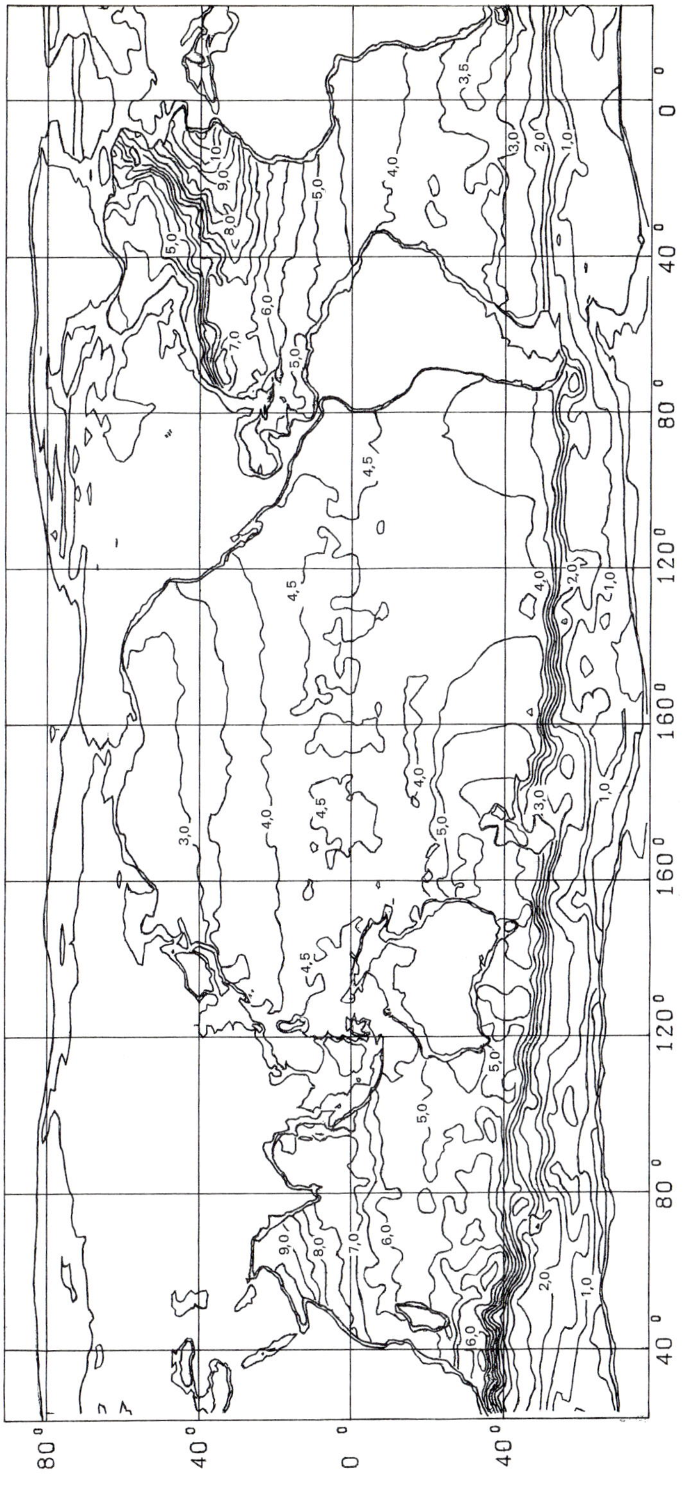

Fig. 8. The temperature at 1000-m depth

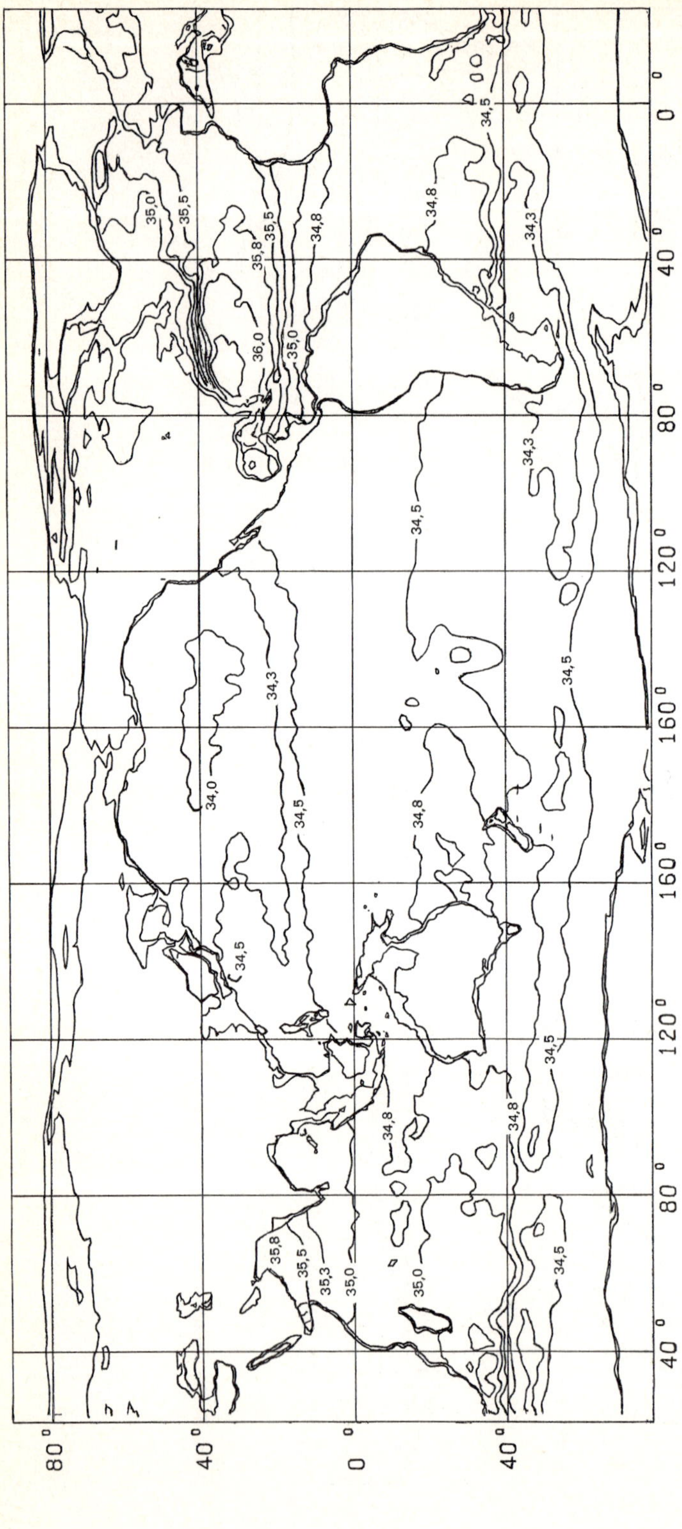

Fig. 9. The salinity at 500-m depth

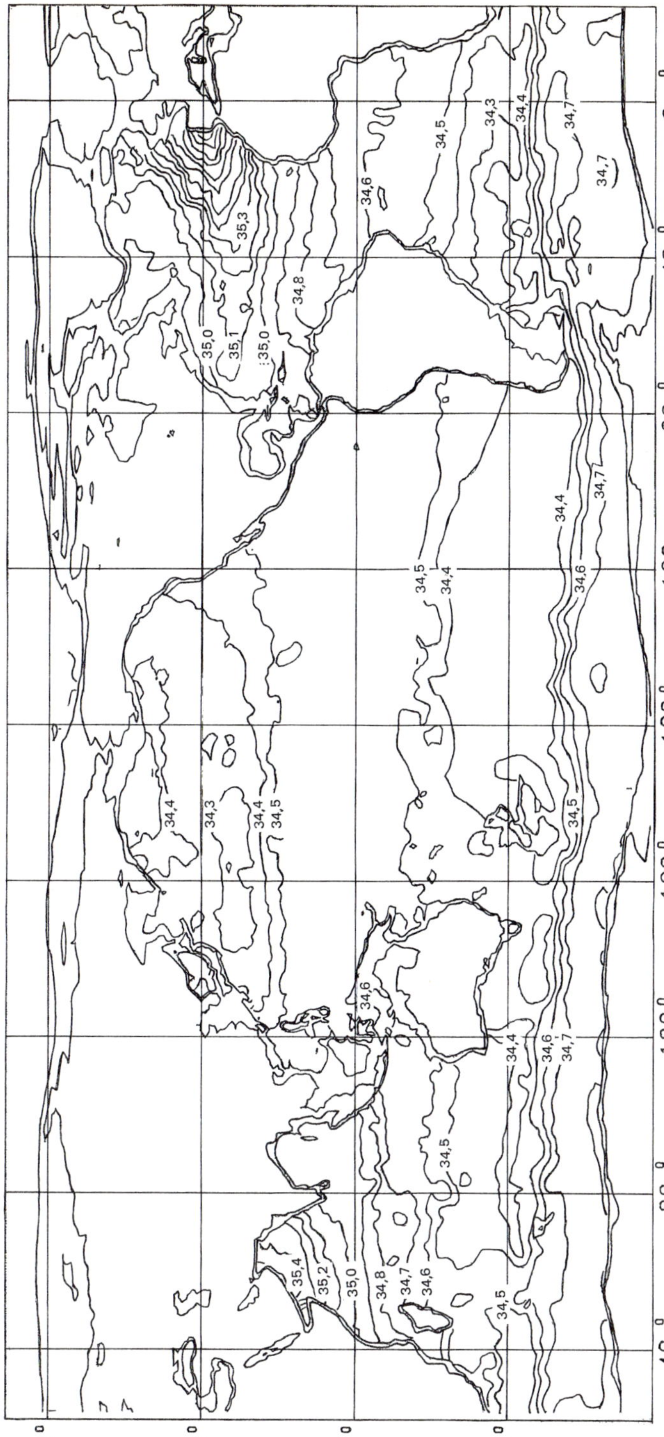

Fig. 10. The salinity at 1000-m depth

temperature reaches 10°–11°, then drops to the south and reaches 0° near the Antarctica coasts. The temperature gradients are weakest over the vast Pacific water basin. Maximum values are observed in the subtropical maximum east of Australia. The area of higher values (5°–5.50°) is located in the equatorial region. In general, it links with the subtropical maximum in the Southern Hemisphere. The subtropical zone of maximum temperatures is preserved in the Indian Ocean near the African extremity in the intermediate water. The spreading of Red Sea water is less pronounced compared with the Mediterranean water in the Atlantic. In the Antarctic zone at a depth of 1000 m all oceans are characterized by a continuous zone of the subtropical front, where strong temperature gradients are preserved. These peculiarities in temperature distribution in intermediate water are generally in agreement with those in Burkov (1980) and Stepanov (1974).

So, salinity and temperature distributions reflect an important peculiarity of the subtropics in the Northern Hemisphere, which is related to the intensive salting due to evaporation and significant downwelling. The development of these maxima in the Southern Hemisphere is restricted by the subsurface intermediate waters. Cold and fresh water transport from the Antarctic areas northward prevails in the deep sea. Temperature and salinity distribution is indicative of a monotonous decrease in values from north to south at a depth of 4000 m.

Let us dwell on the calculation technique. The model used is inapplicable for calculating equatorial currents, so when solving boundary value problems for the sea surface topography, the count procedure of "leaping" over the equator is used (Demin 1975b; Sarkisyan 1977b). Furthermore, as in the case of sea surface topography and velocity component calculation, φ is assumed to be const $= 5°$ in the belt from 5° N to 5° S. This makes it possible to smooth the equatorial current (otherwise the velocities are strongly overestimated). A similar assumption was made for the neighbourhood of the North Pole: φ = const $= 85°$ in the belt from 89.5° to 85° N.

Because of technical difficulties due to a large grid numbers (the grid net has about 2×10^6 points) and the existence of many islands, the sea surface topography at the boundary was calculated with the help of a quasi-dynamic method according to formula (1.103). The formula, as it was mentioned in Chapter 1, is deduced from the boundary relations (1.90), (1.91), whereby the surface wind stress is ignored and the water flux is specified in these formulas by the quasi-dynamic method. The applied simplification results inevitably in some errors, although as numerous investigations (Sarkisyan 1977a, b) have indicated earlier, they are insignificant and decrease with the distance from the boundary.

In diagnostic problems the right-hand side (especially baroclinic summands) rather than the boundary conditions contributes much in their solution. The technique of calculating the right-hand side of the equation for the sea surface topography proved to be much more significant. Thus, the attempt to remove the derivatives of density from the integrals, which makes it possible to significantly simplify the problem, and is therefore often used in practice, has resulted in a drastic distortion of the solution, especially in coastal regions with a very inhomogeneous bottom relief. This procedure is allowable only in the case of a rather smooth bottom relief, therefore it is no longer used. The differences in the solutions obtained when using these two methods of calculating the right-hand side of the equation are more significant than those in solutions obtained through the schemes for the approximation of the first and second orders (Sarkisyan et al. 1980). The advantage in using integrals from the derivatives is also that such an approach eliminates the problem of constructing density field anomalies since density values are used when calculating flow velocity and sea surface topography within the area. The anomaly of the density field is needed only to calculate the sea surface topography at the boundary with the help of a quasi-dynamic method. Since in calculating internal points of the region one should calculate derivatives from the density by horizontal coordinates, it is natural to subtract from the boundary density values its standard mean $\rho(z)$ to obtain the correspondence between the calculated values on the contour and inside the basin. Furthermore, for the closed seas (at 1° resolution), e.g. Caspian, Baltic, Black and Mediterranean, the average density and anomaly fields were calculated separately.

The equation of sea surface topography was approximated by the scheme of directed differences. The solution of the finite-difference problem was sought by the Gauss-Zeidel iteration method. The difference formula for the sea surface topography is similar to Eq. (2.25).

To enhance the accuracy of the numerical calculation the first derivatives from the density on the right-hand side of the equation for the sea surface topography were approximated with differences directed similarly to the corresponding derivatives from the sea surface topography of the left-hand side (Sarkisyan 1977a, b). If the next point in the direction of this difference is beyond the basin boundary, the difference is taken in the direction where the density is determined. The velocity components were calculated with the help of a shifted grid method, i.e. the velocity was calculated in the centres of the elementary grid cells from the wind stress values in the four nearest points and from the calculated pressure values. It is of great importance when calculating narrow jet streams, in particular, coastal ones.

All calculation results were presented in maps with a plotter "Calcomp" without any subjective changes. For simplicity and visualization the wind vector rows in all given maps of the World Ocean currents are given at 2°. If the velocity was less than a specific value (given in each map), the corresponding rows were not drawn at all. However, for the analysis both the simplified maps given here and complete maps like those given in Bulatov et al. (1980) were used. In view of the poor quality of data in the Indian archipelago region, the latter was excluded from consideration. For the same reason the schemes for the Arctic Ocean currents are of low reliability.

The organization of computations is briefly summarized.

1. Temperature, salinity, bottom relief and atmospheric pressure fields are recorded in arrays suitable for use in computers with an allowance for the specificity of the algorithm and calculation program (all programs are in "FORTRAN" language).
2. Sea surface wind stress components τ_θ, τ_λ are calculated by the Ackerblom formula (1.150) and interpolated into the 1° grid set.
3. The density is calculated by formula (5.1).
4. Integrals $\int_0^H \frac{\partial \rho}{\partial \lambda} dz$, $\int_0^H \frac{\partial \rho}{\partial \theta} dz$, $\int_0^H \Delta\rho dz$, $\int_0^H z \frac{\partial \rho}{\partial \lambda} dz$ and the right-hand side of the equation are calculated.
5. The sea surface topography is calculated.

6. Flow velocity components v_λ, v_θ, v_z are calculated from the calculated sea surface topography and integrals from density and given wind stress.
7. The results are plotted with the help of a plotter in cylindric, equi-spaced (rectangular) projection.

5.2.3 The Peculiarities of the World Ocean Surface Topography and the Surface Gradient Currents for the Summer Season

The available data on the World Ocean climatic surface topography and gradient currents structure are mainly based on calculation results from the density field specified in a 5° grid set. As a rule, these are calculations according to the dynamic method or its modifications, performed either for a multi-year average or for separate seasons.

The calculations performed according to a diagnostic model are also done only for the yearly mean characteristics and on a grid set of 5° size. The main disadvantage of all these calculations is a crude resolution of jet like streams. More detailed study is carried out only for certain sections of the World Ocean.

The calculated World Ocean climatic surface topography in the summer season is given in Fig. 11. Qualitatively the scheme agrees well with the modern idea about the World Ocean circulation (Wyrtki 1975). The anti-cyclonic gyres (ACG) are much pronounced in the subtropical areas, and cyclonic gyres (CG) in subpolar and tropical latitudes. The level drop in the Gulf Stream and Kuroshio regions is about 1 m, in the Antarctic circumpolar current (ACC) region it is almost 1.5 m. The principal, qualitative difference of the scheme given above from previous ones is that all the large gyres of the World Ocean happen to be closed near the western coasts, which is pronounced even at a rather large isoline interval, e.g. 20 cm, taken for simplicity of presentation. The advantage in calculations is evidently connected with a much higher resolution of the coastal currents when calculating on the 1° size grid set.

Let us consider surface gradient currents for the summer season and compare them with the previous investigation results. Characteristic values of gradient velocities for the major currents of the World Ocean are given in Table 1a and the currents map shown in Fig. 12.

The largest gyres in the Atlantic ocean are the northern and southern subtropical ACG. In the Northern Hemisphere the ACG cover the space between 15° and 50° N. In the north it is limited by the North Atlantic current, in the west by the Florida current and Gulf Stream. Its eastern boundary is located near the African coasts, approximately at 30° W (though some jets penetrate even into the Canary upwelling region), whereas the southern boundary is not distinctly determined (in the total currents field it is formed by the North Equatorial current which is mainly of a drift origin). Within the region 40° N and 60° W we can observe that the Gulf Stream is transformed into three branches. Its northern branch gives start to the North Atlantic current. The middle branch has an eastward direction and in the region 30°–40° W it turns southward, forming the eastern boundary of the main planetary ACG, mentioned above. The southern branch (the recirculation) is directed opposite to the Gulf Stream, forming together with it the internal, narrow ACG.

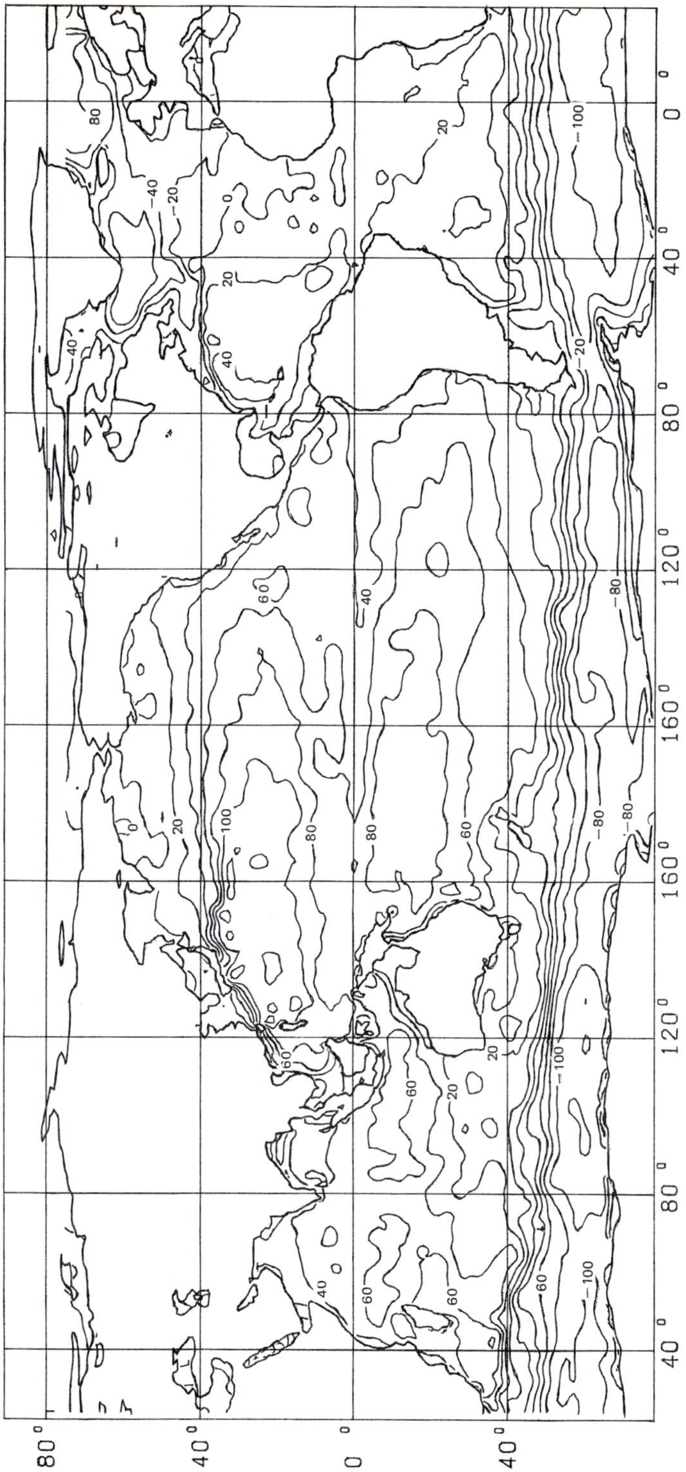

Fig. 11. The transformed sea-surface topography (SST) in the summer season (cm)

Table 1a. Characteristic values of the Atlantic ocean surface gradient currents for the summer and winter seasons (in cm/s)

Current	Summer	Winter
1. Antarctic circumpolar	10	10
2. Antilles	8	10
3. Benguela	10	10
4. Brazil	20	15
5. East-Greenland	10	13
6. East wind drift	10	15
7. Guiana	40	55
8. Guinea	30	50
9. Gulf Stream[a]	32	35
10. West-Greenland	15	20
11. Irminger	10	15
12. Canarian	8	5
13. Labrador	12	15
14. Equatorial countercurrent	30	20
15. Norwegian	12	15
16. Portuguese	10	8
17. North Atlantic	10	10
18. North Equatorial	5	5
19. Florida	20	20
20. Falkland	8	10
21. South Atlantic (west wind drift)	8	5
22. South Equatorial	10	10

[a] In Section 5.4 it is shown that in sophisticated models the flow velocity in the Gulfstream reaches 80 cm/s

In the region 5°N and 40°W the eastward equatorial countercurrent and a flow directed westward and north-westward along the South American coasts, which corresponds to the Antilles and Guiana currents, are formed.

The general nature of circulation in the region is cyclonic, and near the African coast the gradient currents are mostly directed northward, unlike the Canary current (the latter is pronounced in the total currents field). This indicates that the surface circulation in the region is principally of a drift origin. The North Equatorial current is also poorly pronounced both in the SST isoline field and gradient currents scheme which is also attributed to its drift origin (in contrast, the South Equatorial current is well pronounced in the gradient currents field as well).

In the Southern Hemisphere a similar ACG is located in the region 20°–40°S. Both the northern and southern ACG are displaced: the first towards the north-west, and the second to the south, compared to their location in the total currents scheme in the summer season.

The northern, subpolar CG is located in the northern Atlantic, in the region from 50°N to the Greenland coasts. The Norwegian and North Seas are also covered with a cyclonic gyre. The velocities in the western Norwegian Sea near the Greenland coasts exceed substantially [15–50(20)cm s^{-1}][8] that of the Norwegian current.

[8] The velocity range is given. In the parentheses the characteristic velocity is given.

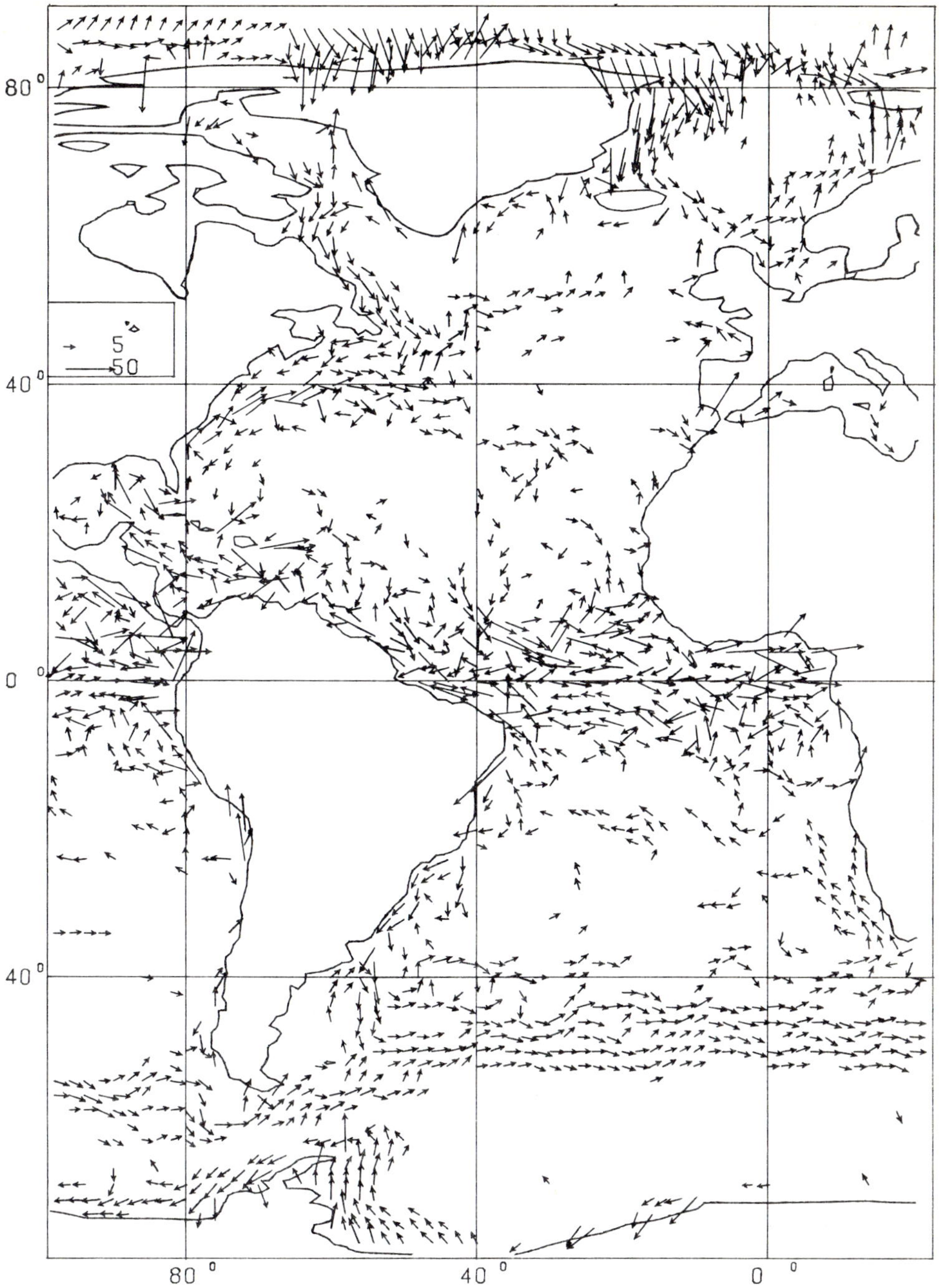

Fig. 12. Sea-surface gradient currents of the Atlantic ocean in summer season, in cm s^{-1}

Table 1b. Characteristic values of the Pacific ocean surface gradient currents for the summer and winter seasons (in cm/s)

Current	Summer	Winter
1. Aleutian	5	5
2. Alaska	8	8
3. Antarctic circumpolar	10	10
4. East Australia	50	50
5. East wind drift	10	15
6. California	20	20
7. Kamchatka	10	13
8. Kuroshio	50	50
9. Mindanao	40	45
10. Cape Horn	10	10
11. New-Guinean	90	60
12. Oyashio	8	8
13. Peru	50	55
14. North Pacific	12	15
15. North Equatorial	10	15
16. Formosa	30	50
17. South Equatorial	25	30
18. South Pacific	6	5

The largest gyres in the Pacific, as in the Atlantic are the northern and southern subtropical ACGs. The northern ACG has huge dimensions and covers the space from the western to the eastern oceanic coasts within 6°–40° N. The currents which form it, i.e. North Equatorial, California, Formosa, Kuroshio and North Pacific, have rather high velocities (Table 1b). The southern ACG is somewhat smaller than the northern one. It covers the space from the western to the eastern coasts of the southern ocean within 10°–30° S.

The CG and ACG are located in the tropical and equatorial zones of the northern Pacific, as in the Atlantic, along the equator. The Equatorial counter-current located between them spreads out almost from 140° W to the eastern oceanic coasts. Certain meanders are characteristic of it.

The largest gyre in the southern Indian Ocean is the subtropical ACG. It is a bit smaller on the gradient currents map compared with that on the total currents map. The gyre is displaced to the north-west and covers the space between 10°–30° S and 50°–80° E. In all three oceans the subtropical ACG are characteristic of a common physical nature related to available planetary anti-cyclonic gyres in the atmosphere and intensive processes of heating and evaporation. These gyres are asymmetric and their centres are displaced considerably westward; the currents in the western parts are stronger than in the eastern because of the known western intensification.

In the northern Indian Ocean the circulation has a variable monsoon nature. The Somalian current directed to the east produces the monsoon flow in the Arabian Sea and has a velocity of more than 100 cm s^{-1} (Table 1c).

South of 40° S the sea level decreases towards Antarctica by 120–180 cm. The CG which is a southern boundary of the ACG in each ocean of the Southern Hemisphere coordinates with the slope. The East Wind drift current passes along the Antarctica coasts as a narrow current directed eastward. Separate branches of

Table 1c. Characteristic values of the Indian ocean surface gradient currents for the summer and winter seasons (in cm/s)

Current	Summer	Winter
1. Agulhas	60	60
2. Antarctic circumpolar	10	10
3. East wind drift	10	10
4. West Australian	25	25
5. Equatorial countercurrent	30	30
6. Mozambique	45	45
7. Monsoon (in Arabian Sea)	10	15
8. Somali	100	60
9. South Equatorial	20	30
10. South Indian ocean	10	10

the current, together with the ACC, form three cyclonic gyres. Their centres are located at 110° W, 110° E and 0° longitude (Figs. 11, 12).

The maps of the World Ocean surface dynamic topography issued earlier refer, as a rule, to the multi-year average fields. So, in principle, only a qualitative comparison is possible. In general, the issued maps are similar to ours. However, there are a number of differences. So, in the Pacific the southern tropical CG in Fig. 11 is located east of 160° W to the South American coasts. In Burkov (1980) for example, the gyre mentioned above starts only east of 150°W. Furthermore, the ACG near the western coasts in Burkov (1980) are not closed. In the equatorial and tropical latitudes of the Atlantic and Pacific in Burkov (1980), ACG and CG are observed. In previous diagnostic investigations (Bulatov and Pojarkov 1975; Bulatov et al. 1980), where the grid size was 5°, it was rather difficult to single out these formations since they have a small meridional length. In Burkov (1980) the mentioned gyres extend latitudinally and consist of small gyres. In the present calculations the distribution of the gyres is more complicated and does not have a simple latitudinal character.

The differences in the Indian Ocean are more significant, i.e. seasonal currents are of a monsoon nature. So, the major differences are observed in the northern ocean.

Significant differences are not observed in the Atlantic Ocean. The schemes for the Indian Ocean surface dynamic topography in the summer and winter seasons are given in Neuman (1970). The correspondence of the circulations in the summer schemes is quite good. The calculations performed on the 1° grid set provide a more detailed picture of the circulation and represent smaller gyres.

5.3 The Large-Scale Circulation and Seasonal Variation of the World Ocean Waters

5.3.1 The Circulation in the World Ocean Surface and Intermediate Layers in the Summer Season

The maps of surface currents (Figs. 13–15) indicate all known gyres in the World Ocean. The northern and southern subtropical anti-cyclonic gyres (ACG) are the largest in the Atlantic.

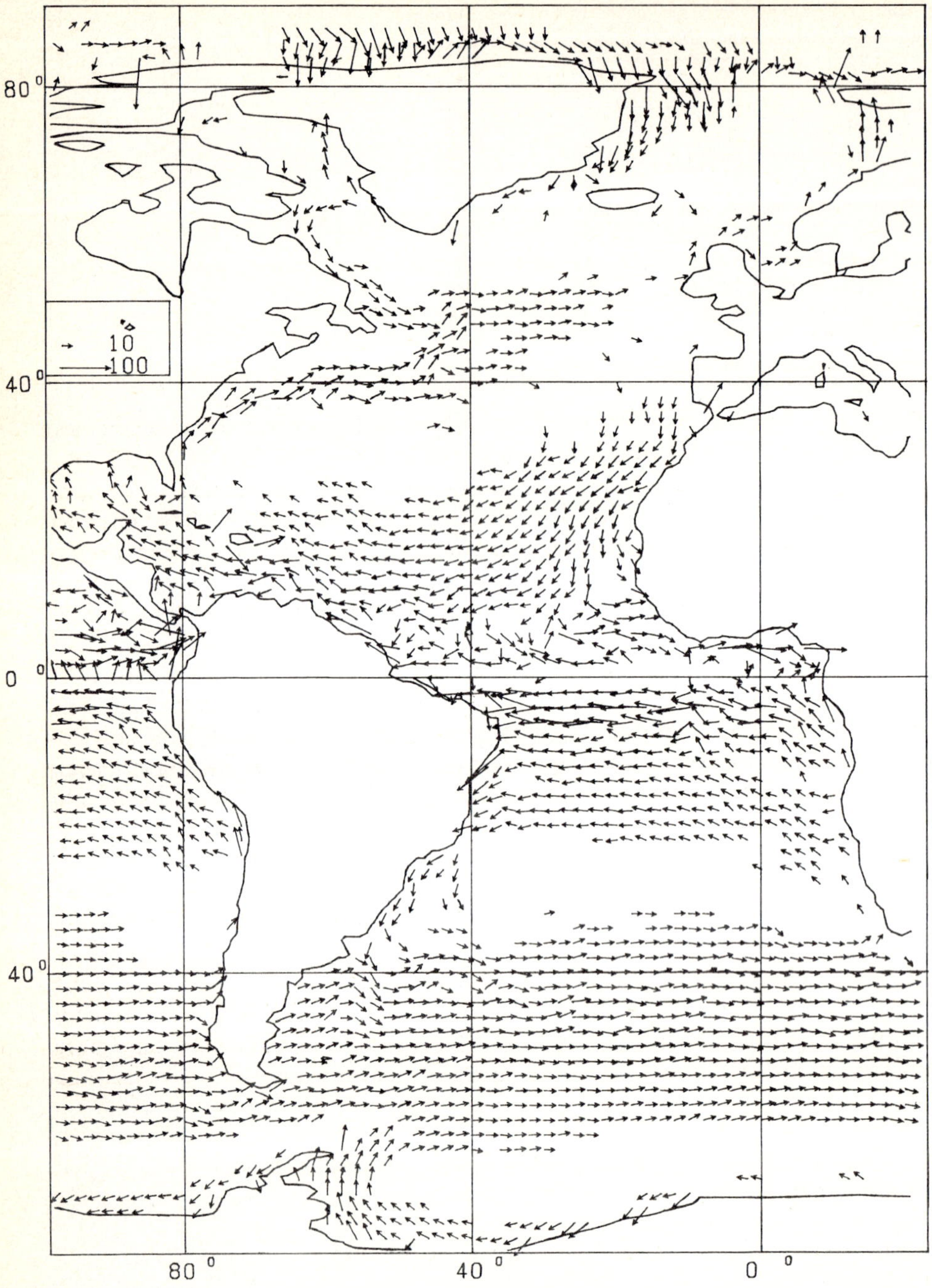

Fig. 13. Sea-surface currents of the Atlantic Ocean in the summer season

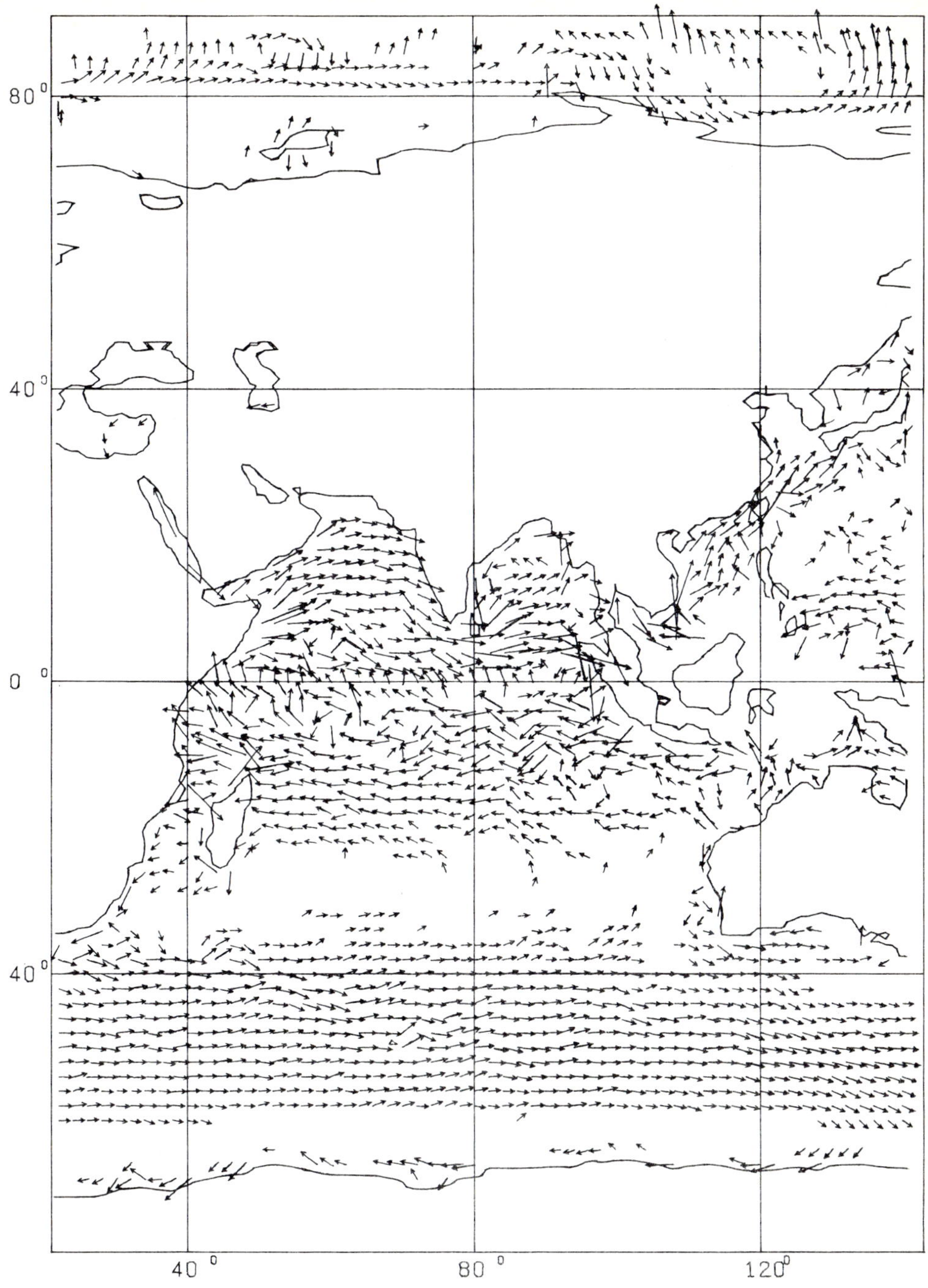

Fig. 14. Sea-surface currents of the Indian Ocean in the summer season

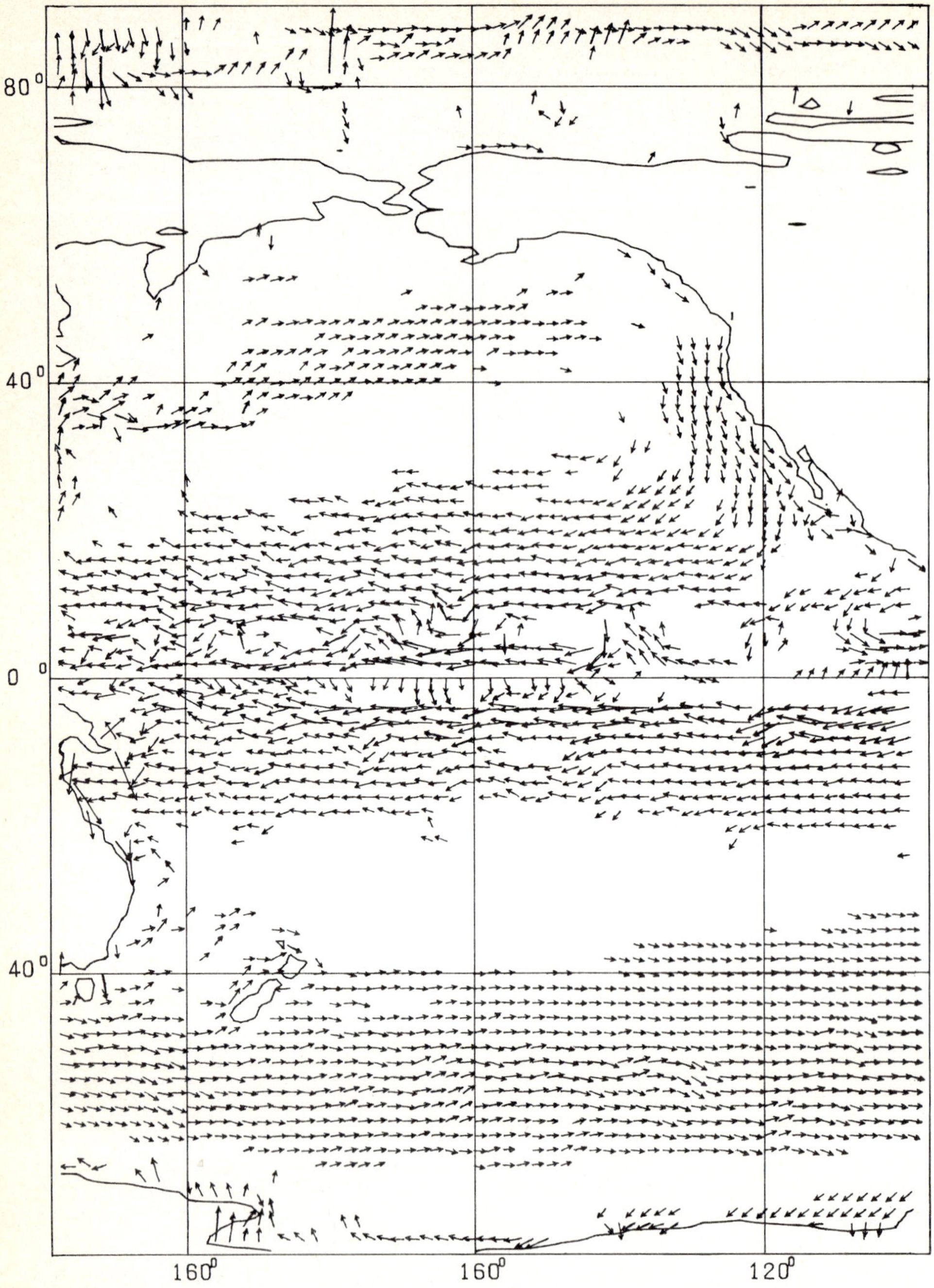

Fig. 15. Sea-surface currents of the Pacific Ocean in the summer season

Tropical ACG are observed in tropical and equatorial zones of the North and South Atlantic. In the Northern Hemisphere such a gyre is displaced eastward (40°–17° W), and is formed partially by the North Equatorial current [10–30(25)] and Equatorial countercurrent [10–45(30)].

The Equatorial countercurrent is observed east of 40° W. In the Southern Hemisphere such a gyre is much less pronounced. It is formed by the South Equatorial current [10–50(30)] and a slightly pronounced Equatorial countercurrent (10).

The northern tropical ACG is located in the eastern equatorial zone of the Northern Hemisphere. It extends from 35°–40° W to the western African coasts, and from 0° to 5° N. It is formed by the Equatorial countercurrent, Guiana [20–75(50)] and South Equatorial currents. Both in the Atlantic (Fig. 15) and Pacific Oceans the major space is occupied by the northern and southern subtropical ACG. Both northern and southern tropical CGs are distinctly pronounced in both hemispheres. The gyre in the Northern Hemisphere occupies a small space in the eastern ocean (80°–115° W, 0°–15° N).

In the Southern Hemisphere such a gyre extends along the equator almost across the whole ocean (160° E–105° W, 0°–4° S), and includes the South Equatorial current [10–50(30)] and Equatorial countercurrent [0–30(10)].

In the Indian Ocean (Fig. 14) the major space is occupied by the southern subtropical ACG. Its width ranges over 5°–40° S. The southern tropical CG formed by the South Equatorial current and the Equatorial countercurrent [10–35(20)] is located north of the gyre up to the equator. The Equatorial countercurrent occupies a narrow belt within 60°–95° E and 0°–2° S. A part of the Indian Ocean located north of the equator is under the influence of the summer monsoon which governs the water circulation in this area. North of the equator 0°–70° N the flow across the whole ocean is directed to the east and its velocities increase from west to east within 10–100 cm s^{-1}.

The Somalian current is directed to the north-east and its velocities reach 100 cm s^{-1}. In the Benguela Bay the cyclonic circulation [20–75(30)] is observed in its southern part and the anti-cyclonic one, in the northern part.

The Antarctic circumpolar current (ACC) directed from west to east incorporates all the oceans from the south. The current velocities are stable and are 10–35(20) cm s^{-1}. Its width ranges between 40°–60° S. The East Wind drift is directed along the Antarctica from east to west.

The major oceanic fronts in the World Ocean coincide with the mid-streams of main currents, which separate one gyre from the other. The North Polar front in the Atlantic formed by the Gulf Stream and North Atlantic current is located between the Florida and England coasts from 30° to 50° N. In the Pacific Ocean it is formed by the Kuroshio and North Pacific currents and is directed from the Japanese coasts to those of Alaska (35°–55° N). The southern polar front coincides in all oceans with the southern boundary of the ACC and is located at 60° S. The northern tropical front in the Atlantic and Pacific Oceans coincides with the North Equatorial currents. In the Atlantic Ocean it is located in the interval 15°–20° N, in the Pacific 10°–15° N. The southern tropical front coincides with the mid-stream of the South Equatorial currents in the Atlantic Ocean (10° S) in the Pacific Ocean (10° S) and in the Indian ocean (10° S) and is located almost parallel to the equator.

The subtropical front in all oceans coincides with the northern boundary of the ACC and is located approximately at 40°S.

The corresponding divergences and convergences are located inside all cyclonic and anticyclonic gyres. The subpolar divergence in the Atlantic is located at 50°–55°N. The divergence in the Norwegian Sea is its continuation. In the Pacific Ocean the location of such a divergence is approximately 60°N. The northern subtropical convergence in both oceans is located between the western and eastern coasts within 30°–40°N. The southern divergences in the Atlantic, Pacific and Indian Oceans are located aong the equator, crossing practically all three oceans. In the Atlantic Ocean the divergence extends from the western to eastern oceanic coasts in the interval 1°–2°S; in the Pacific, 110°W–160°E and 2°S; in the Indian Ocean, 60°–90°E and 1°–3°S. The southern subtropical convergence crosses all three oceans approximately at 30°S. The Antarctic divergences (60°–70°S) are located between ACC and the East Wind drift near the Antarctica coasts. As a whole, due to the effect of the drift component, the circulation becomes more arranged and close to the classic concept. In many respects this is attributed to the smoothness of the used atmospheric pressure field. The drift current velocity is also closely related to the phenomenological factor of the vertical turbulent momentum exchange. Based on numerical experiments we take $v = 100\,\mathrm{cm}^2\,\mathrm{s}^{-1}$. In the majority of ocean areas, this leads to acceptable results, although using $v = \mathrm{const}$ for the entire area under consideration is a rather rough approximation.

At the 50-m level major gyres, almost all gyre-generating currents and their locations are conserved. The nature of circulation in the major planetary gyres is complicated due to smaller gyres which reflect peculiarities of the specified temperature and salinity fields. The current structure becomes more complicated, the drift component attenuates and current velocities decrease considerably. The circulation becomes similar to that of the sea surface gradient.

Substantial decrease in current velocities is observed almost everywhere. The circulation at 100 m depth is shown in Figs. 16–18. There are no significant changes compared with the 50-m level.

At a depth of 250 m (Fig. 19) in the Atlantic Ocean, the northern subtropical ACG decreases in size and displaces to the north-west. It is replaced by the broadening, tropical *CG*.

The current velocities decrease considerably. The maximum decrease is observed in the East Australian [3–42(15)], Guianan [3–24(10)], Peruvian [15–65(30)] and Somalian [5–40(15)] currents.

The circulation at the 500-m level (Figs. 20–22) does not differ essentially from the previous level. Specific changes in the Atlantic Ocean are not observed. The northern subtropical ACG in the Pacific Ocean is reduced in the meridional dimensions, and has retreated northward. The region of tropical and equatorial latitudes is covered by small gyres. Further weakness of the circulation takes place, however, it is lower compared with that at the 250-m level.

The zonal flows at a depth of 1000 m (Figs. 23–25) undergo further destruction. The meridional circulation progresses further. There is an increase in a number of small cyclonic and anti-cyclonic gyres the location of which is often related to the bottom topography.

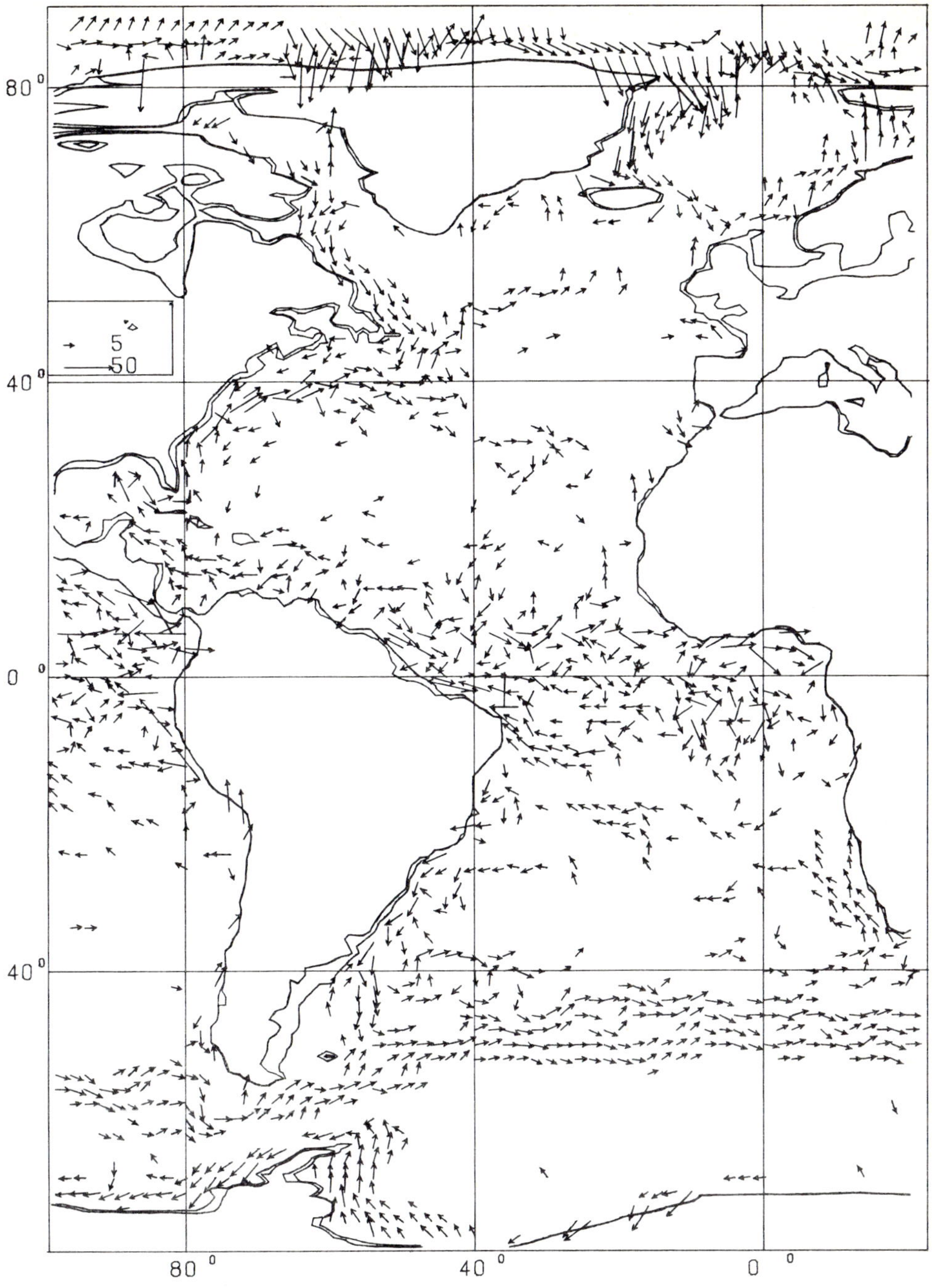

Fig. 16. The currents of the Atlantic Ocean at 100-m depth in the summer season

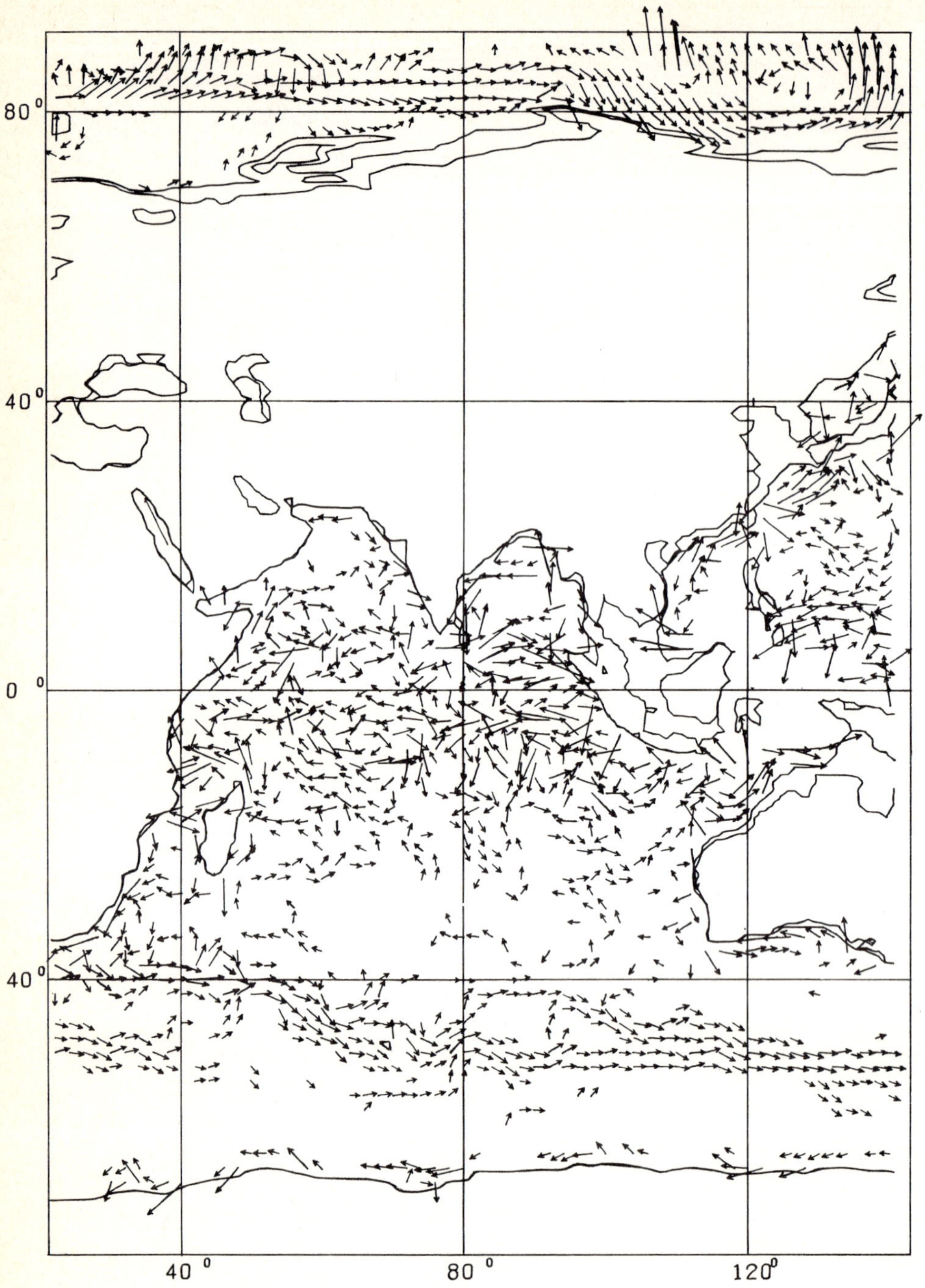

Fig. 17. The currents of the Indian Ocean at 100 m depth in the summer season

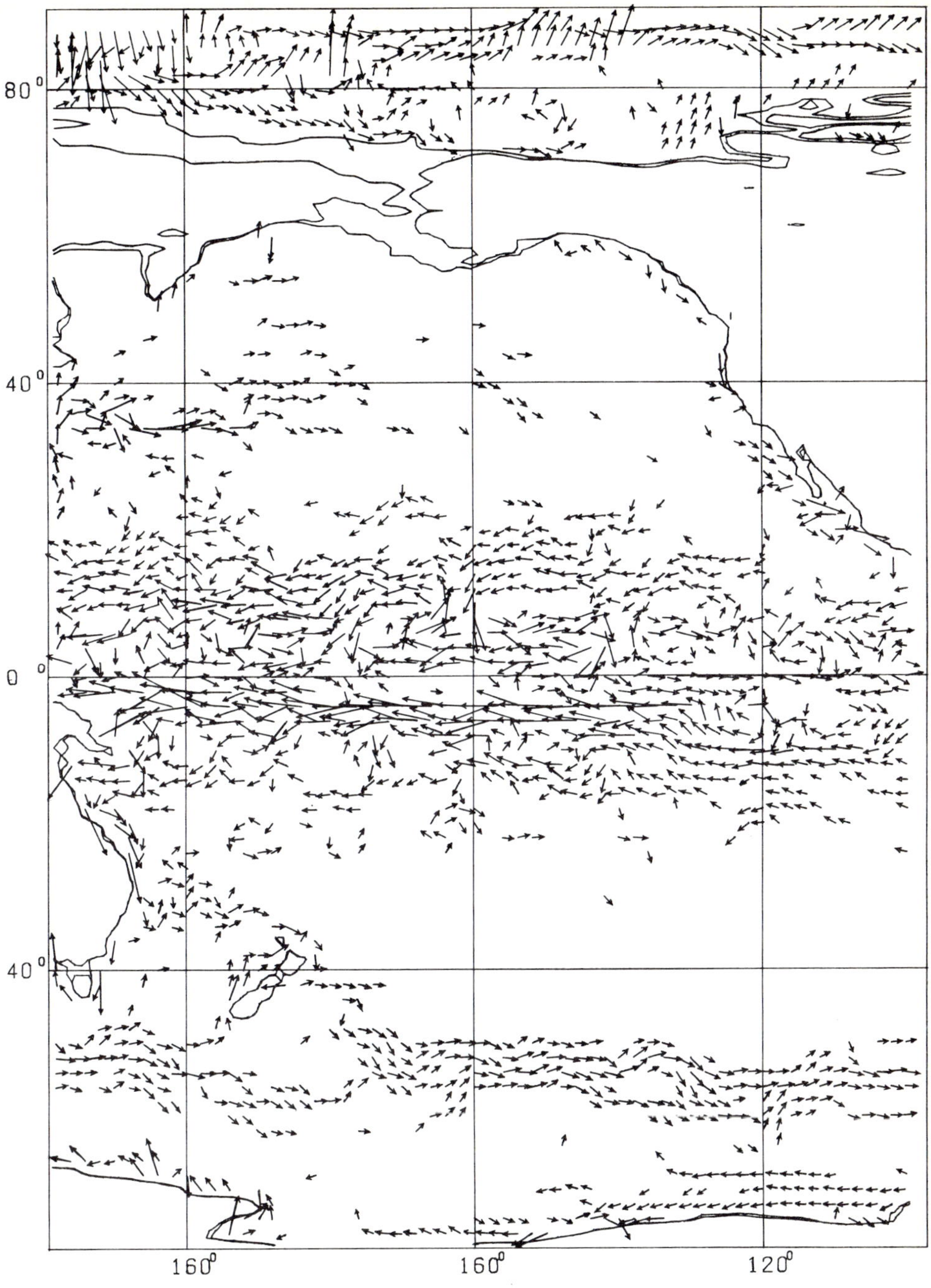

Fig. 18. The currents of the Pacific Ocean at 100 m depth in the summer season

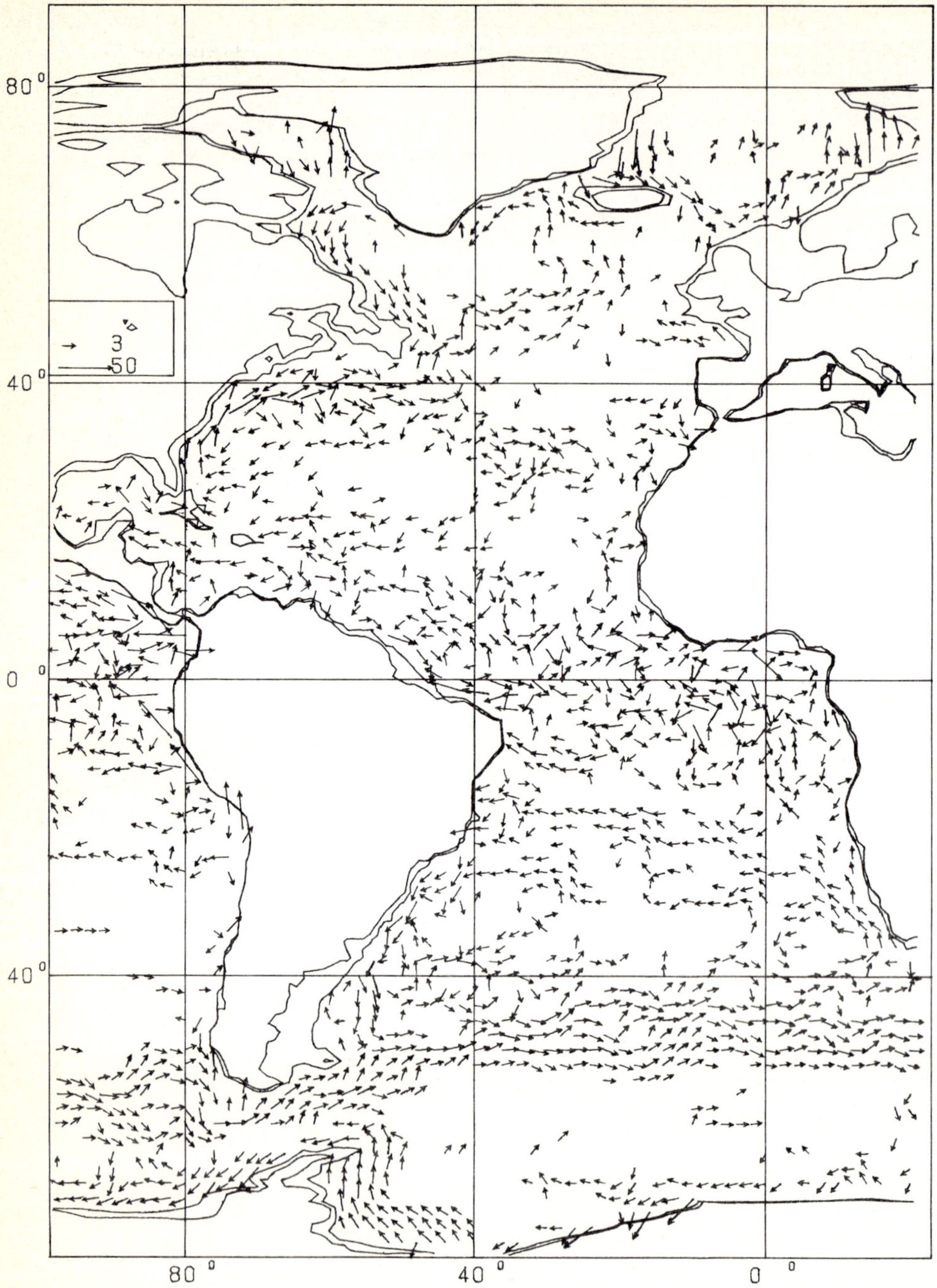

Fig. 19. The currents of the Atlantic Ocean at 250 m depth in the summer season

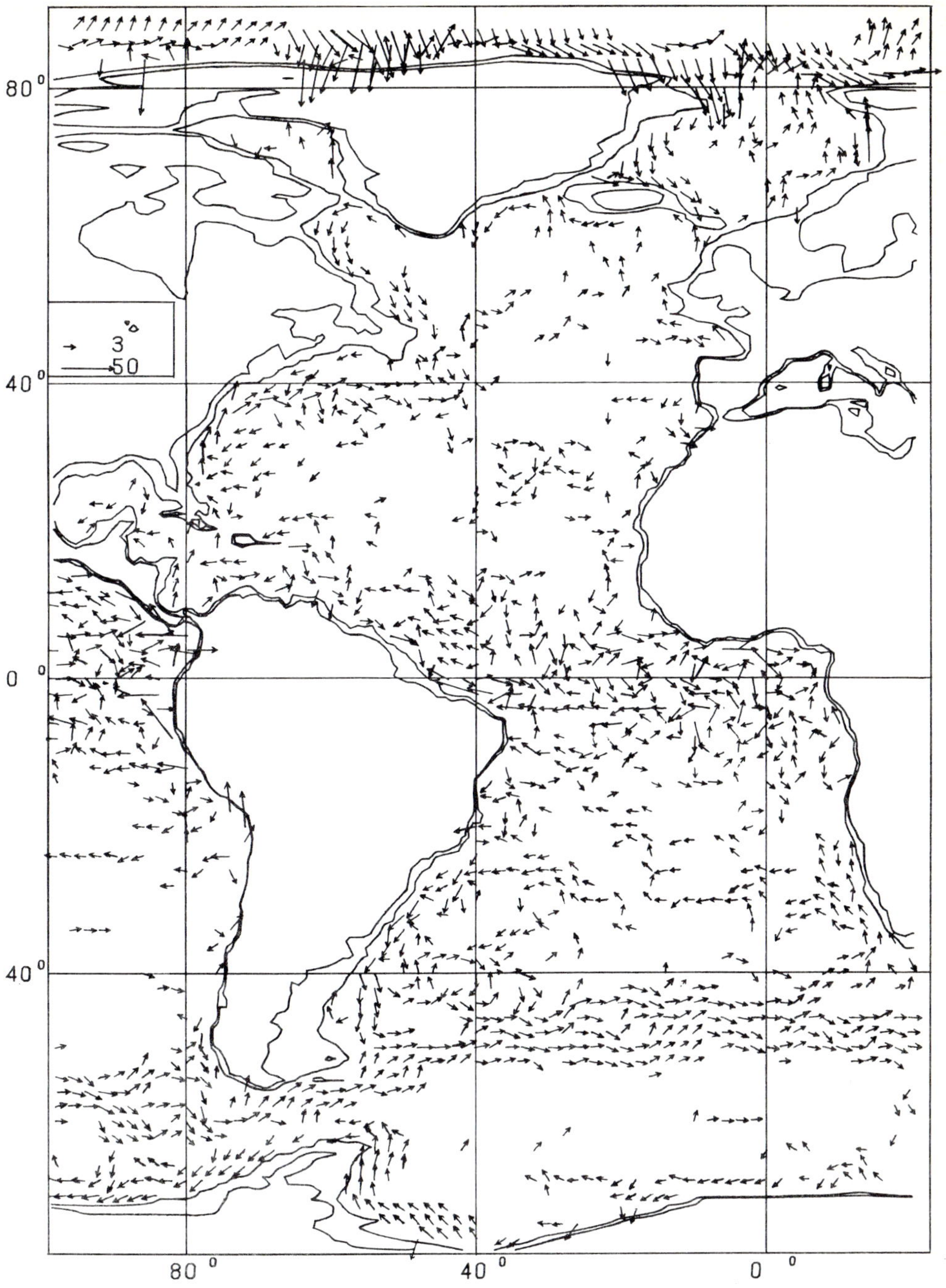

Fig. 20. The currents of the Atlantic Ocean at 500-m depth

Fig. 21. The currents of the Indian Ocean at 500-m depth

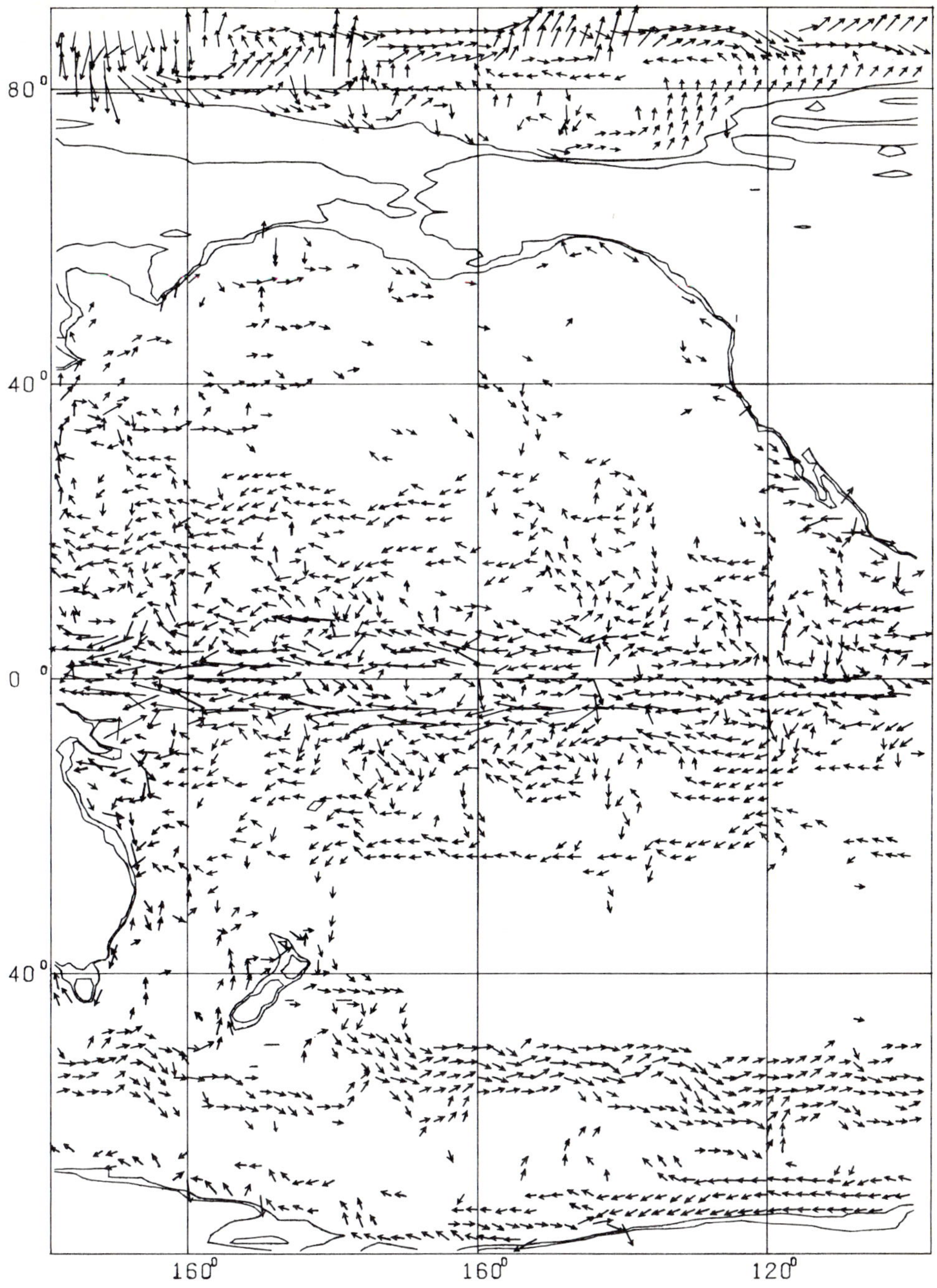

Fig. 22. The currents of the Pacific Ocean at 500-m depth

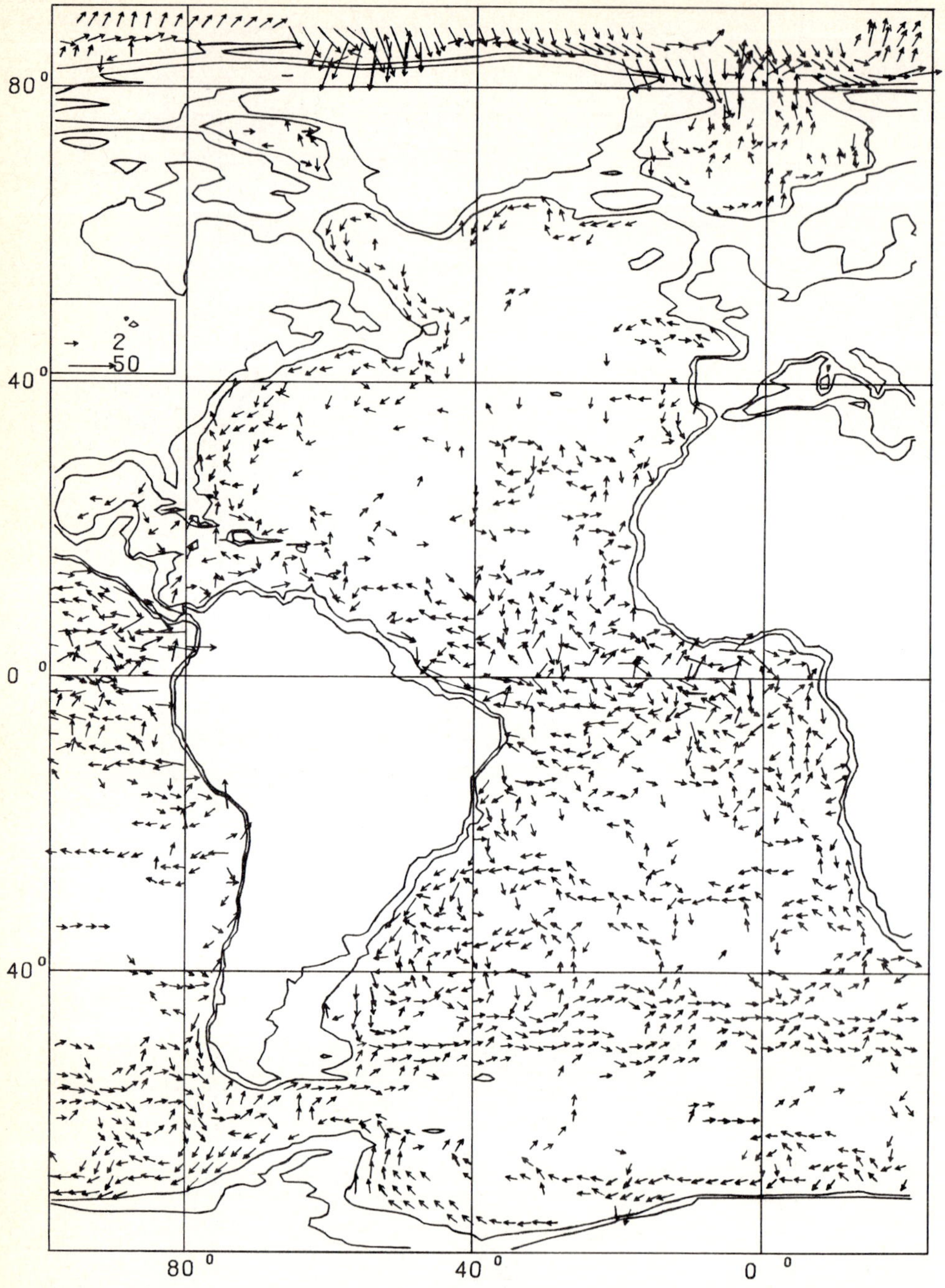

Fig. 23. The currents of the Atlantic Ocean at 1000-m depth

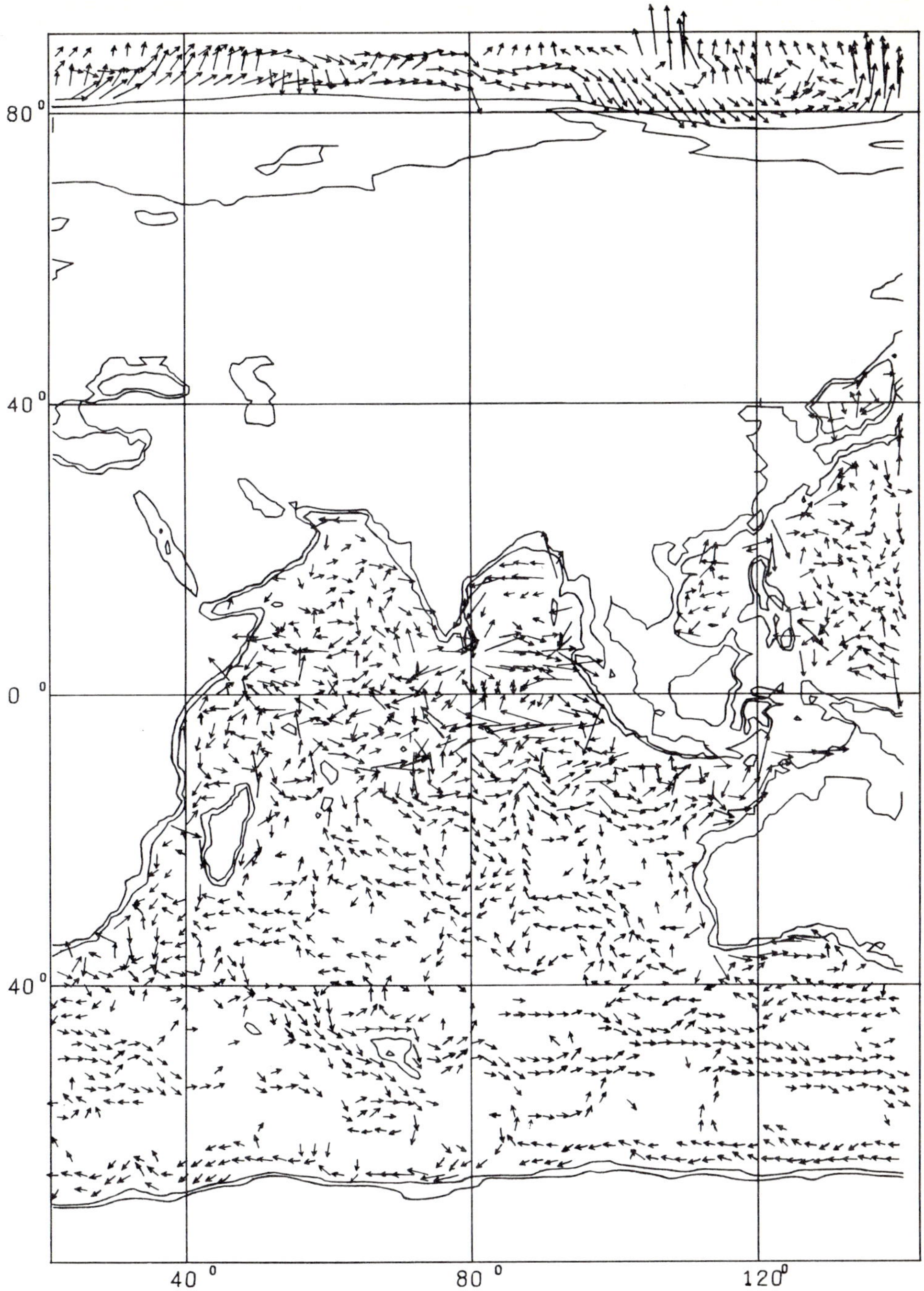

Fig. 24. The currents of the Indian Ocean at 1000-m depth

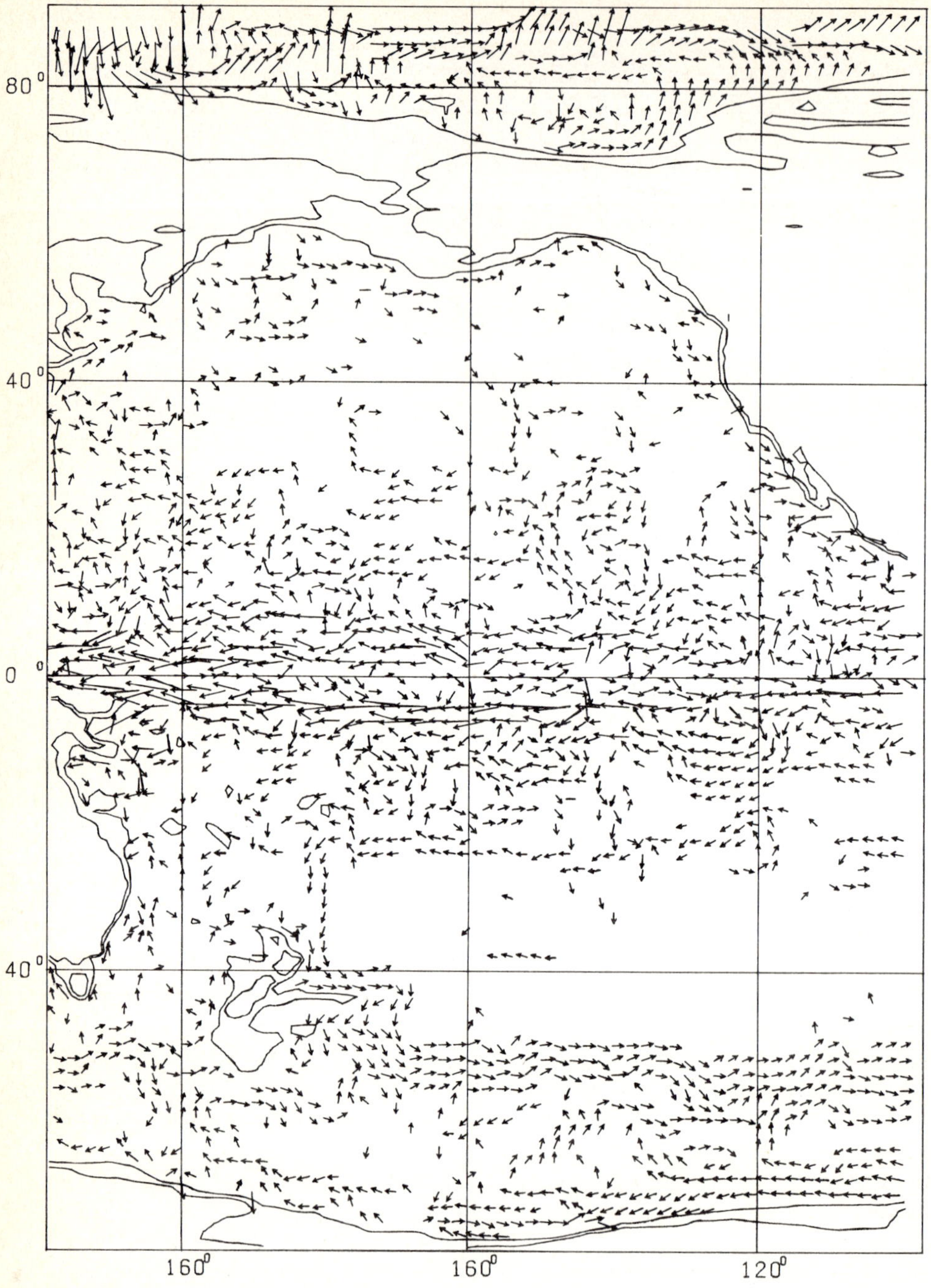

Fig. 25. The currents of the Pacific Ocean at 1000-m depth

In the Atlantic and Pacific Oceans the northern ACG tend to further reduction and decomposition into a small gyre system. The current direction changes partially in the opposite direction. Two, cyclonic and anti-cyclonic, gyres are generated in the central Gulf Stream. At 40° S the Gulf Stream changed its direction to the west, and in the region of Gatteras, to the west and south-west. The current, opposite to the Gulf Stream, suppressed the north-western flow which is now indicated by several arrows. This united (together with the East Greenland and West Greenland currents and that of Labrador) flow transports the arctic waters southward, which is consistent with the general concept of abyssal circulation (Stommel 1965).

So, the circulation at levels from 0 to 1000 m (surface, subsurface and intermediate waters) is generally a large gyre system which attenuates with depth. The current velocities decrease with depth, more abruptly at depths ~250 m and at lower depths, the changes are less significant. Both northern and southern ACG, while attenuating, displace northward and southward respectively, approximately by 10° N and S. The direction of a number of currents changes at a water depth of 0-1000 m, e.g. Antillean, Gulf Stream, Canarian, Portuguese currents and the deeper part of the Somalian current. Significant changes are observed at a level of 1000 m. The subtropical ACG are destroyed, being substituted by small gyres.

In all oceans separate areas under the trade-wind currents are observed. Beginning with 500 m the currents closing southern subtropical ACG (South Atlantic, South Pacific and South Indian Ocean) disappeared altogether. ACC and northern subpolar CG remain invariable in location and dimensions although they attenuate with depth.

5.3.2 The World Ocean Deep and Bottom Layer Water Circulation

In the previous section it has been shown that already at a level of 1000 m the circulation changes considerably compared with the higher levels, and loses its resemblance to the surface circulation. At depths of 1000 to 2000 m (Figs. 26–28) the circulation is still more characterized by the small gyres of a cyclonic and anti-cyclonic nature. The countercurrents generated at the 1000-m level are preserved. The current under the trade wind in the North Atlantic occupies the same position as at a depth of 1000 m, but becomes sinuous and is related to the small gyre system. In the South Atlantic in width it occupies a larger space (15°–20° S). In the Gulf Stream region two cyclonic and two anti-cyclonic gyres are formed. In the Pacific Ocean the subsurface countercurrents maintain the same location and dimensions as at 1000 m. A similar current in the Indian Ocean crosses practically the whole ocean between 15°–22° N. In the eastern Indian Ocean, in the band 20°–30° S, the current is directed westward. Up to that depth the current changed its direction twice, generating three layers of circulation. The current at the 0–50 m layer is directed westward; from 50 to 100 m, eastward; and from 250 to 2000 m, westward again. The North Pacific current still keeps its westward direction, covering the same area (30°–40° N, 150°–180° E).

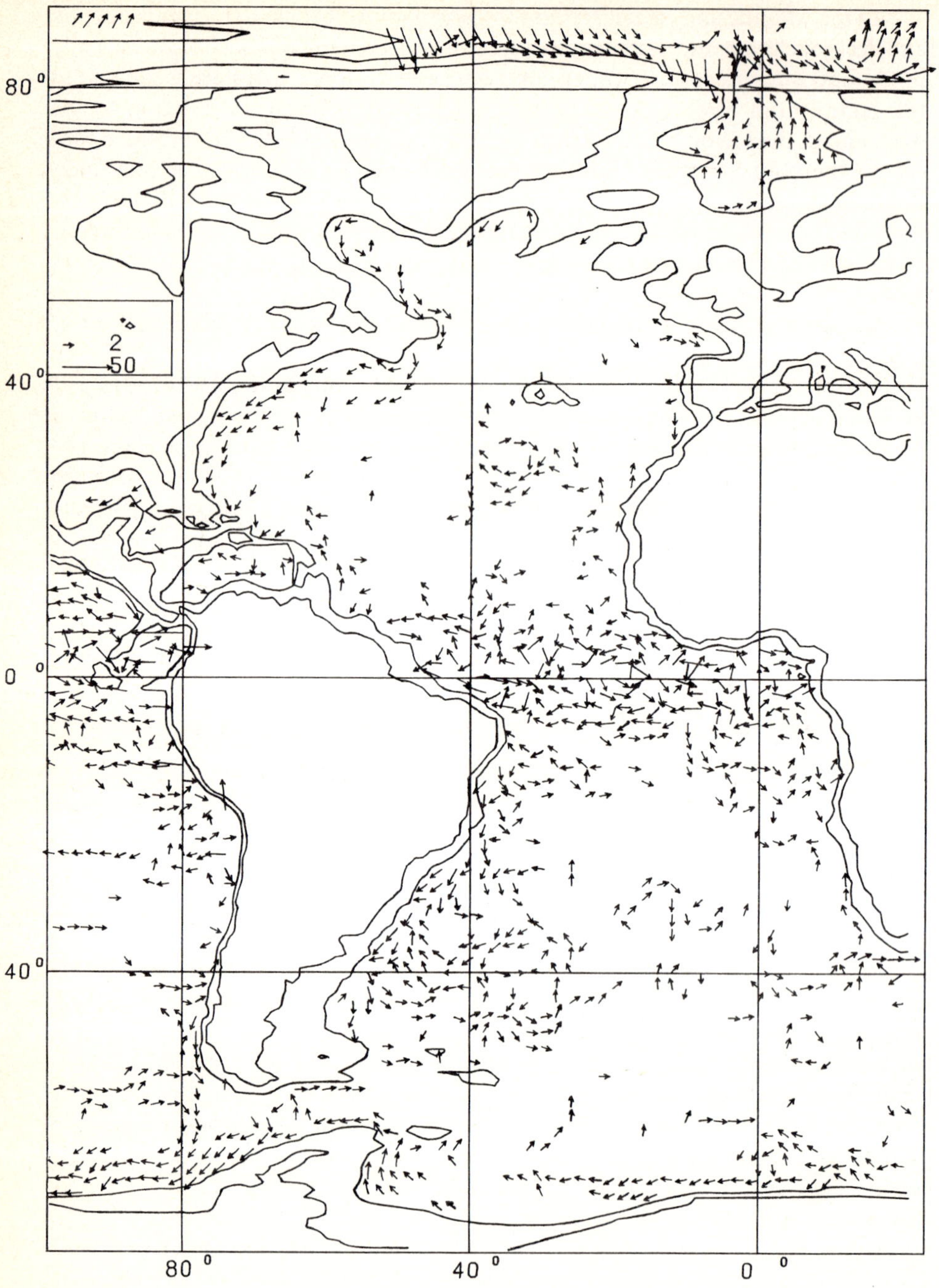

Fig. 26. The currents of the Atlantic Ocean at 2000-m depth

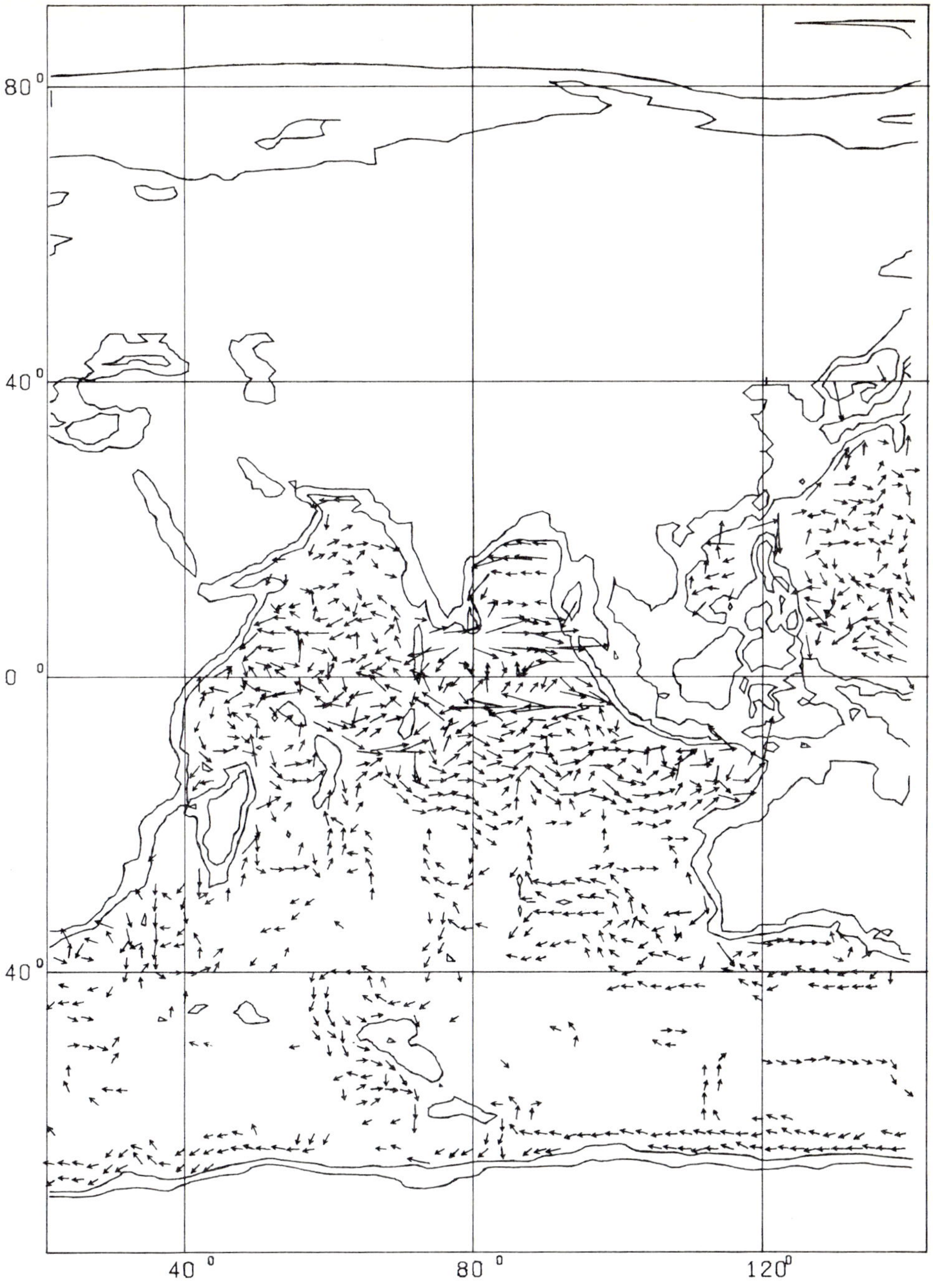

Fig. 27. The currents of the Indian Ocean at 2000-m depth

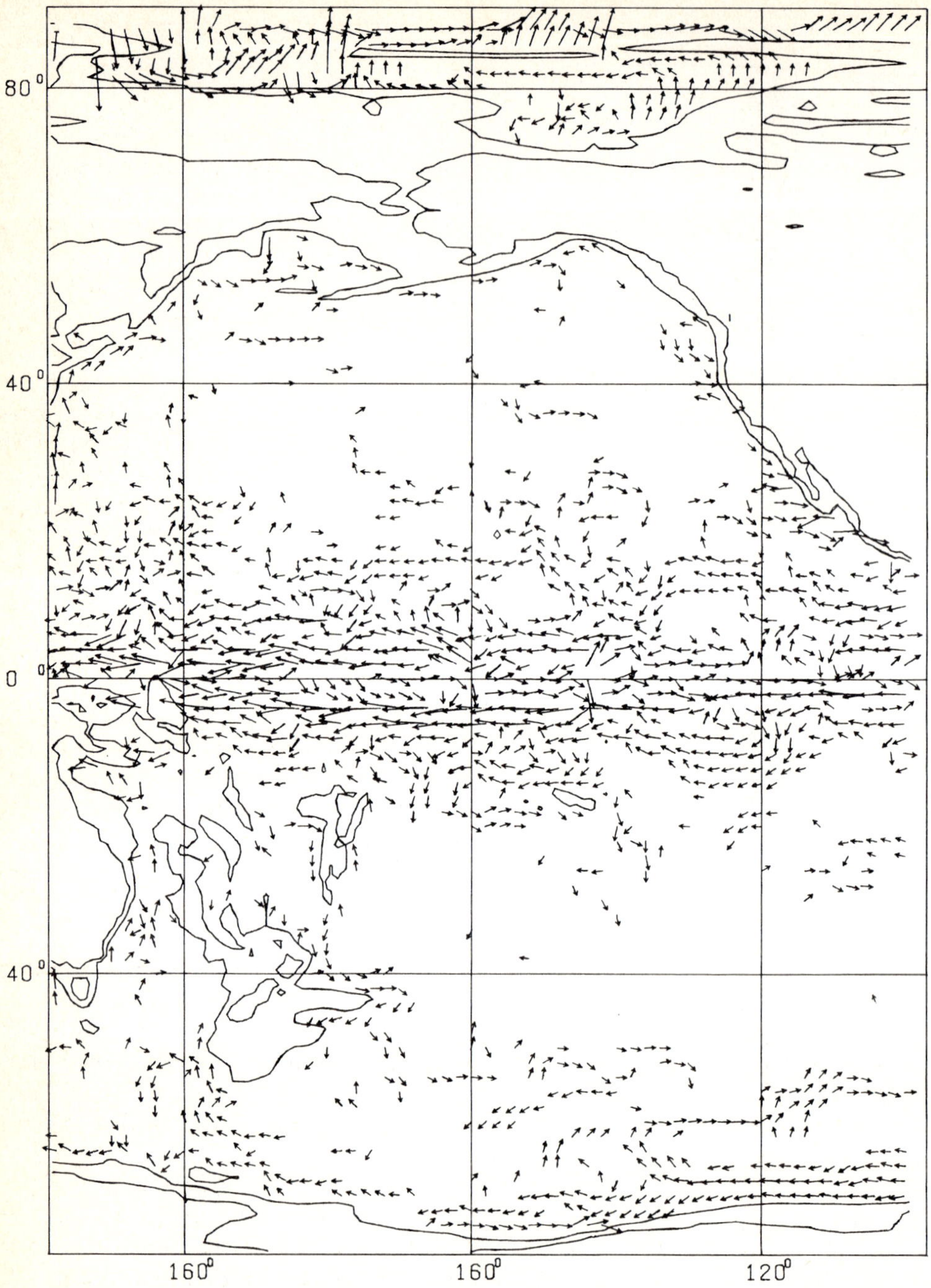

Fig. 28. The currents of the Pacific Ocean at 2000-m depth

The ACC becomes more sinuous, presenting a small gyre system. So, the circulation at a level of 1000–2000 m undergoes strong reconstruction, and the effect of the bottom relief becomes more intensified.

At a depth of 3000 m the general circulation scheme is similar to the current field at 2000 m. However, there is no total similarity. The subpolar CG, though somewhat distorted, is still preserved in the North Atlantic. The Gulf Stream region is still characterized by two anti-cyclonic and two cyclonic gyres. Two cyclonic gyres are located in the western and eastern North Atlantic separated by the Mid-Atlantic ridge. The deep countercurrent is much pronounced under the Gulf Stream region. Its velocity values are rather high (10–20 cm s^{-1}). The flow forms a western branch in a general cyclonic gyre in the western oceanic abyssal plain, and is observed at a depth of 1000 m. The Mid-Atlantic ridge divides the ocean according to its circulation structure into two parts: western and eastern. In the eastern North Atlantic one can observe an intensive cyclonic gyre within 10°–30° N. In the South Atlantic an anti-cyclonic gyre still exists, which due to the effect of the Mid-Atlantic ridge, splits into two gyres. In the region of 40–50° S, particularly in the Argentinian abyssal plain, the anti-cyclonic circulation nature is also considerably pronounced. It is represented by a chain of gyres. Their southern chain consists of an ACC flow, which at this level meanders with much bending around the bottom rise. In the tropical latitudes of the Atlantic the circulation changes little compared with 1000 m. Near North America, within the belt 0°–10° S, the cyclonic gyre is pronounced. Its coastal chain corresponds to the Antilles-Guiana countercurrent. The cyclonic gyre is located at the same latitudes in the eastern half of the ocean. The eastern flow of the equatorial undercurrent in the range 0°–2° S forms the southern chain of the gyre. The western flow is observed in the band 2°–10° S. The flow forms the northern chain of the anti-cyclonic gyre in the tropical region. In the Northern Hemisphere the anti-cyclonic circulation generates round the bottom rise in the region of the mid-ocean ridges, while the cyclonic one is generated in the Southern Hemisphere. The World Ocean deep waters, as it is known, are formed at the surface in the Labrador Sea, Norwegian Sea, Greenland and Antarctic Seas as a result of a winter convective displacement and distribution in deep layers in accordance with density. This transport is carried out by the countercurrent under the Gulf Stream and the western chain of the cyclonic gyre in the eastern abyssal plain of the North Atlantic. The Mediterranean waters of high density, transported from the Straight of Gibraltar (Demin and Mikhailichenko 1981), considerably effect the formation of the deep waters of the Atlantic. Then the abyssal waters circulate in the tropical and equatorial latitudes and through the Atlantic gyres in the Southern Hemisphere they are transported to the southern part from where they are taken by the ACC and transported to the Indian and Pacific Oceans. The Arctic Ocean and Antarctic Sea waters serve as a source for bottom layer waters. They penetrate into the Atlantic Ocean through the Faroes-Iceland cascade.

The waters in the Indian Ocean move northward, east of Madagascar island. These waters might have come from the Atlantic and Antarctic coastal seas. South of 20° S, because of the bottom rise, the meridional current velocity components prevail. These currents are principally directed northward which is indicative of water spreading from the Antarctic throughout the ocean. Numerous bottom rises impede the spreading of ACC, and make its movement difficult here. North of 20° S the circulation changes little compared with the level at 1000 m; the eastern flow passes within 10°–20° S, then to the north, up to 2°–3° S; the flow direction changes to the west, and near the equator the eastern equatorial undercurrent passes again. These currents form the system of cyclonic and anti-cyclonic gyres. On both sides of the Central-Indian Ridge there is a chain of small cyclonic gyres.

The same situation of circulation as a whole as that at the level of 1000 m is observed in the tropical latitudes of the Pacific Ocean: the westward flow prevails between 20° N and 20° S, and the eastward equatorial undercurrent passes within 1°–5° S. Two anti-cyclonic gyres are located in the northern ocean along the Aleutian and Kuril Islands, being centred at about 40° S. The southern Pacific is mainly characterized by the cyclonic circulation which performs the interlatitudinal exchange. The location of large gyres is related to the influence of the bottom relief. The ACC is much pronounced at a level of 3000 m as well. It now undergoes a still more intense effect of the bottom topography and generates a number of cyclonic gyres.

In the Southern Hemisphere the nature of circulation remains as a whole similar to that at a level of 1000 m. This is attributed to the fact that the stable ACC serves as a major circulation element. The meanders penetrating far to the north and their branches are of a topogenic nature and are also stable.

The circulation at a depth of 4000 m is under the impact of a strong bottom relief. In the North Atlantic, on both sides of the Mid-Atlantic Ridge in the region of the Brazilian abyssal plain, the anti-cyclonic circulation is preserved. It includes waters overpassing the equator. The anti-cyclonic gyre, which is fed by the ACC, is located in the western Atlantic with a centre at 40° S. Eastward at the same latitude near the southern Africa coasts a whole system of small, mostly anti-cyclonic, gyres is located.

The Indian Ocean at this depth is considerably cut by the bottom relief. So, in the western ocean water, coming from the north across the equator, does not mix with that coming from the south. In the eastern ocean, from north to south throughout the whole ocean, a whole system of cyclonic gyres is located.

In the southern Pacific, from 10° to 40° S there is a cyclonic circulation, which performs the interlatitudinal exchange. South of it, from 40° to 60°, the anti-cyclonic circulation is marked, which promotes the exchange between the ACC and the anti-cyclonic gyre north of it. In the northern Pacific the circulation differs slightly from that in the upper level.

The bottom layer water circulation in the Atlantic Ocean is shown in Fig. 29. There are three sources of the bottom layer waters: the North Atlantic (east and south-east of Greenland), Antarctic (near the Antarctica coasts) and arctic (in the Greenland Sea). The North Atlantic waters take an active part in the bottom layer water circulation while motions of the arctic water are hindered by the Faroes Iceland Cascade and Greenland underwater ridges. The motion of the water is pronounced (Fig. 29) where the depth is close to 3000 m. The Antarctic waters are submerged and spread, together with the ACC, which meanders, bending round the bottom rise and penetrating far to the north. It has branches which transport the Antarctic waters to tropical latitudes.

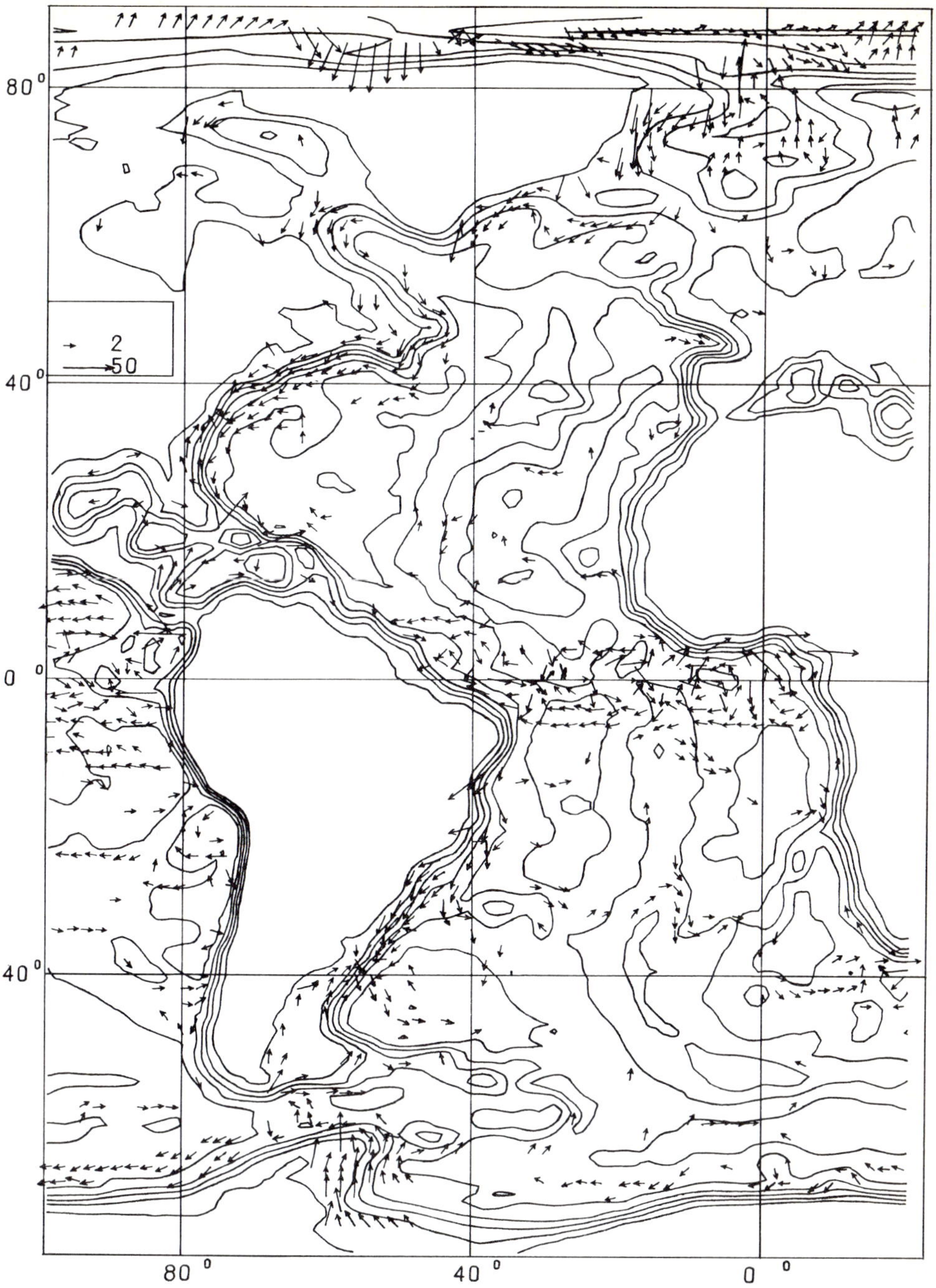

Fig. 29. The bottom circulation of the Atlantic Ocean waters. The isobaths are drawn to the scale of 1 km

The ACC preserves its general eastward direction. In the Atlantic and Pacific Oceans the Antarctic abyssal layer water transported by the ACC branches to the north through abyssal plains and trenches. In all the abyssal plains of Argentina, Brazil, Angola, etc., the anti-cyclonic circulation is generated. The Antarctic water crossing the equator at three points (10°, 18°–20°, 30° W) extends to the north in North America and to the Canarian abyssal plains to 30°–40° N.

In the Indian Ocean the Antarctic waters move to the north, west of the Central Indian Ridge through Madagascar and Mascarene abyssal plains. A cyclonic circulation is observed in the Madagascar abyssal plain, while in the Mascarene an anti-cyclonic one is observed. In the eastern ocean the Antarctic waters penetrated into the West Australian, Cocos and Central abyssal plains from the south. The circulation there is primarily cyclonic.

In the southern Pacific the Antarctic water moves northward along the South American coasts. Beginning with 20° S it deviates westward and then southward, generating the anti-cyclonic circulation. The second northward flow passes along the Kermadec and Tonga trenches.

At the near-equator, southern Pacific, from 0° to 4° S, the flow is directed to the east along the equator (160° E–120° W) and south of it; in the latitudinal range 4°–10°, practically across the whole ocean, a powerful flow passes in a westerly direction. Near the western oceanic coasts the current splits into two parts. One part is directed to the south along the Australian coasts, and the other one to the Japanese coasts. These flows are generally observed at all levels, from surface to bottom.

The Antarctic bottom layer waters penetrate into the northern Pacific across the equator in the region 170°–145° W and 120°–105° W.

As a whole the bottom layer water circulation is similar to the deep circulation but with an increased bottom relief impact. Briefly the character of circulation at a level of 0–1000 m will be mentioned. The circulation in the upper layer represents an ordered system of oceanic gyres which is common for all oceans, while in the lower layer this is a set of relatively small-sized gyres without a common system. The gyre sizes are related to the bottom relief. The lower layer current direction is, as a rule, opposite to that of the upper layer current. The available western boundary currents are typical of all levels, although current directions are not similar for all oceans. So, in the Atlantic Ocean it is directed from north to south, across the whole ocean, and in the Indian and Pacific Oceans, to the north. The World Ocean deep and bottom layer waters are formed at high latitudes in the process of convective mixing at cooling and salting. Then these waters spread from high latitudes to low. In the Atlantic Ocean the deep waters are transported from the northern seas and Arctic Ocean primarily through the Faroes-Iceland Cascade and spread along the western coast to the southern area where they mix with the ACC waters. Other oceans are not connected with the Arctic ocean, and deep waters come here mainly from the Antarctic seas. After circulating in the zonal tropical flows, waters from high latitudes enter the other hemisphere across the equator. Maximum changes in circulation compared with the upper (from 0 to 1000 m) water layer take place in the Atlantic Ocean. In the Pacific and Indian Oceans the deep water circulation is not particularly transformed in comparison with the upper levels.

Let us consider briefly the comparison of the results obtained with those published earlier.

The maps of the dynamic topography of the Pacific sea surface and at levels of 200, 500, 1000, 2000 and 3000 db, obtained with the help of a dynamic method, are given in Arsenyev and Galerkin (1976); and Wyrtki (1975); 4000 db is taken as a reference level, the grid size is $5^\circ \times 5^\circ$. The calculations are performed for a multi-year mean. The general circulation schemes, both at the sea surface and at the deep levels, are qualitatively alike. At the sea surface the currents, according to these calculations, are of a more zonal nature, particularly the ACC. The northern and southern subtropical ACGs in paper of Brechovskich et al. (1983) are not closed near the western oceanic coast. The differences in the equatorial and tropical zones are pronounced. Tropical gyres, according to our calculations, are of less dimensions. The southern tropical CG has a somewhat different configuration: according to these calculations, it is flatter and is located strictly along the equator. Partial closing of the subtropical ACG is pronounced at a level of 500 m, although the southern tropical CG still remains narrower. In accordance with the calculations, it is located exactly along the equator. The equatorial undercurrent also moves along the equator within 2°–3° S. In the Northern Hemisphere, according to calculations, the tropical CG and ACG occupy, as before, less area. There is almost no qualitative difference at levels of 1000, 2000 and 3000 m, excluding a narrow current (3°–8° S) along the equator which is shown in our maps. It is directed westward and is a southern branch of the southern tropical CG.

The oceanic Atlases (1974, 1977) include the maps of the World Ocean water circulation at the surface and at levels of 100, 200 and 500 m. The sea surface circulation is given for February and August of the multi-year mean. The general surface circulation scheme in August is close to that obtained from the calculations for the summer season. However, the map of currents in the Atlas is considerably smoothed, so it contains only major, large gyres and currents. Even large, southern tropical, cyclonic gyres of the Atlantic and Pacific are absent from the Atlas. In the northern Pacific the Equatorial countercurrent in the Atlas crosses the whole ocean, in contrast to our maps, where it is located from 110° W to the American coasts. Correspondingly, the tropical CG in the Atlas as well crosses the whole ocean. The comparison of current velocity values indicates that most currents (two-thirds) in the Atlas have velocities within the limits obtained by us.

At levels of 100, 200, and 500 m the Atlas includes only the multi-year mean circulation, and only at a level of 100 m is there a scheme for the northern Indian Ocean circulation for February and August of the multi-year mean. The large size of the calculating grid set resulted in a considerably smoothed picture of circulation. The circulation at a level of 100 m is, as a whole, similar to that obtained by us. The absence in the Atlas of the circulation as a chain of gyres in the tropical area of the Atlantic, then the absence of change in the current direction in the region of the Canarian and Portuguese currents, a small length of the southern tropical CG in the Pacific, which was not observed at the surface, and the presence of only one gyre (anti-cyclonic) in the Benguela Bay of the Indian Ocean can be considered as significant differences. The circulation at a depth of 200 m differs little from that at 100 m, and the differences remain almost the same. As a new significant characteristic one can point out the absence in the Atlas of the western boundary currents in the northern Pacific (excluding Kuroshio), i.e. the North Equatorial current abuts upon the coast. The circulation at a depth of 500 m is qualitatively also similar to that obtained by us. The still unchanged direction of the Canarian and Portuguese currents, the presence of only cyclonic circulation in the Arabian Sea, the absence of western boundary currents from the northern Pacific and the considerably smoothed field of the currents in the region of 20° N to 20° S can be noted as differences of importance.

The current calculations for the multi-year mean were performed according to a modified version of the dynamic method on a grid set with a size of 5° × 5° (Burkov 1980). The results are given as schemes for dynamic horizontals. The general schemes for circulation are qualitatively alike. A number of sea surface currents in Burkov (1980) appeared to be suppressed (West Australian, East Australian, Canarian, Mindanao, Guinean, Portuguese). The tropical ACG and CG in the North Atlantic were not obtained in Burkov (1980).

One more distinct characteristic of the calculations is that the main transformation of the upper layer water circulation to that of the lower layer is already pronounced at a depth of 1000 m, whereas in Burkov (1980) it is 2000 m.

5.3.3 The Vertical Structure of the World Ocean Currents

The vertical structure of currents will be studied according to the change first of the zonal currents with depth, and then to meridional currents. The zonal flows play the major role in the oceanic surface layer water circulation for distant from the coastal layers. At greater depths, this role is played by meridional flows.

The ACC, which zonates the whole globe and is directed eastward, represents the largest zonal flow. It changes neither direction nor location (40°–60° S) from surface to bottom.

All South Equatorial currents are flows of a zonal direction. At the surface they are located approximately identically in all oceans: in the Atlantic — within 3°–25° S, in the Indian Ocean — 2°–25° S, in the Pacific — 3°–20° S. The South Equatorial currents displace southward with depth. At a depth of 1000 m in the Atlantic and Indian Oceans the equatorial currents are located within 20°–35° S. At depths of 1000–2000 m in all oceans the currents are transformed. The equatorial western currents are interrupted by separate flows directed eastward. At these depths a strong effect of the bottom relief is observed, which results in the change of the current direction.

South Atlantic, South Pacific, South Indian Ocean currents present weak east flows, which disappear already at depths of 250, 50 and 100 m respectively.

Narrow zonal currents pass in the southern, near-equatorial latitudes in all oceans (0°–5° S). In the Atlantic Ocean along the equator (0°–2° S) from the surface to a depth of 3000 m, there exists an eastward current. At a latitude of 2°–5° S, at the same levels, a west flow is already observed. From 3000 m to the bottom there is one westward flow instead of two.

In the same region of the Pacific Ocean, from surface to bottom, two opposite flows exist: eastward (0°–3° S) and westward (3°–5° S).

In the Indian Ocean the situation of currents in the near-equatorial belt is more mixed. First and foremost, it is divided into two parts by the meridian 80° E. West of it in the width of 0°–5° S, the westward current is observed from the surface to a depth of 1000 m. At a level of 2000 m the strong influence of the bottom relief is pronounced. Eastward flows are observed near the underwater ridges. East of 80° E two narrow flows, eastward (0°–2° S) and westward (2°–5° S), are observed from surface to bottom.

In the northern Indian Ocean, from equator to the northern banks of the Arabian Sea, the surface of the ocean is covered with a continuous eastward flow generated by the summer monsoon. A narrow westward flow, covering the space 1°–2°, is observed at a depth of 50 m along the northern banks of the Arabian Sea. South of it, up to 10° N, an eastward flow is observed, which is part of the anti-cyclonic gyre well pronounced at a depth of 500 m. From 10° N to the equator cyclonic gyres are located. At 10° N they generate the westward flow, and closer to the equator (0°–3° N), an eastward flow. Such a circulation with significantly pronounced cyclonic and anti-cyclonic gyres is preserved to the bottom.

In the northern Atlantic the zonal flows include North Atlantic, East Greenland and West Greenland currents. The North Atlantic current at a depth of 0–2000 m is transformed step by step into a system of small gyres which are under the impact of a distinct bottom relief. The East Greenland current has a western direction from surface to bottom. The West Greenland current has a western direction from the surface to a depth of 500 m, and an eastern direction from lower levels to the bottom.

In the Pacific Ocean the zonal flows include the North Pacific and Aleutian currents. The North Pacific current in the 0–1000 m layer is directed eastward and beginning with 1000 m the current direction changes to the western one.

The meridional currents joint the zonal currents, producing circulation systems in the ocean.

In the Atlantic Ocean the meridional flows include the Brazilian, Benguela, Falkland, Gulf Stream, Canarian, Portuguese, Labradorian and Norwegian currents. The Gulf Stream, at a level of 1000 m, changes to a southern and south-western direction, forming a system of gyres at this depth. The Canarian and Portuguese currents are wind drift by nature, and at a level of 50 m they change to a northern direction. Labradorian, Norwegian and Benguela currents keep their direction to the bottom. The Brazilian current up to a depth of 3000 m is directed mainly southward. At a depth of 4000 m, due to the significant impact of the bottom relief, it was involved in two anti-cyclonic gyres with southward branches. The Falkland current, directed north-eastward at the sea surface, changes its direction to the north at a depth of 50 m. At a depth of 2000 m it is completely "removed" by the bottom relief, and only a southward flow remains.

In the Pacific the meridional flows include Kuroshio, Mindanao, Formosa, Oyashio, Kamchatka, Californian, New Guinea, East Australian and Peruvian currents. The Peruvian current, directed northward in the upper layer, changes its direction to the south at a depth of 1000 m. The Kuroshio current up to 2000 m has northern and north-eastern directions. At a depth of 2000 m, under the influence of the bottom relief, a southern and south-eastern current component is observed. The East Australian current keeps its southern direction to a depth of 1000–2000 m, then its order is disturbed, and water motion becomes very complicated. The rest currents preserve their direction to the bottom (Mindanao, Oyashio, Kamchatka, California).

In the northern Indian Ocean the meridional flows include the Somalian current. Its main direction at the surface is northern. Already at a depth of 50 m the flow in its southern part has a southern direction, and at 250 m within 0°–8°N the southern direction prevails and is maintained to the bottom. The Mozambique current in its upper layer is directed southward. At a depth of 2000 m some sections of northern directions are observed. At 3000 m the current system is considerably transformed by the impact of the bottom relief, and so it includes both northern and southern flows, which are preserved to the bottom. The Agulhas current of a southern and south-western direction beginning with a depth of 1000 m becomes less ordered and its northern part at a depth of 2000 m acquires a northern direction. The West Australian current splits into northern and southern flows from surface to bottom.

Thus, the structure of zonal and meridional flows indicates that central parts of the oceans are characterized by the eastern transport up to subtropical convergences, then the transport changes to a western one, and nearer to the equator it is of an eastern nature. Furthermore, the general circulation structure of the World Ocean is characterized by a complicated multi-layer. In general, it can be said that a one-layer current structure prevails in subpolar latitudes, a two-layer structure prevails in tropical latitudes, and in subtropical and equatorial latitudes a more complicated structure prevails. The current interface of the upper and lower layers is most often at a depth of 1000–2000 m.

5.3.4 Seasonal Variations of the World Ocean Surface Topography and the Surface Gradient Currents

In Section 5.2.3, the surface topography of the World Ocean and the structure of the surface gradient current in the summer season were considered. The surface topography and surface gradient currents are poorly related to the direct force of the wind, being almost completely related to density and bottom relief fields. They characterize subsurface circulation rather than the surface circulation. Furthermore, they are much more suitable for comparison with results of previous calculations mostly performed with the help of a dynamic method, which is based on the density field and which ignores the direct impact of the wind.

To estimate the seasonal variation the surface topography and surface gradient currents were calculated for all four seasons of the year. The surface topography in the winter season is shown in Fig. 30. From the comparison with a similar map for the summer season (Fig. 11), it is clear that even in the extreme seasons the surface topographies are very similar to each other in global, "large-scale" characteristics. Significant differences are characteristic of the areas of monsoon currents. In general, only details and quantitative characteristics are under change. Compared with the calculations performed earlier on a 5° grid set, the schemes shown are more

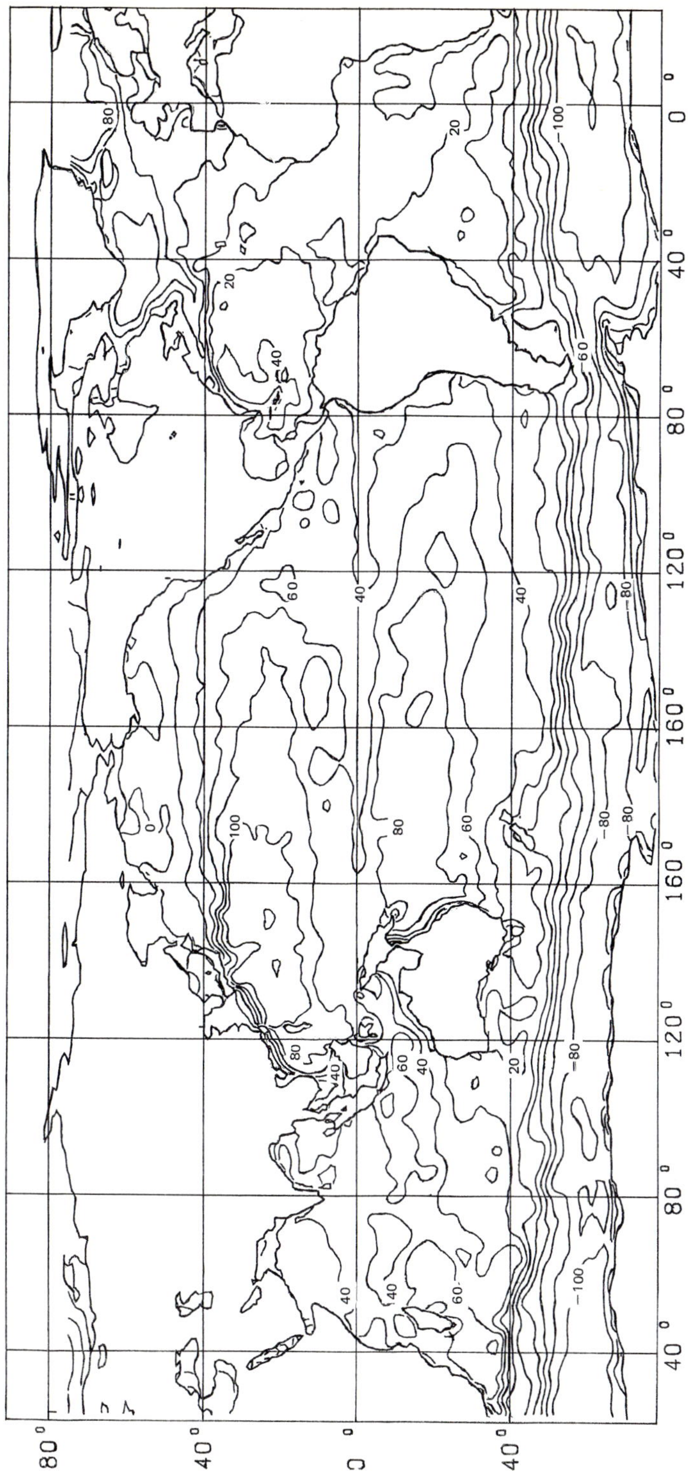

Fig. 30. The transformed SST in the winter season (cm)

detailed (despite the large-scale interval of isolines, whereby 20 cm is taken for demonstration), the jetlike currents are better pronounced, all northern, subtropical ACG are closed, the subpolar CG is more markedly expressed and tropical CG are observed in the Atlantic Ocean.

The seasonal variations of the density-driven currents can be considered in more detail in the maps of surface gradient currents (cf. Figs. 12 and 31).

In general, the surface gradient currents as well as the ocean surface topography are qualitatively similar in all four seasons. All major currents are preserved, being only slightly displaced over the ocean water area and changed in intensity from season to season. Significant seasonal climatic reconstruction in density-driven currents is observed only in a few areas, for example in the Arabian Sea, the Somalian current and in some other areas of monsoon and trade-wind nature (Brekhovskikh et al. 1983).

Let us first compare fields of seasonal extremes, summer and winter, and then compare results obtained with those from the previous study. Characteristic velocities of major World Ocean currents for summer and winter seasons are given in Tables 1 (a, b, c). It is noteworthy that the smoothed, specified fields made lessening of the synoptic "noise" level possible with simultaneous, significant decreasing of velocities and broadening of jetlike currents.

The qualitative difference between summer and winter seasons is insignificant. However, the area of monsoon currents in the northern Indian Ocean is of interest. In the winter season the circulation here changes significantly compared with that in the summer season, which is quite natural. The eastern flow in the Arabian Sea is turned into a cyclonic circulation, and nearer to the equator (0°–3° N) two small CG are turned into the east flow. The velocities also change quantitatively: in the winter season in the Arabian Sea, it is somewhat higher [5–35(15)], and in the near-equator zone it is somewhat lower [5–30(15)]. In the Bengal Bay the nature of circulation is generally preserved, while velocity increases slightly [5–55(25)]. The Somalian current is under the impact of strong monsoon winds. In the winter season in the direct vicinity of the current, a small CG is generated. Part of it is directed southward, forming the main flow in the Somalian current. Only a very narrow, though very fast, flow along the coast preserves its northern direction.

The velocities of a number of currents changed only quantitatively. Thus, the velocity of the Equatorial countercurrent in the North Atlantic dropped substantially, whereas in the northern Pacific it increased slightly.

The velocity of northern trade-wind currents changed. The velocity range in the Atlantic Ocean was reduced, thus maintaining the prevailing values. The prevailing velocity in the Pacific Ocean increased. The velocity of South Equatorial currents in the Indian Ocean increased significantly, whereas in the Pacific Ocean its increase was insignificant.

The velocity of Guinean and Guianan currents increased and that of the Brazilian current decreased slightly. The velocity of the other currents remained either the same or changed insignificantly.

So, the most substantial seasonal variations in the surface gradient currents are observed in the northern Indian Ocean, in the equatorial and tropical zones of the Atlantic (Equatorial countercurrent, North Equatorial, Guinean and Guianan

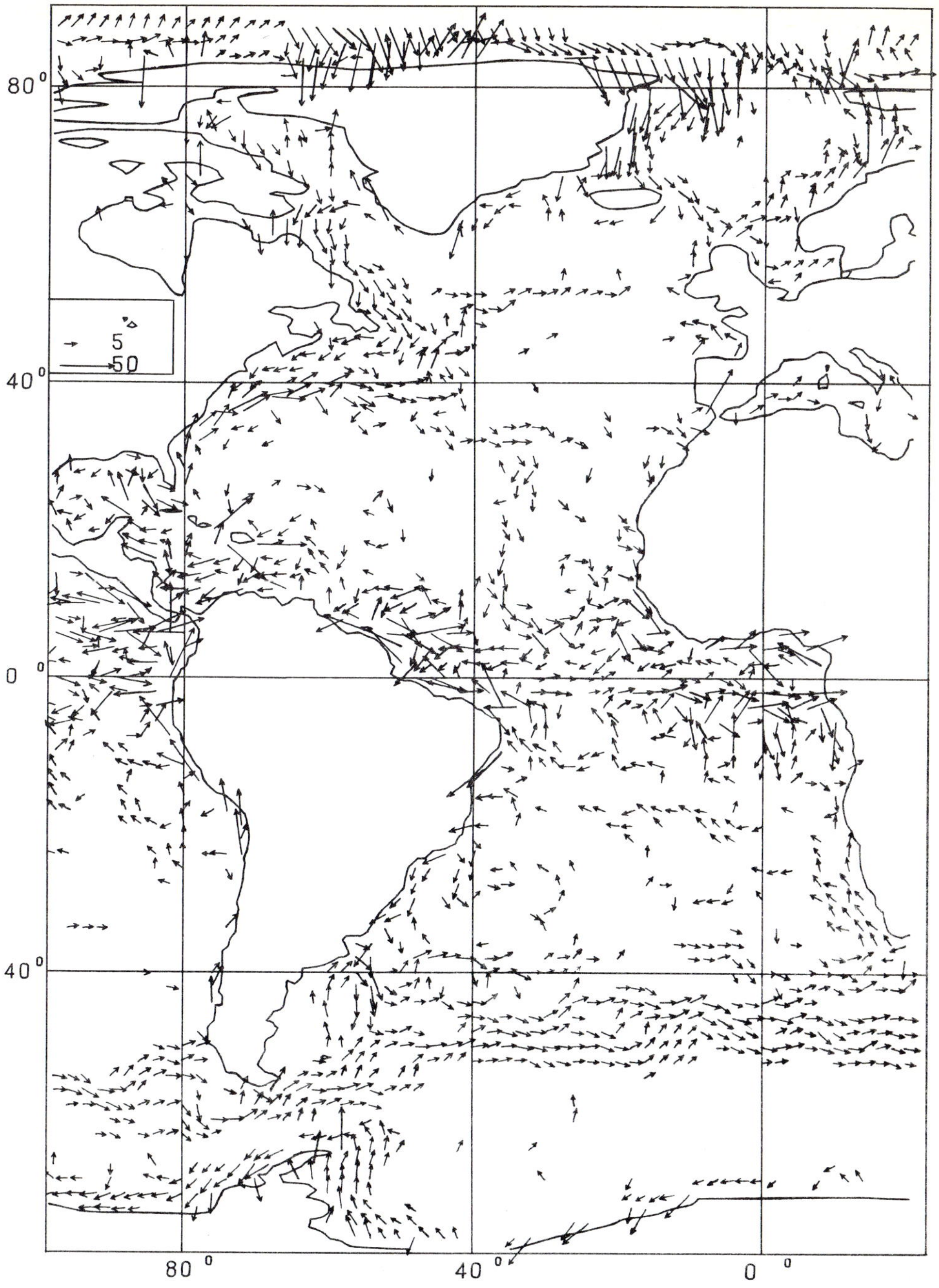

Fig. 31. The sea-surface gradient currents of the Atlantic Ocean in the winter season

currents) and in the trade currents in all oceans. The seasonal variation of total surface currents, related to the seasonal variation of both density and wind fields, which is pronounced in the drift current field, is more substantial. This issue will be discussed in the next section.

The comparison of the results obtained with those published earlier for the summer season and multi-year mean was made in the previous section.

The differences in the Indian Ocean are most significant, since seasonal maps are of a monsoon nature. The major discrepancies are observed in the northern ocean. Thus, the dynamic topography of the multi-year mean annual of the Arabian Sea (Burkov 1980) is more similar to our maps for the summer season. Nearer to the equator (0°–5° N) the current scheme in Burkov (1980) as a whole becomes more similar to that for the winter season, although the Somalian current is directed northward as in the summer season. In the Bengal Bay the general situation is closer to that observed in winter.

The calculations for the winter season are very poor. The maps of the dynamic surface topography of the Indian Ocean in summer and winter seasons have been published in Neuman (1970). The comparison with the summer season was performed in Section 5.2. A number of differences are observed in maps for the winter season. According to Neuman (1970), there is only an anti-cyclonic circulation in the Bengal Bay, while in our maps there are anti-cyclonic (19°–24° N), cyclonic (5°–19° N) and anti-cyclonic (2°–5° N) circulations. The differences in the southern ocean are less important and attributed to our detailed maps.

The map of the geostrophic water circulation at the World Ocean surface for the winter season, shown in Stepanov (1974), agrees generally with our results.

5.3.5 Seasonal Variations of the Upper Layer Water Circulation of the World Ocean

The seasonal variations of the surface topography and gradient currents of the World Ocean, i.e. pure density-driven currents, were discussed in the previous section. It was shown that the climatic variation of these fields is significant only in areas with a monsoon circulation regime. The applied density fields are of a seasonal nature only in the upper 200-m layer, below it — annual average fields. So, the seasonal variation of currents (both gradient and total, with allowance for the direct force of wind) decreases rapidly with depth, being much pronounced in the upper layer. It is very characteristic of the total currents, which are strongly related to the effect of the wind-drift component, which depends upon the wind field and is of significant importance only in the thin, upper layer. The seasonal variations of the total currents appeared as a whole more substantial. The concept of the sea surface current structure for all four seasons can be based on Figs. 13, 32–38.

Let us compare these results with those published earlier.

The map of the surface water circulation of the Atlantic Ocean for the winter season is given in Bulatov (1971). Almost all known gyres are shown in this map. The difference to Fig. 32 is that gyres in tropical and equatorial zones extend over longer distances (from the African coasts to South America).

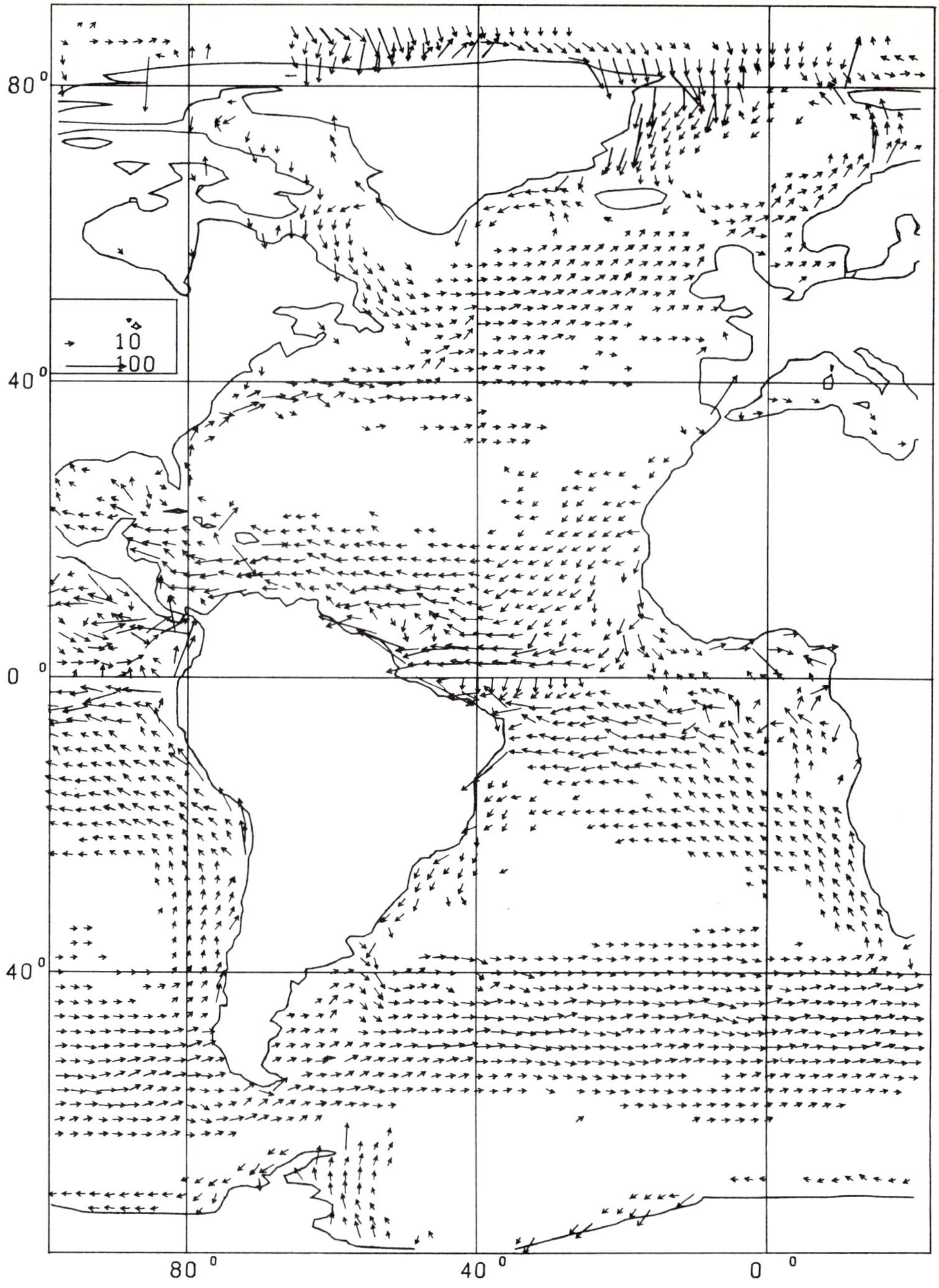

Fig. 32. The sea-surface currents of the Atlantic Ocean in the winter season

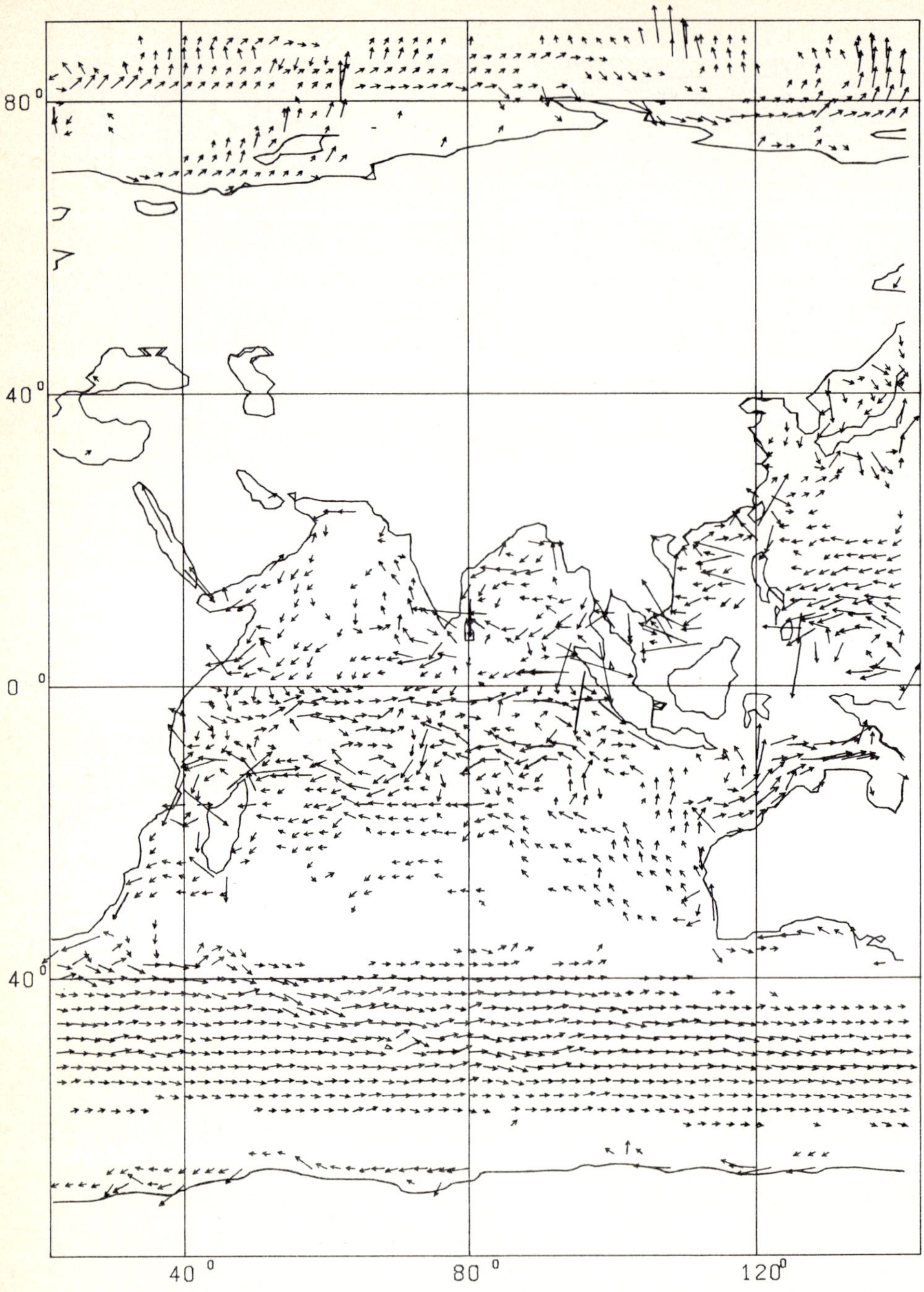

Fig. 33. The sea-surface currents of the Indian Ocean in the winter season

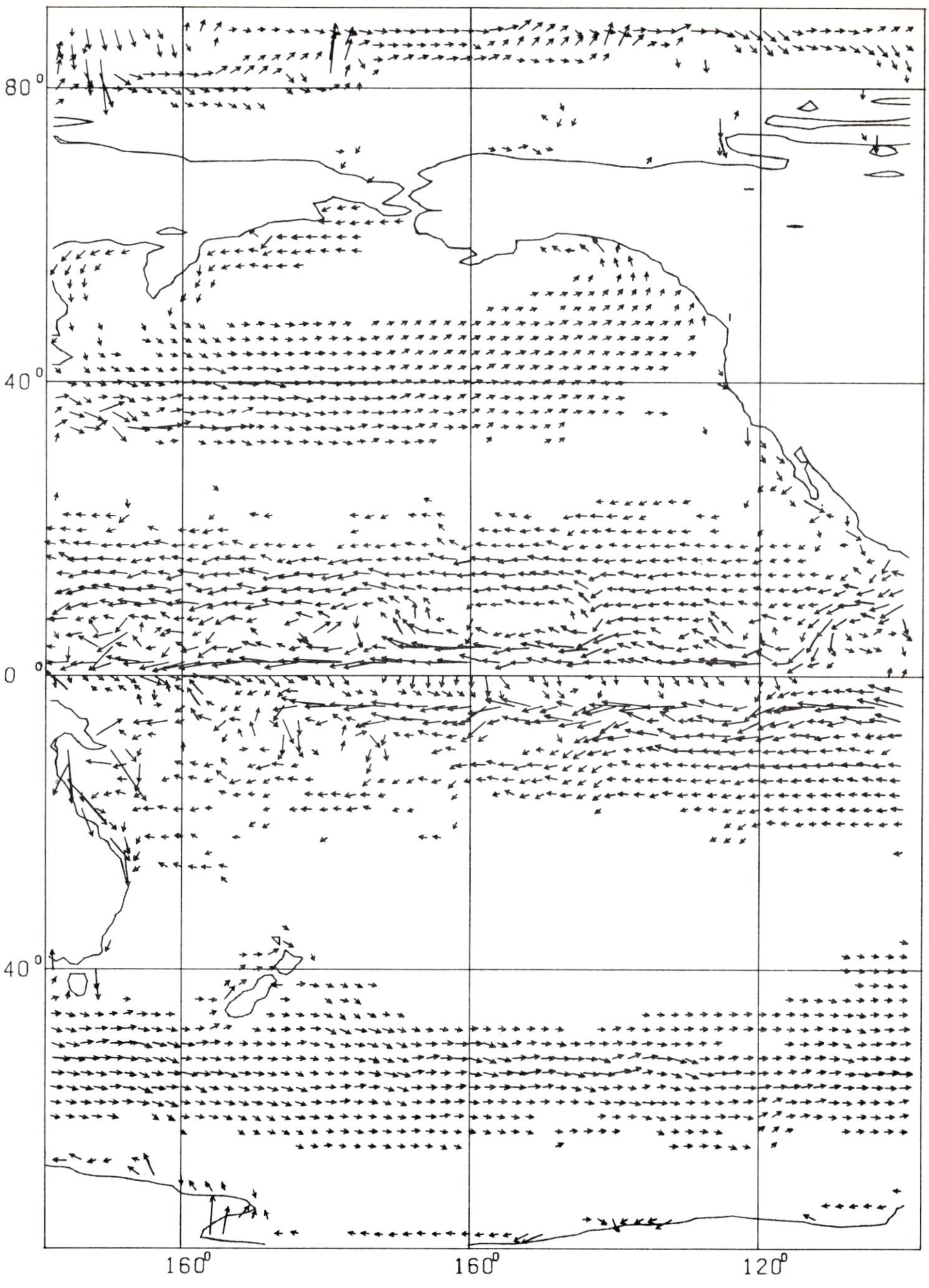

Fig. 34. The sea-surface currents of the Pacific Ocean in the winter season

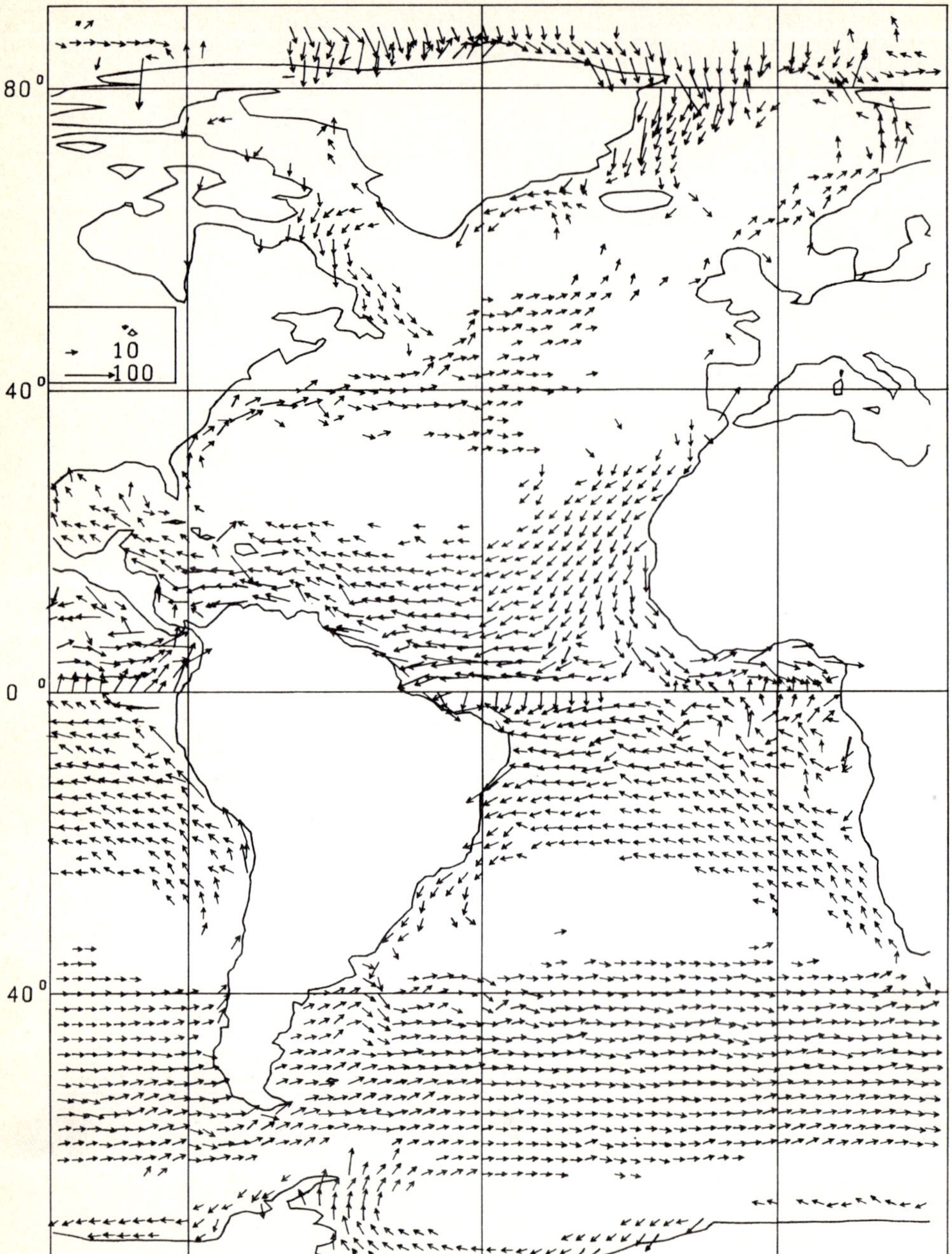

Fig. 35. The sea-surface currents of the Atlantic Ocean in the spring season

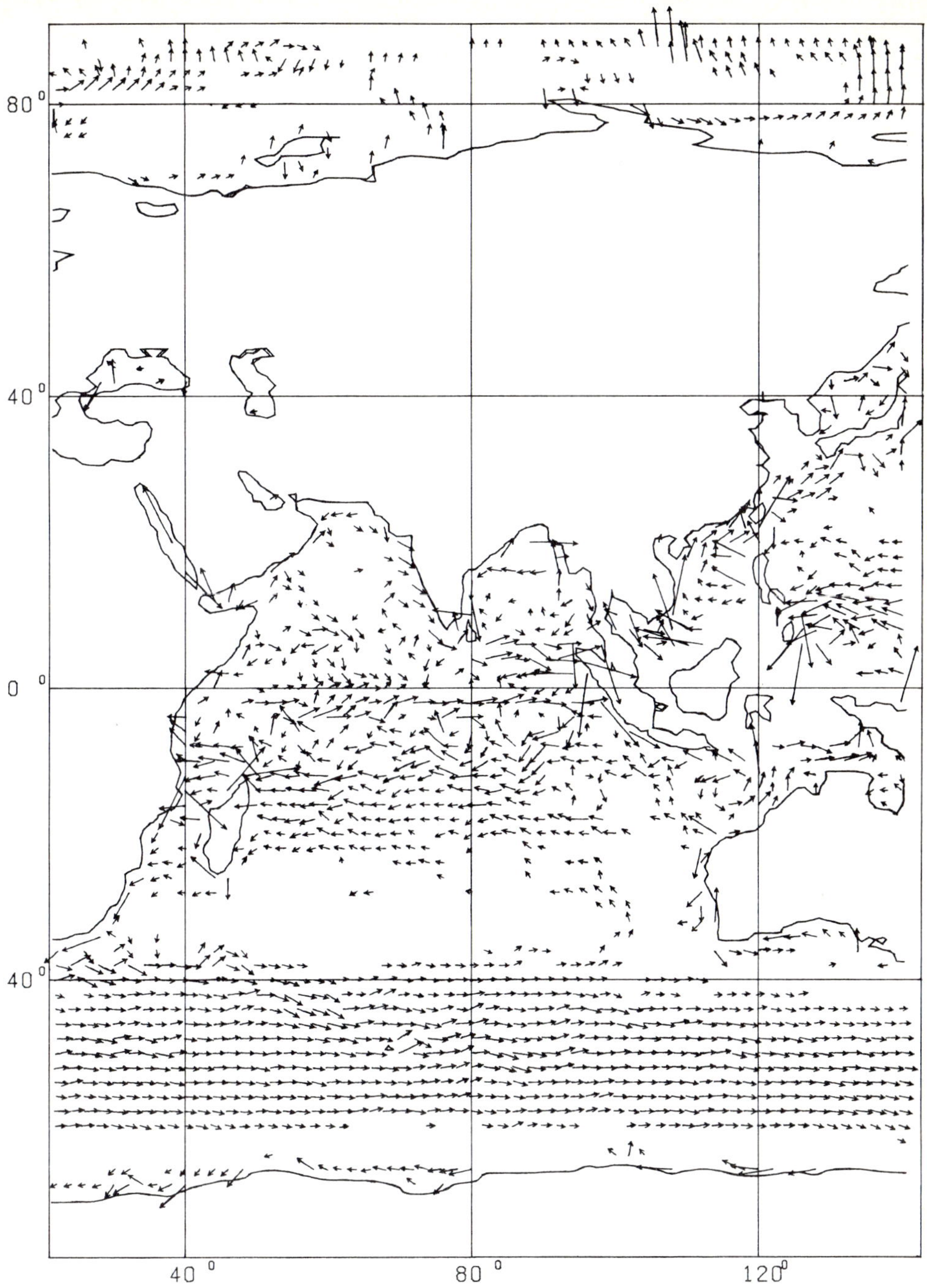

Fig. 36. The sea-surface currents of the Indian Ocean in the spring season

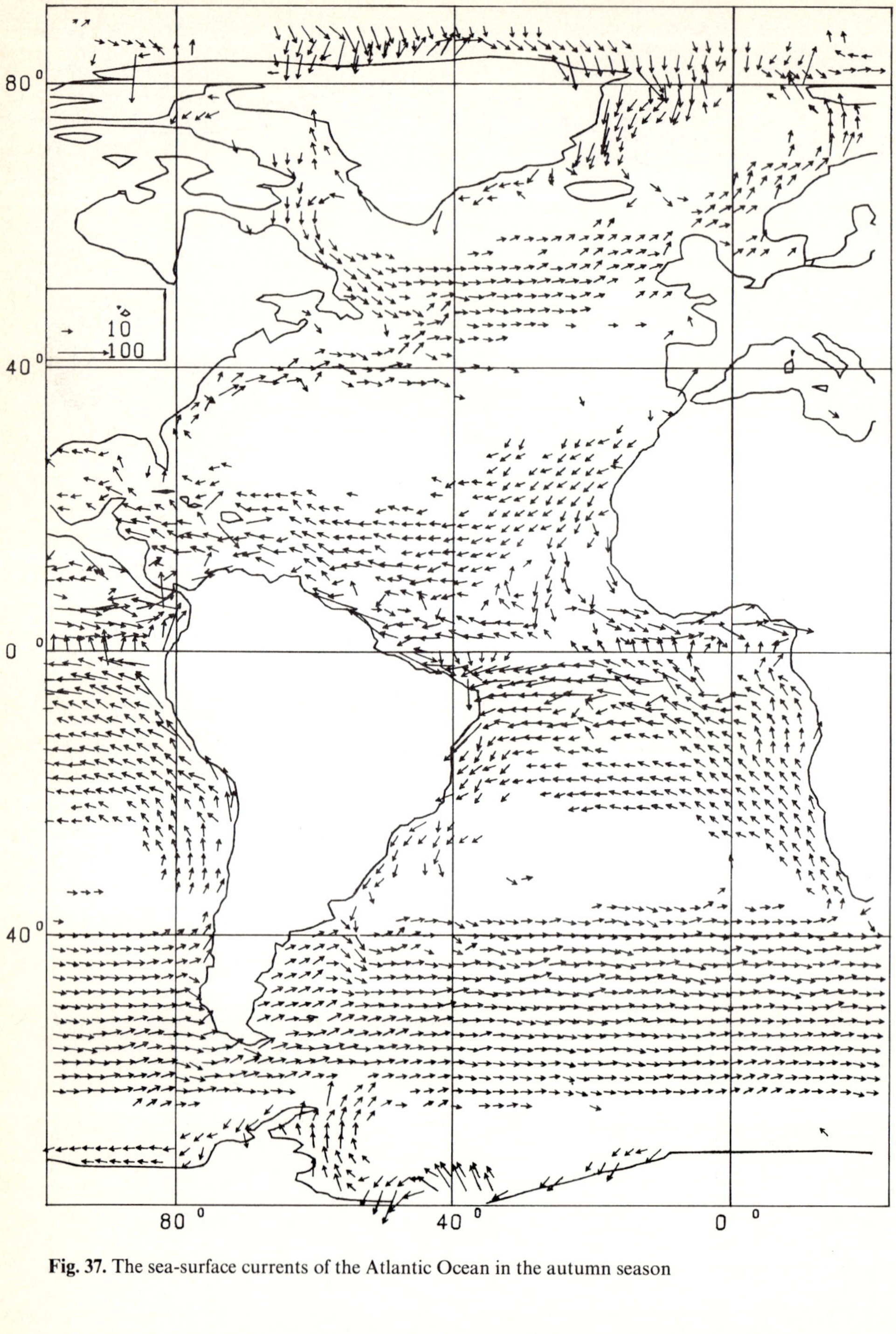

Fig. 37. The sea-surface currents of the Atlantic Ocean in the autumn season

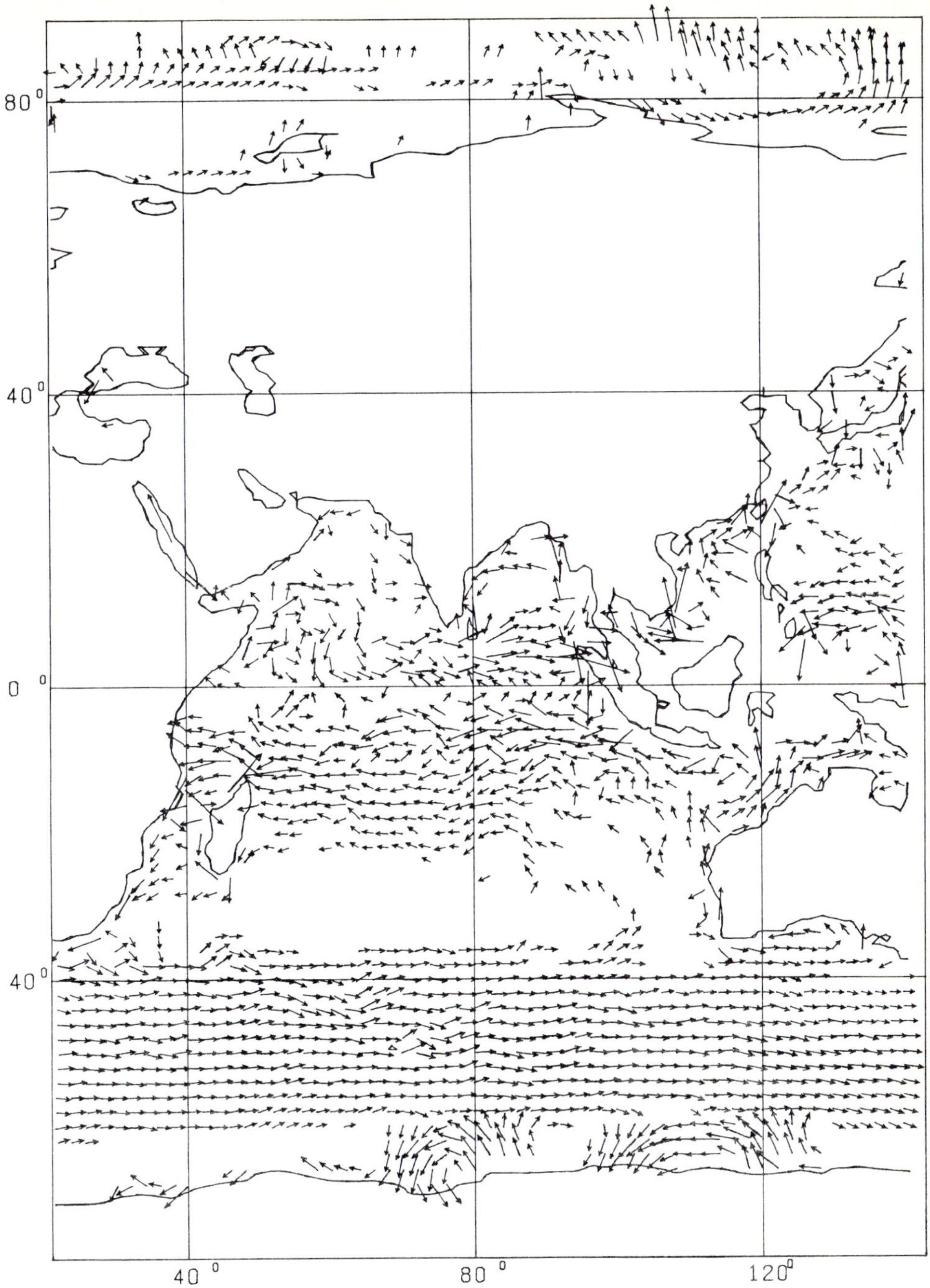

Fig. 38. The sea-surface currents of the Indian Ocean in the autumn season

Furthermore, the northern tropical CG is located much further to the north (by 10°). The location of vergences in these gyres differ correspondingly. Labrador, Norwegian and Falkland currents are absent in Bulatov (1971).

The map of the surface current of the Pacific Ocean for February is given in Burkov (1968). This map, as well as Fig. 34, indicate all known major currents and gyres, although three of them are in different locations. The northern tropical CG is much longer compared with that in Fig. 34. It extends almost from the eastern oceanic coasts to the western coasts. The Equatorial countercurrent has the same location. The northern tropical ACG is extended: it stretches from the coast of America to 140°W. And the southern tropical CG, in contrast, is reduced in size and covers only the western ocean up to 120°W. It is natural that the length of the corresponding vergences of each gyre differs. Besides, the Bering Sea currents (The North Bering Sea and Kamchatka) are absent in Burkov (1968) and the East Australian current exists only in its very southern edge. In general, the velocity values are similar to those obtained by the authors.

The map of the surface currents of the Indian Ocean for the winter season is shown in Neuman (1970). As a whole, the map is similar to Fig. 33. However, some differences do exist. So, in the Northern Hemisphere unlike Neuman (1970), the monsoon current only partially changes its direction to the north near the Maldive Ridge. The rest of the current moves still westward until joining the Somalian current. According to Neuman (1970), the Bengal Bay has ACG, which covers the whole bay. In Fig. 33 it occupies only the northern bay, the southern part is occupied by the flow directed westward. The Equatorial countercurrent, according to our data, is more powerful and longer. According to Neuman (1970), the South Indian Ocean current crosses the ocean from southern Madagascar island to northern Australia. In Fig. 33 this current is separated from the ACG in the region 90°–100°E and 37°S and flows to the north. The location of the southern subtropical convergence differs correspondingly. The Mozambique current in Neuman (1970) flows northward, while in Fig. 33, southward. The velocity values are generally close, excluding areas of the Mozambique and Somalian currents, as well as that of the Bangal Bay, whereas velocities in Neuman (1970) are much lower.

The map of the water surface circulation of the World Ocean in the winter season is given in Stepanov (1974). Qualitatively, the map is also similar to Figs. 32–34. However, there are some differences in details. The northern tropical CG in the Atlantic is not very pronounced in Stepanov (1974); the northern subpolar CG is not closed, i.e. the currents responsible for its generation are absent (East-Greenland and West Greenland currents); the Antilles current is directed eastward. The situation of the Indian Ocean circulation is very similar to that shown in Neuman (1970). The discrepancy with Neuman (1970) was already discussed above. The northern tropical CG and ACG in the Pacific Ocean in Stepanov (1974) occupy a much larger space and extend over the whole ocean. The southern tropical CG is in contrast reduced, and occupies the space between 120°W and the North American coasts. There is no East Australian current directed to the south along the eastern Australian coasts.

All maps cited above, compiled with the help of a dynamic method (or its modified versions, Burkov 1968), are based on the multi-year mean density fields.

Many disadvantages characteristic of them are attributed to large-scale averaging and drawbacks of the dynamic method.

In the spring season (Figs. 35 and 36) all major gyres and currents responsible for their generation are preserved. Sometimes changes in velocity are observed. The velocity increased in the North Equatorial current of the Atlantic [10–60(35)], Mindanao [30–100(75)] California [15–50(25)], Canary [10–27(20)], Gulf Stream [10–50(40)] and to the south of Bengal Bay [15–100(45)]. A decrease in velocity is observed in the North Bering Sea current, in the deeper parts of the Somalian current, directed southward [10–20(17)], in Bengal [10–20(15)], East Australian [10–100(45)], East Greenland [10–47(25)], West Australian [10–50(20)], West Greenland [0–47(20)], New Guinean [30–90(60)], Norwegian [10–23(20)], Oyashio [0–13(10)], North Atlantic [10–30(15)], Kamchatka [0–15(10)], South Atlantic [10–20(15)] and South Equatorial current of the Atlantic Ocean [10–45(20)].

The change in the current direction is primarily observed in the northern Indian Ocean, which is under the effect of summer and winter monsoon (spring and autumn are periods of reconstruction). In winter the monsoon current in the Arabian Sea is directed northward and north-westward along the coasts of India, forming part of the CG. At that time the major part of the Somalian current (located further in the sea) is directed southward, and only its narrow part flows along the coast northward. In spring the monsoon current is directed south-eastward along the Indian coasts, forming part of the ACG. In the Somalian current there is still a fast but narrow flow along the coast directed northward, although velocities of the flow directed southward decrease significantly. In the Bengal Bay in winter the major flow is directed westward and a very narrow flow to the south of the Bay is directed eastward. In spring the eastward flow becomes the major one, and its velocity increases substantially.

The Oyashio (Kuril) current in the Pacific changed its south-east direction to north and north-east.

The position of front divergences and convergences remained without significant changes.

At present for the spring season there are no maps of currents for either the World Ocean as a whole or for individual oceans.

Almost all major gyres and currents (Figs. 13–15) are preserved in the summer season. The North Bering Sea (transverse) current is not observed. A change in velocity is observed in many currents. The velocity dropped in Alaska [0–20(10)], Bengal [10–20(10)], Mindanao [25–75(50)], North Bering Sea (10), South Atlantic [10–15(10)] the North Equatorial current of the Atlantic [10–15(25)] and in the East Greenland current of the Norwegian Sea [10–60(35)]. The velocity increased in the Somalian current (50), Florida current [15–45(25)], in the current of the northern Arabian Sea [10–45(25)], in the South Equatorial current of the Atlantic [10–50(30)], Kuroshio [15–100(48)], New Guinean [30–100(70)] and North Atlantic [10–30(20)] currents.

The seasonal variations in the current direction are only observed in the northern Indian Ocean. In summer the south-west monsoon prevails here, which determines in principle the direction of the current. The whole northern Indian Ocean to the equator is covered by the easterly directed flow. The Somalian current is directed north-eastward.

The seasonal variation of currents in the tropical zones of other oceans is rather significant. The Equatorial countercurrent strengthens substantially in the Atlantic ocean in the summer season. Width and length of the flow (in winter it is observed east of 20° W and in summer, 40° W) as well as current velocity (from 50 cm s^{-1} in winter to 70 cm s^{-1} in summer) are enhanced. The intensity and size of the tropical CG increase correspondingly. The North Equatorial current, and, particularly, the Canarian current, the jets of which penetrate into the Guinean current in the summer period, thus supplying the Guinean current, are also intensified. The South Equatorial current becomes more powerful, which is considerably pronounced in the eastern ocean, near the equator. The Bengal current and the western flow passing at 2°–4° N in the western ocean become weaker.

In the tropical Pacific the Equatorial countercurrent is intensified in summer, the tropical CG is considerably pronounced (it is practically unobserved in winter). The North Equatorial, especially the Californian current, is also reinforced. The summer intensification of the South Equatorial current is considerably pronounced in the western ocean, where it is observed as a system of gyres unlike the ordered flow in the summer season. The current near the coast of Chile is substantially intensified, as well as west flows passing at 2°–4° N and 2°–4° S.

The variation in the position of fronts, divergences and convergences in the summer season are not significant except for a few cases. So, in the Atlantic Ocean the northern tropical CG covered a larger space compared with that in winter and spring (beginning with 40° W) and drove the subtropical ACG northward. The northern tropical front in the central Atlantic ocean displaced northward (from 8° N to 15° N). The length of the northern tropical ACG and the northern tropical convergence increased correspondingly. The southern tropical CG became narrower and more pressed to the equator. The corresponding divergence is observed along the whole equator in the area of 2°–3° S. The northern tropical CG in the Pacific Ocean increased insignificantly. Its extent to the west increased approximately by 5°. In the Indian Ocean the southern tropical divergence displaced to the north closer to the equator (2°–5° S).

The comparison with results published earlier for the summer season was given above in Section 5.3.1.

All major gyres and currents are also preserved in the autumn season (Figs. 37–38). From all known currents only the Portuguese current is absent from the map (its velocity is less than 10 cm s^{-1}). The velocity of many currents changed. Almost in all cases the velocities became similar to those in the winter season. The velocity increased in the Alaskan [10–15(10)], Bengal [10–28(20)], West Greenland [10–48(25)], Cape Horn [15–30(25)] currents, Equatorial countercurrent of the North Atlantic [10–50(35)] and in the Southern Pacific [10–35(20)] and Indian [10–30(25)] Oceans as well as in the South Equatorial currents of the Atlantic [10–75(40)], Pacific [10–60(40)] and Indian [10–50(35)] Oceans. The velocity of the East Greenland current of the Norwegian Sea [15–60(40)] increased. The velocity of the Californian [10–35(20)], Irminger [10–26(10)], Gulf Stream [10–47(30)], Canarian [10–20(17)], Kuroshio [10–75(40)], Florida [10–40(15)], Formosan [30–100(40)] currents and in the North Equatorial countercurrent of the Atlantic [10–30(20)] decreased. The velocity in the northern Arabian Sea

[10–90(18)] and in the deeper part of the Somalian current decreased as well (45 cm s^{-1}).

The changes in the position of fronts, divergences and convergences in the autumn season are insignificant. The sizes of the northern tropical CG changed. Now they cover almost the same position as in the winter season (from 20° W to the African coasts). The length of the northern tropical divergence of the gyre decreased. The sizes of the northern tropical ACG and the northern tropical convergence located in it lessened as well.

At present, there are no published maps of the World Ocean currents or individual oceans for the autumn season.

In these calculations, as it was mentioned in Section 5.2, the sea surface wind stress fields, calculated from the atmospheric pressure fields with the help of the Ackerblom linear model, are used. The input parameters were selected in such a way that the wind stress means were close to those given in the known Hellerman Tables (1967), first used for calculating currents in the summer season (Brekhovskikh 1980) (model density and bottom topography fields are similar to ours). The advantage of the Hellerman Tables is that they are compiled on the basis of real data on wind velocity and direction. The disadvantages, with respect to these calculations, have already been mentioned earlier and they were first and foremost manifested in the coastal currents. So, according to Brekhovskikh (1980), unlike the results given, Peruvian and Californian currents in the Pacific Ocean are very weak and are not expressed as jetlike currents; the Falkland current in the Atlantic Ocean penetrates far into the north (north of 50° S); the direction of the Florida current is distorted and that of the Canarian is very weak without its jets penetrating into the Guinea Bay; the Agulhas current is absent in the Indian Ocean. In all the oceans the Antarctic circumpolar current (ACC) has, according to Brekhovskikh (1980), a strong meridional component close to the zonal one, and the circulation near the Antarctica is distorted due to lack of wind data. In low latitudes the major characteristics of the circulation are the same, although there are a number of differences. According to Brekhovskikh (1980), the equatorial divergence of the currents in the Pacific and Atlantic Oceans is much more pronounced, and at the same time the Equatorial countercurrent is much less pronounced, having a very strong meridional component. In the Indian Ocean the currents are on the contrary almost of a zonal nature. Furthermore, the tropical CG near the eastern Pacific coast, and the Equatorial (South Equatorial) countercurrent in the Indian Ocean are absent according to Brekhovskikh (1980). So, the scheme for currents in these calculations have become more realistic also at low latitudes, although the application of the wind stress, calculated from the atmospheric pressure with the help of the Ackerblom model, results inevitably in certain errors at these latitudes. It is obvious that errors at low latitudes are also due to the nature of the diagnostic model (the absence of the non-linear summands) and very large-scale averaging of the wind fields.

The analysis of the maps obtained is indicative of the available seasonal variations in the climatic circulation of the World Ocean. In most World Ocean currents they are only of a quantitative nature. Some currents have maximum velocities in the summer season (Gulf Stream, Californian, Canarian, Kuroshio, Florida, Equatorial countercurrent of the North Atlantic and the current in the

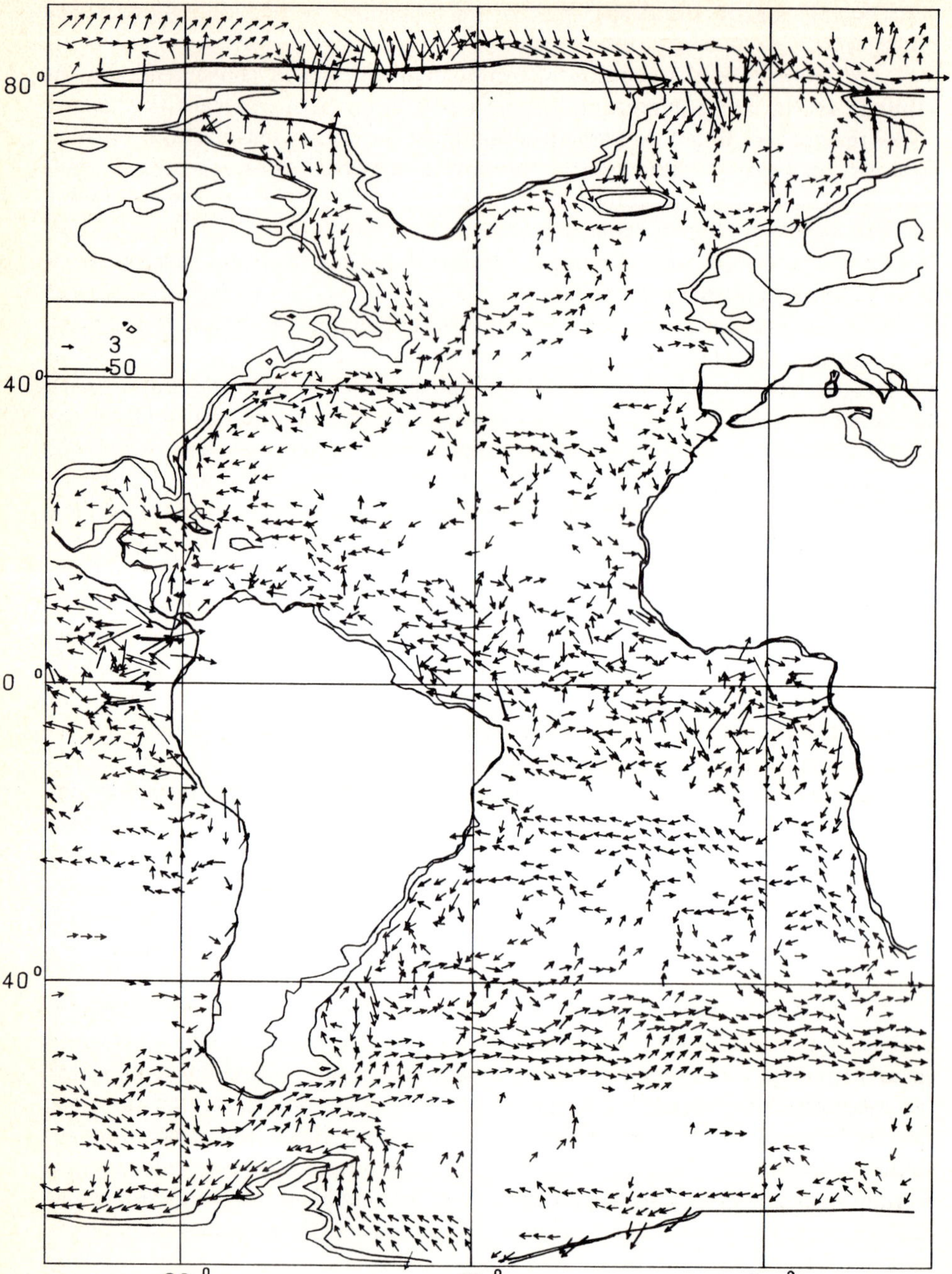

Fig. 39. The currents of the Atlantic Ocean at 250 m depth in the winter season

northern Arabian Sea); others, in the winter season (Bengal, West Australian, West Greenland, Irminger, New Guinean, Norwegian, Oyashio, North Bering Sea, Kamchatka, South Atlantic); and in some currents the velocity remains practically constant throughout the year.

Qualitative seasonal variations are primarily characteristic of the northern Indian Ocean, which is under the influence of a strong monsoon.

The subtropical ACG and tropical CG in the Northern Hemisphere gain in size and displace northward in the summer season in comparison with the winter season. The northern tropical front displaces in the central Atlantic by 7° to the north (from 8° to 15° N), and in the Pacific Ocean by 5° (from 9° to 14° N). The southern tropical CG in the summer season displaces towards the equator, and the corresponding divergence displaces in the Atlantic Ocean by 3° (2°–5° S), in the Pacific by 1° (2–3° S), in the Indian Ocean by 4° (6–10° S) to the west of 70° E and by 2° (3°–5° S) to the east of 70° E.

The seasonal variations of the surface (total) climatic currents of the World Ocean were considered above, and the previous section dealt with the surface gradient currents. The latter characterizes the structure and variations of the subsurface climatic circulation. The seasonal variations in the climatic currents attenuate very fast with depth, and at 250 m (Figs. 19, 39) they are not very pronounced. Even the absolute values of the velocity are similar. This refers not only to winter and summer seasons, but to the transient seasons spring and autumn. To a certain extent such a structure is stipulated by the fact that the used density fields are of a seasonal nature only in a narrow upper layer of 0–200 m, if the specified fields are below the multi-year mean. The wind impact, the most variable characteristic in time, weakens rapidly with depth and the temporal variation can be observed only through the barotropic velocity component, related to the ocean free surface topography. The time variation of the latter is considerably pronounced only in the northern Indian Ocean, in other areas it is insignificant. The same can be noted of the circulation at a depth of 50 m, i.e. weak seasonal variations in climatic currents of the subsurface and intermediate layers are related to both a relatively small thickness of the upper seasonal layer of the density and to weak seasonal variations of the density inside the upper layer.

The velocity of most currents at 200 m decreases significantly both in the winter and summer seasons. The Alaskan, Peruvian, Kamchatka, West and East Greenland currents as well as the Equatorial countercurrent in the Indian Ocean, where velocities drop insignificantly, are exceptions. In most currents the velocity does not exceed 20 cm s^{-1}. The velocities of New Guinea (50), Peru (45), Mindanao (35), Guinea (30), the East Greenland current of the Norwegian Sea (30), Formosa (25), Gulf Stream (25), Kuroshio (25) remain the most significant.

5.3.6 The Structure and Seasonal Variations of the Vertical Circulation of the World Ocean Waters

When studying the general circulation of ocean waters, of great importance are the directions and intensity of vertical motions. Despite the low value of the vertical component of flow velocity, its impact on transport of momentum, heat and salt is quite essential.

The vertical water circulation is of special importance for the removal of nutrient salts from deep waters to the surface layer, and when chemical substances and radioactive contaminants penetrate into deep water. At present, because of difficulties in determining the vertical velocity component with the instrumental method, there are no experimental data to study the vertical velocity fields in the ocean. So, numerical oceanic circulation models are still more important for studying vertical motions than for horizontal motions. The application of the diagnostic model (Sarkisyan 1977b) made it possible to calculate the vertical velocity field over the whole thickness of the World Ocean waters, from surface to bottom (Bulatov et al. 1975). The major peculiarities of the vertical circulation are the following. First, there is a strong increase of the vertical velocities in the 10° band on either side of the equator, where they are one to two orders higher than over the rest of the ocean water area. Second, the vertical velocities increase with depth, which are one to two orders higher in deep waters than in the surface waters. Third, there is a strong inhomogeneity of the vertical velocity fields in deep waters, which is manifested by the elongated (in the meridional direction) bands with the alternating up- and downwellings stipulated by the bottom relief.

Let us first and foremost consider the structure of the vertical climatic circulation obtained in these calculations for the summer season. At a level of 50 m (Fig. 40) the drift component prevails in the vertical velocity. So, the zones of up- and downwelling alternate as in Bulatov et al. (1975), in accordance with the large-scale zones of the atmospheric circulation. In the moderate and tropical region in both hemispheres, where there is cyclonic wind circulation and precipitation surpasses evaporation, the upwelling predominates. At subtropical latitudes, where anti-cyclonic wind circulation and evaporation predominates, the downwelling of warm and salt water is characteristic. The water rise is manifested in the major areas of the coastal upwelling. A narrow band of water lowering, corresponding to the equatorial anti-cyclonic gyres, extends to north of the equator.

The gradients component dominates in the subsurface and intermediate layers, considerably changing the vertical velocity field structure. If at a level of 50 m the area of water descending predominates, then the area of water rise increases with depth. A continuous band of water rise is preserved in the Antarctic zone. The areas of water descending wedge into the zones of water rise in the moderate and tropical latitudes, and "tongues" of water rise in the areas of water descending.

At a level of 1000 m and in the deeper layers the areas of water rising and descending alternate regularly, extending in the meridional direction in strips of 2°–10° width. It is stipulated by the bottom relief, which is characterized by the alternate abyssal plains and ridges extending meridionally. The water inside the abyssal plains circulates like gyres extending from north to south.

Below 1000 m the area of the water rise increases relatively, in agreement with the results given in Bulatov et al. (1975). In the near-bottom layer the oceans are cut by the ridges and continents to individual abyssal plains. Both cyclonic and anti-cyclonic circulation takes place in these abyssal plains (the cyclonic circulation

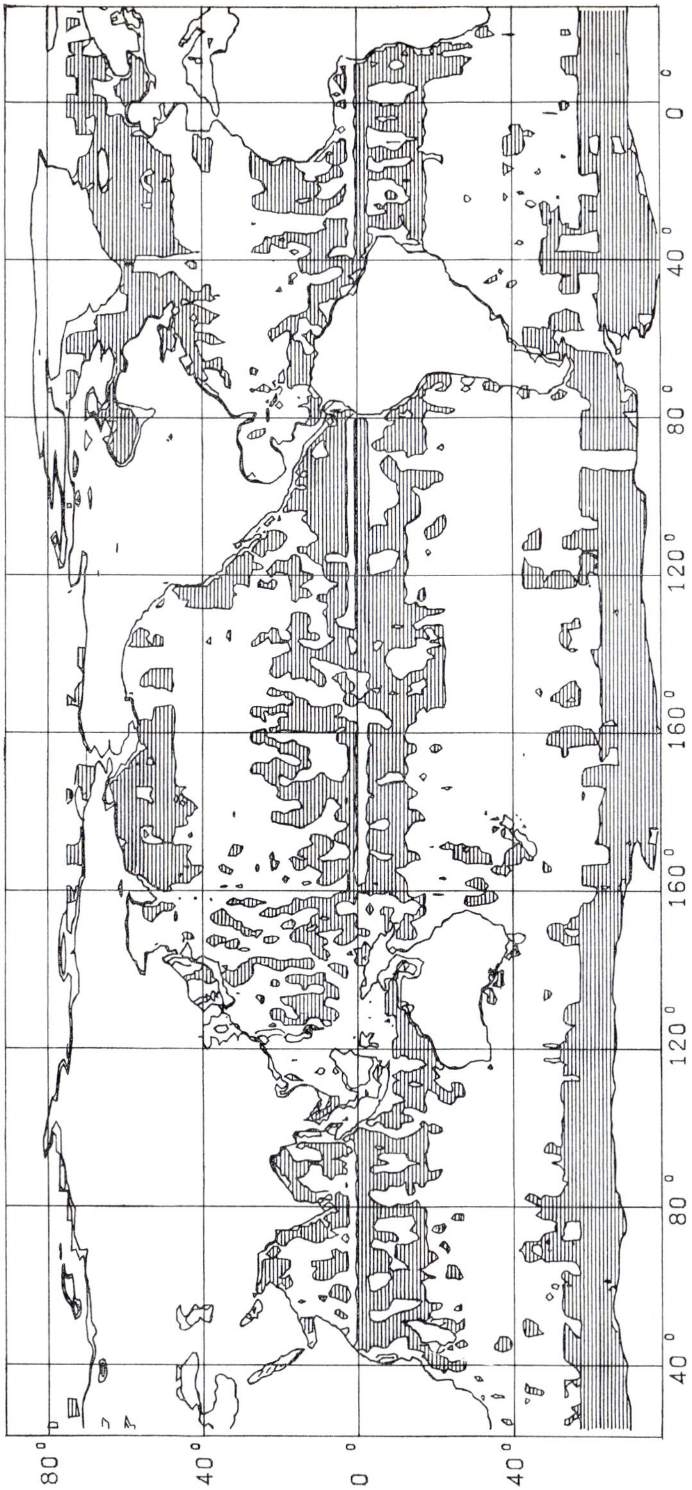

Fig. 40. The vertical velocity at 50 m depth in the summer season. The upwelling areas are *shaded*

prevails). Thus, the wind conditions significantly effect the formation of the vertical velocity field in the surface layers, and beginning with the intermediate and particularly in the abyssal and near-bottom layers, the effect of macroforms of the bottom relief prevails.

In the geostrophic approximation the vertical velocity can be given as:

$$w = \frac{\beta}{l} \int_0^z v d\xi \,. \tag{5.8}$$

It follows that the vertical velocity reflects the flow of the meridional component in the layer from the surface to the level under the study. So, narrow meridional bands of water rising and descending in the field w indicate the sequence of the meridional flow zones of different sign with longitude.

Let us consider the seasonal structure of the vertical circulation. In the winter season (compared with summer) the zones of water rise in the Northern Hemisphere at a level of 50 m broaden to the south, which is related to the increased cyclonic circulation of the wind and lowered heating and evaporation at subtropical latitudes. At tropical latitudes the boundary of the water rise zone displaces also to the south, particularly in the Pacific and Indian Oceans. The zone of the Antarctic water rise is slightly reduced. The vertical circulation in the northern Indian ocean endures considerable changes.

Seasonal differences subside rapidly with depth, and at 250 m the seasonal differences are insignificant. Only the northern Indian Ocean and areas of subpolar cyclonic gyres in the Pacific and Atlantic Oceans differ considerably. In the Pacific Ocean to the north of 40° N, in the Atlantic to the north of 50° N, these areas in the winter season are practically continuous areas of water rise.

At a level of 1000 m there are essentially no differences between the seasons. So, seasonal variations of the climatic fields of the vertical velocity are observed only in the surface layer and are mainly attributed to the atmospheric pressure variations.

5.4 The Hydrodynamic Adjustment of the Ocean Temperature, Salinity, Density and Flow Fields

The results of the diagnostic calculations of climatic currents are most reliable in the upper 1000-m oceanic layer. The amount of the available temperature and salinity data reduces drastically with depth, particularly below 1000 m (which is manifested by the climatic TS curves), so the results of the diagnostic calculations are less reliable. The situation is worsened since the density gradients diminish with depth as well, which results in the further increase of relative errors. In deep layers these errors can, in addition to quantitative changes, lead even to qualitative changes in the velocity fields. And as a result important integral characteristics, such as meridional heat and mass transport, are rather inaccurate. So the quality of the diagnostic calculations is in many aspects related to the quality of the density fields used. The requirements for the density field depend greatly on the hydrological peculiarities of the area. For instance, in the equatorial zone, an area with rather weak density gradients with strong flows, the requirements for the density field should be higher than at mid-latitudes. The problem is of great urgency when calculating on a grid set with high resolution (1° and less), needed for the description of jetlike currents, the most important elements of the ocean circulation. The available climatic temperature and salinity fields with such an averaging scale are statistically inadequate and are characteristic of a high "noise" level which is attributed to both available synoptic and other small-scale disturbances in the fields and to erroneous observations. To filter out the "noise" when performing the diagnostic calculation of the World Ocean currents (Sect. 5.2) the cosine filter was used. However, it only partially resolves the problem. Alongside the filtration of the noise, the jetlike currents are considerably smoothed. The issue on the field correlation is still open.

A substantial disadvantage of the diagnostic calculations is that the density fields (temperature and salinity) obtained from the observations do not correlate with wind and bottom topography fields. The situation mostly affects deep currents, where the bottom relief effect is of great importance, and even relatively small errors are significant, as well as coastal currents with a very irregular bottom relief.

To adjust the fields hydrodynamically some diagnostic models use supplementary equations of property conservation. In the models of Stommel-Schott (1977) (β-spiral model) this is the condition of the current isopycnicity. However, almost all known models of this type are, in principle, modified dynamic methods. The work of Mellor et al. (1982), presents an interesting diagnostic calculation.

The robust diagnostic method proposed by Sarmiento and Bryan (1982) might be the most complete approach of all known methods. The system of equations of diagnostic models incorporating equations of motion, hydrostatics and continuity (as well as an equation for an integral stream function) includes as well an equation of heat and salt transport with corrective sources. The equations are as follows:

$$\frac{df}{dt} = \nu \frac{\partial^2 f}{\partial z^2} + \mu \Delta f + \gamma(f - f^*), \tag{5.9}$$

where f, f^* are the calculated and observed temperature and salinity respectively, γ is the proportionality factor, selected on the basis of numerical experiments. In the extreme cases when $\gamma = 0$ we have the prognostic model, and when $\gamma = \infty$, the diagnostic model. The disadvantage of this approach is the uncertainty in the selection of the coefficient γ.

The models based on the primitive equations of motion are the most universal diagnostic models. However, for calculations of the World Ocean currents in the rather small grids needed for the reliable description of the jetlike currents, these models are uneconomic, since they make very high demands on the operative memory of the computer and are very time-consuming. On the other hand, when calculating on a large grid set (with a size of several degrees), the account of the non-linear effects, excluding the equatorial zone, will be principally of a formal character. To lessen the requirements for the computer power it is expedient first to use an irregular grid set. Thus, to describe large-scale currents in the open sea it might be quite sufficient to have a size of 100 km in the equatorial zone, and

25–50 km in the areas of jetlike coastal currents; in the zones of the coastal upwelling, 10–20 km. Second, it is known that while calculating the large-scale currents of the World Ocean, the account of the non-linear effects is far from being necessary everywhere. In the open sea, i.e. the major part of the World Ocean, they can be ignored without any damage. So, the numerical experiments performed on the 1° grid set for the Gulf Stream area (Allakhverdova and Demin 1985) and the Mozambique strait (Demin and Usychenko 1984) indicated that non-linear effects in such scales are of little importance, despite the fact that the Mozambique strait is for example, characterized exclusively by a complex and "mixed" current system. Furthermore, the calculations of currents in the coastal upwelling zone, done on a grid set with a rather small size equal to 10′, indicated that the role of the non-linear effects in this case is insignificant as well (Demin et al. 1983). Only the more detailed calculations performed for the Black Sea coastal zone with a grid size smaller than 5 miles, revealed the importance of the non-geostrophic effects, though only directly in the narrow (relative to the shallow zone) shelf sea zone. In the abyssal part of the sea the balance was quasi-geostrophic. The importance of the non-linear effects is somehow related to the extent of the inhomogeneity of the used density field. On the whole, however, based on the numerical experiments performed at present, the idea is that the non-linear effects should be taken into account when performing the diagnostic calculations of large-scale currents only in the equatorial zone. Bearing this in mind, one can use for diagnostic calculations of large-scale currents of the World Ocean the economic technique based on the application of two models: (1) a relatively simple one to use on the computer, the quasi-geostrophic model for the extra-equatorial areas of the World Ocean, and (2) a model based on the primitive equations of motion for the equatorial zone (and if necessary for other oceanic areas complicated from the dynamic viewpoint). It is desirable that both models be based on one and the same integral function (in the calculations considered here, it is the sea surface topography). The solutions for the extra-equatorial and equatorial subregions are adjusted through the boundary conditions on the overlapping grid sets. It is expedient to use an irregular grid: the grid size in the equatorial zone as well as in other dynamically complicated areas should be much less compared with that in the open ocean of mid-latitudes.

General comments on the calculation technique were made above. Let us return directly to the problem of the hydrodynamic adjustment. So, the diagnostic calculations are characteristic of two significant disadvantages. First, the input density fields (temperature and salinity) practically always contain "noise" effects of non-stationary processes, motions of small subgrid scales, in particular synoptic. This leads naturally to distortions in the fields of the climatic currents to be calculated. Second, the specified density fields do not correlate with the bottom topography and wind fields. This is also a source of error. Use of such hydrodynamically mismatched fields as an initial state in the tasks of monitoring or hydrodynamic weather forecast in the ocean results in a "mixture" of two processes: (1) the process of hydrodynamic model adjustment of specified fields (including filtering of noise of subgrid scales), and (2) model response to the time variations of the boundary conditions. As a result there is a need to design numerical models and to develop a technique which would provide hydrological fields, which, on the one hand, are similar to the observed data and, on the other hand, well adjusted hydrodynamically without small-scale "noise".

The approach to field adjustment, which is here under consideration, is to combine the major (from the practical viewpoint) advantages of the diagnostic models (based on the observed data and so being of direct practical significance) and prognostic models, providing hydrodynamically adjusted filtered fields, and at the same time eliminating such disadvantages as the mismathching of fileds and the presence of "noise" (in the diagnostic models) and idealization and weak gradients of the fields (in the prognistic models). The approach is based on the consecutive application of the diagnostic and prognostic models. The specified temperature and salinity fields (based on observations) and flow fields (calculated by the diagnostic model) are used as the initial approximation in the prognostic model. The integration with the model continues only until the fields fast reconstruction takes place, without reaching the steady state. The process is controlled by fast waves, primarily, by gravitational waves. This stage is called an adjustment. The advective and diffusive processes are of much less importance at this stage. They cannot be yet fully distinguished since the adjustment process is very fast and ends after some days of the model time. The adjustment problem is an initial value problem (Cauchy problem) rather than a boundary value problem. If the integration time is very long, the solution will be "spoiled", differing considerably from the specified data and approaching the prognostic model regime, i.e. the integration time should be limited from the top. To some extent the limit depends on the model characteristics (for instance, viscosity) and basin sizes. The solution can be "spoiled" by being either extremely smoothed or "losing" qualitative peculiarities available in the specified fields. The integration time should be limited from the bottom also, being related to the velocity of major waves, which provide the field correlation, as well as to the basin sizes.

Some criteria are certainly needed to control the adjustment processes. Several possible adjustment criteria were tested in the process of the numerical experiments. The control over the adjustment by the temporal variations of the kinetic energy was the most efficient and simplest one. At the first stage of the integration, i.e. the adjustment stage, the kinetic energy is characterized by the wavelike, fast oscillations, then comes the stage of slower and regular changes of its quasi-monotonous nature.

It seems useful to control the kinetic energy of fast wave motions separately as well (determining them as deviations from mean values for some characteristic time period compared with the time of adjustment, i.e. for several days). The energy of the wave motions, providing the adjustment of mean fields, should tend in time towards zero (or towards a minimum). The control of the adjustment process with the temporal variations of the kinetic energy, total and wave energy, should be supplemented by the control of the total energy of the system, i.e. kinetic and available potential, which should be preserved in time.

As an adjustment criterion one can use criteria used in the prognostic models, in particular tracking the process by the behaviour of the normalized residuals for temperature, salinity and density, e.g.

$$\varepsilon = \min \varphi(t) ,$$

$$\varphi(t) = \frac{f^{n+1} - f^n}{\sqrt{\left(\frac{\partial f}{\partial x}\right)^2 + \left(\frac{\partial f}{\partial y}\right)^2}} \tag{5.10}$$

where φ is the temperature, salinity or density, n is the number of the time step. The criterion is based on the assumption that at the first stage of the integration, i.e. the adjustment stage, the numerator decreases much faster than the denominator (horizontal gradient of the characteristic). At the second stage the numerator does not decrease so rapidly. Its decrease becomes almost the same as that of the denominator. As a result the character of the temporal variation of the curve $\varphi(t)$ (or the envelope of the curve in the case of the wave disturbances) changes in such a way that either some minimum of $\varphi(t)$ or strong variations of its behaviour are achieved. This means that some optimum state is achieved when fields are principally adjusted, and property gradients are not too smoothed by the dissipative factors (physical and computer calculation). Experience in calculations indicates that in the case of a long-term integration, more and more smoothed and weak gradient fields are obtained.

One of the possible criteria of this type of adjustment will be given here. The most suitable, specific type of the criterion which indicates distinctly the transition from the first stage of adjustment (with fast field adjustment and slow smoothing of gradients) to the second one (with relatively slower adjustment of the fields and relatively faster smoothing of gradients) can be found from the numerical experiments. The possible variations can refer either to the nature of the averaging of the function $\varphi(t)$, or to a specific type of denominator. Instead of the horizontal density gradient, the horizontal velocity gradient or a gradient combination can be used. The averaging can be performed over each level rather than over the whole thickness. In this case the section of transition of each level will correspond to a different integration time and will be more noticeable. The tracking of the adjustment process by the temporal variation in the energy and normalized residuals should be supplemented by the reference calculations with a visual viewing of the solution, particularly in the first numerical experiments with a selected, specific model.

It is noteworthy that most sea current models use total mass transport functions as an integral function. The models, in which the sea surface topography is an integral function, are not used so often and, as a rule, only for the diagnostic calculations. Examples of using such a model for solving the adjustment problem are given below. The calculations are performed on rather different grid sets: with a size of 100-8 km in the horizontal and from 100 to 5 m in the vertical. Most of calculations are performed with a model based on the primitive equations of motion.

In the adjustment problems the system of equations of the diagnostic models is supplemented by the equations for temperature and salinity or the equation of the density diffusion. In this case either a Dirichlet problem is solved, when the distribution of characteristics is specified at all boundaries or a mixed boundary value problem, when the distribution of properties is specified at the surface and in the open lateral boundaries, while at the coastal walls and at the bottom the fluxes are equal to zero. If the grid is rough and the depth at the basin boundary is shallow, it is more expedient to set up the Dirichlet conditions at the lateral boundaries, even if it is a coastal wall. The velocity in the open sections of the boundary is specified from the solution of the diagnostic task for the region, including the area under study (the method of "telescopization"). It is noteworthy once again that unlike the prognostic stationary problems (the problems of reaching the steady state) the solution of which is completely related to the boundary conditions, in the adjustment problems, the boundary conditions are of much less importance, although the prognostic model is used at one of the stages. Furthermore, cases are possible when the adjusted fields inside the region are obviously mismatched with the given (non-changeable) field distribution at the basin boundary.

After the adjustment the structure of the hydrological fields is principally preserved, although substantial hydrodynamic "noise" filtration, lowering of minor and weak peculiarities and a relative (not absolute) increase of major dynamic systems take place. As a rule, the correlation of the hydrological fields with the bottom relief increases noticeably. As a whole, the field structure after the adjustment becomes more precise and regular.

Since fast wave processes (first, gravitation waves, initiated by the specified fields "noise") are a major adjustment mechanism, the adjustment occurs relatively fast and ends after several days of the model time. The adjustment criteria were mentioned above. It is worth stressing that the integration time should not considerably exceed the adjustment time determined by these criteria, otherwise the specified fields can be markedly "spoiled" by the model.

5.5 The Diagnostic Calculations of Flows and the Adjustment of the Hydrological Elements of the North Atlantic

Because of great practical concern the area is given particular attention. Furthermore, it had been used as a testing ground before starting the global diagnostic calculations of the World Ocean flows described previously. This section deals with the results of the diagnostic and adjustment calculations performed both by quasi-geostrophic and non-linear models (Demin and Brekhovskikh 1985; Gurina et al. 1983; Sarkisyan et al. 1980; Sarkisyan and Demin 1983). All calculations were performed for the multi-year mean summer season.

5.5.1 The Diagnostic Calculations of Flows by the Quasi-Geostrophic Model

The calculations were performed by the modified quasi-geostrophic model, reducing it to a boundary value problem for the near-bottom pressure ζ'.

In contrast to Kozlov (1973), the transition to the equation for ζ' is more accurate, the near-bottom water stress effect being preserved. As a result on the right-hand side of the equation the following summand appears:

$$\frac{1}{2\alpha\rho_0}\left[\rho_H \Delta H + 2\left(\frac{\partial H}{\partial\theta}\frac{\partial\rho_H}{\partial\theta} + \frac{1}{\sin^2\theta}\frac{\partial H}{\partial\lambda}\frac{\partial\rho_H}{\partial\lambda}\right)\right]. \tag{5.11}$$

The solution was sought according to a scheme of directed differences on the grid sequence. The calculations were performed in three stages. At the first stage the solution was sought at the specified grid size according to the directed differences scheme, i.e. with the first order of approximation. At the second stage the solution was sought in the same manner, but the grid was half the size. The input data were determined by the linear interpolation of the initial data. And finally, at the third stage the corrector (2.29) was constructed. This procedure of increasing the order of approximation of difference schemes is one version of the methods of embedded grids (Kochergin 1978). It is described in more detail in Chapter 2. The solution of difference problems was carried out by the Gauss-Zeidel method. The vertical turbulent exchange coefficient is specified as $10\,cm^2\,s^{-1}$.

The southern boundary of the area under consideration is located at 3.5°N, and the northern one at 72.5°N. In the eastern part of the basin it extends over the shallow waters in the region of the Denish rapids. The grid size, number of levels and specified density and bottom topography fields are the same as those in the World Ocean flow calculations, though unlike the latter, the fields were not preliminarily filtered. The sea surface wind stress fields are taken from the Hellerman Tables (1967) and interpolated to a 1° grid set.

The sea surface topography field coincides principally with that in Section 5.2, although the number of small eddies increased considerably. The sea surface topography slope is about 1.5 m. One of the interesting peculiarities of the sea surface is the pronounced flow of the slope waters along the American coast, spreading further south-westward up to Cape Gatteras.

Compared with Bulatov and Pojarkov (1975) and Bulatov et al. (1980), where the results of calculations on a 5° grid set were considered, and the sea surface topography difference in the Gulf Stream was 1.5-fold less, the jet is much narrower and the velocity is much higher (up to $80\,cm\,s^{-1}$). The basic circulation characteristics coincide qualitatively, although more detailed peculiarities are missed in Bulatov and Pojarkov (1975), and Bulatov et al. (1980). The main reasons for these differences are certainly the grid size and the detailed density field. However, besides the density field quality the peculiarities of the model and the numerical method are of significant importance. In Sarkisyan (1977a, b), where the quasi-geostrophic model is used and the grid size is approximately the same, the sea surface topography differs significantly. The subpolar cyclonic gyre is poorly pronounced, and the slope water flows are completely absent, i.e. the Gulf Stream covers the whole coastal area. One of the main reasons for these discrepancies is a methodological one: during the realization of the numerical model the "baroclinic" bottom relief effect was strongly overestimated, which led to suppressed "anomolous" coastal flows. The sea surface topography map in Holland and Hirschman (1972) where calculations were performed by the non-linear model on a 1° grid, does not indicate the Labrador waters in flow. The sea surface topography difference across the Gulf Stream jet is (similar to our calculations) about 1 m, although other jetlike currents are poorly pronounced. As a whole the sea surface topography is similar to that obtained on a 5° grid set in Bulatov et al. (1980). To a great extent it is evidently related to a strong smoothing effect of the horizontal viscosity (because of an overestimated value of μ equal to $4 \times 10^8\,cm^2\,s^{-1}$).

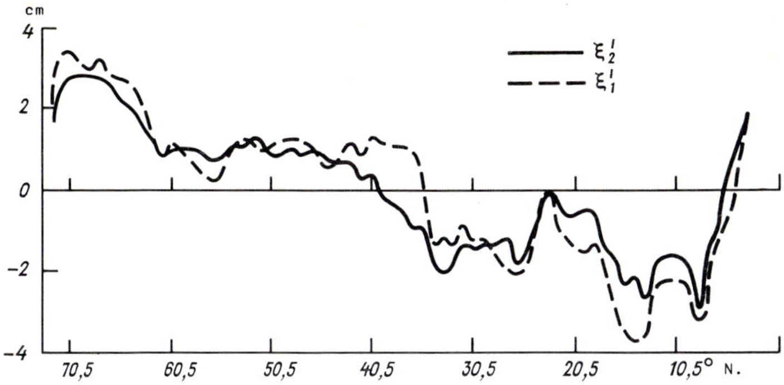

Fig. 41. The zonal distribution of the values of the anomaly of the transformed pressure of the North Atlantic, in cm (*continuous line* denotes the second order of approximation; *dashed line* the first order of approximation)

We have investigated the anomaly of the field of normalized near-bottom pressure ζ' obtained with the help of a corrector. It represents some dynamic characteristic of the abyssal and near-bottom currents. It is of interest that the near-bottom pressure map, unlike the sea surface topography map, does not show the westward intensification. For example, near the African coast the gradient of the near-bottom pressure is much greater than in the Gulf Stream area, i.e. the bottom relief effect competes with the β-effect with success.

The calculations of summand values on the right-hand side of the equation for ζ' indicate that the "weight" of the summand, taken into account in this task and which is indicative of the bottom stress effect, is approximately the same as the "weight" of the wind summand. So, for correct calculations of the near-bottom pressure all three summands should be taken into account.

The solutions for ζ', obtained with the first and second order of accuracy, appeared to be similar qualitatively. However, quantitative corrections are significant (Fig. 41). The differences in the sea surface topography fields are much less pronounced since the major part of the solution, ζ_d, is the same in both versions. It is of interest that the zonal mean profile of the sea surface topography is diametrically opposed to the near-bottom pressure profile: the first is inclined from north to south, and the second from south to north. This illustrates once again that countercurrents are a characteristic peculiarity of the ocean.

When calculating abyssal flows, even slight quantitative corrections are of significance. Flows of the upper 1000–1500 layer undergo only quantitative variations, whereas at 3000 m and lower the discrepancies in some areas are even of a qualitative nature.

The structure of the currents was previously discussed when dealing with the calculations of the World Ocean water circulation. We shall confine ourselves to the area of the Gulf Stream and the Sargasso Sea (Gurina et al. 1983). The velocity was calculated by the quasi-geostrophic relations in the centres of the elementary cells of the specified grid set, i.e. shifted grid points. The procedure is similar to that used in the World Ocean flow calculations, and is of particular importance for narrow jets.

One of the peculiarities of the area is the slope water jet along the American coast. The sea surface topography slope is about 20 cm. The cyclonic system, including several small gyres (the divergence zone), is located between the Gulf Stream flow and the slope water jet. At the same time a number of small anti-cyclonic gyres (the convergence zone) are located at the southern periphery of the Gulf Stream. After the Gulf Stream separates from the coast near the Cape Gatteras and the jet penetrates into the open ocean, it assumes a meandering nature. The most powerful quasi-stationary meander is located at 60° W and might be stipulated by the topogenic effect. The currents reconstruct considerably with depth. Step by step the Labrador jet (slope water) slackens and a recirculation in a north-east direction is observed under it with ever-increasing clarity. The anti-cyclonic gyres generating in the area destruct the divergence zone. The Gulf Stream jet attenuates considerably as well. Its pressure difference decreases almost four times in the intermediate layers. The whole system of the subtropical anti-cyclonic gyre decreases in dimensions and becomes more closed. As a result there is no water flow to the equator as is observed in the subsurface layer. In deep-sea water the anti-cyclonic gyre is destructed and turns into a broad cyclonic system. Its west and north-west peripheries are represented by a flow directed opposite to the Gulf Stream, a markedly pronounced countercurrent. However, its velocity is much lower compared with that of the Gulf Stream and does not exceed 5–10 cm s^{-1}. Only a very weak jet pressed back to the coast maintains its north-eastward direction, i.e. in the direction of the Gulf Stream.

Since the problem was solved without preliminary filtration of the specified fields, the flow fields are more complicated than those considered above, and are characterized by the much higher number of available gyres (which stipulated filtration of the fields before calculating the World Ocean currents). However, all basic elements of the circulation, particularly in the upper 1500-m layer, adjust rather well despite differences in the calculation technique and the extent of smoothing of the specified fields. These calculations can be considered as a supplementary, qualitative test of the global calculations of the World Ocean currents.

The approach to global calculations related to the presentation of ζ as a sum of ζ' and ζ_d did not, however, appear quite adequate. The disadvantage of the approach is attributed to the fact that the sea surface topography, and consequently, the current velocity are related to the derivatives of the integrals of density rather than integrals of the derivatives with respect to density. When the bottom relief is significantly irregular, this results in considerable errors, particularly if the density field is specified over the whole ocean thickness rather than in the upper layer, i.e. it "interacts" with the bottom relief. This drawback was already noted in these calculations, and for the World Ocean it appeared naturally still more pronounced particularly in the front zones characteristic of strong irregularity of both the bottom relief and density.

5.5.2 The Diagnostic Calculations of Flows by the Non-Linear Model

The calculations were performed for the area within 24.5° N and 62.5° N on the same grid and for the same density and bottom relief fields as in the previous section. Nineteen levels are taken vertically: 0, 5, 10, 20, 30, 50, 75, 100, 125, 150, 200, 300, 400, 500, 800, 1100, 1500, 2000 and 3000 m. The semi-implicit scheme of directed differences was used. The time step is equal to 1 h. Based on the numerical experiments and grid scale, values μ and ν are assumed to be 10^7 and $10\,\mathrm{cm^2\,s^{-1}}$.

To compare solutions involving the constant and variable coefficients of the vertical turbulent exchange, the numerical experiment with parameterizaton of the upper mixed layer was carried out by the technique of Marchuk et al. (1976a) and Obukhov (1946), according to which ν is determined from the relation:

$$\nu = (0.05\,h_z)^2 \sqrt{\left(\frac{\partial u}{\partial z}\right)^2 + \left(\frac{\partial v}{\partial z}\right)^2 - \frac{g}{\rho_0}\frac{\partial \rho}{\partial z}}\,. \tag{5.12}$$

The thickness of the upper mixed layer h_z was based on the first estimated point z_k, where the requirement was met:

$$\tilde{\nu} = (0.05\,\Delta z_k)^2 \sqrt{\left(\frac{\partial u}{\partial z}\right)^2 + \left(\frac{\partial v}{\partial z}\right)^2 - \frac{g}{\rho_0}\frac{\partial \rho}{\partial z}} \leqslant \min \nu = \nu_0\,, \tag{5.13}$$

where ν_0 was assumed to be $1\,\mathrm{cm^2\,s^{-1}}$.

The calculations were indicative of a strong response of ν to inversions in the upper layer density fields. In such cases the model indicates a much greater depth of the turbulent penetration in comparison, for example, to the thickness of the upper mixed layer, calculated from the temperature fields as the depth, at which the water temperature does not differ from the sea surface temperature more than 1°. Except for the inversion points, the thickness of the upper mixed layer when using these two techniques is adequate. The most significant factor, however, is that flow fields did not change markedly when introducing the variable ν.

The force balance in the model is similar to the geostrophic, and as a consequence, the fields of the sea surface topography and horizontal flows are similar to those obtained by the quasi-geostrophic model. The vertical velocity field in the upper layer coincides qualitatively with that obtained by the quasi-geostrophic model. The area of the subtropical anti-cyclonic gyre is characteristic of the downwelling zone, while that of the subpolar cyclonic gyre is characteristic of the upwelling zone. The complication of the vertical velocity field increases progressively with depth. The adequacy of the quasi-geostrophic model deteriorates substantially. The value w increases from 10^{-4}–$10^{-5}\,\mathrm{cm\,s^{-1}}$ in the upper layer to 10^{-2}–$10^{-3}\,\mathrm{cm\,s^{-1}}$ at a depth of 3000 m.

So, the non-linear and quasi-geostrophic models, used for the diagnostic calculations of the sea surface topography and horizontal flows on the 1° grid set, provide similar results. Principal discrepancies are observed in the solutions for the vertical velocity, when the accuracy of the quasi-geostrophic approximation is evidently inadequate.

5.5.3 The Adjustment of the Temperature, Salinity, Density and Flow Fields by the Quasi-Geostrophic Model

The adjustment calculations according to the quasi-geostrophic model were carried out for the North Atlantic within 3.5°N and 72.5°N on the 1° grid set for 31 levels by height. In the first numerical experiments the climatic density and flow fields

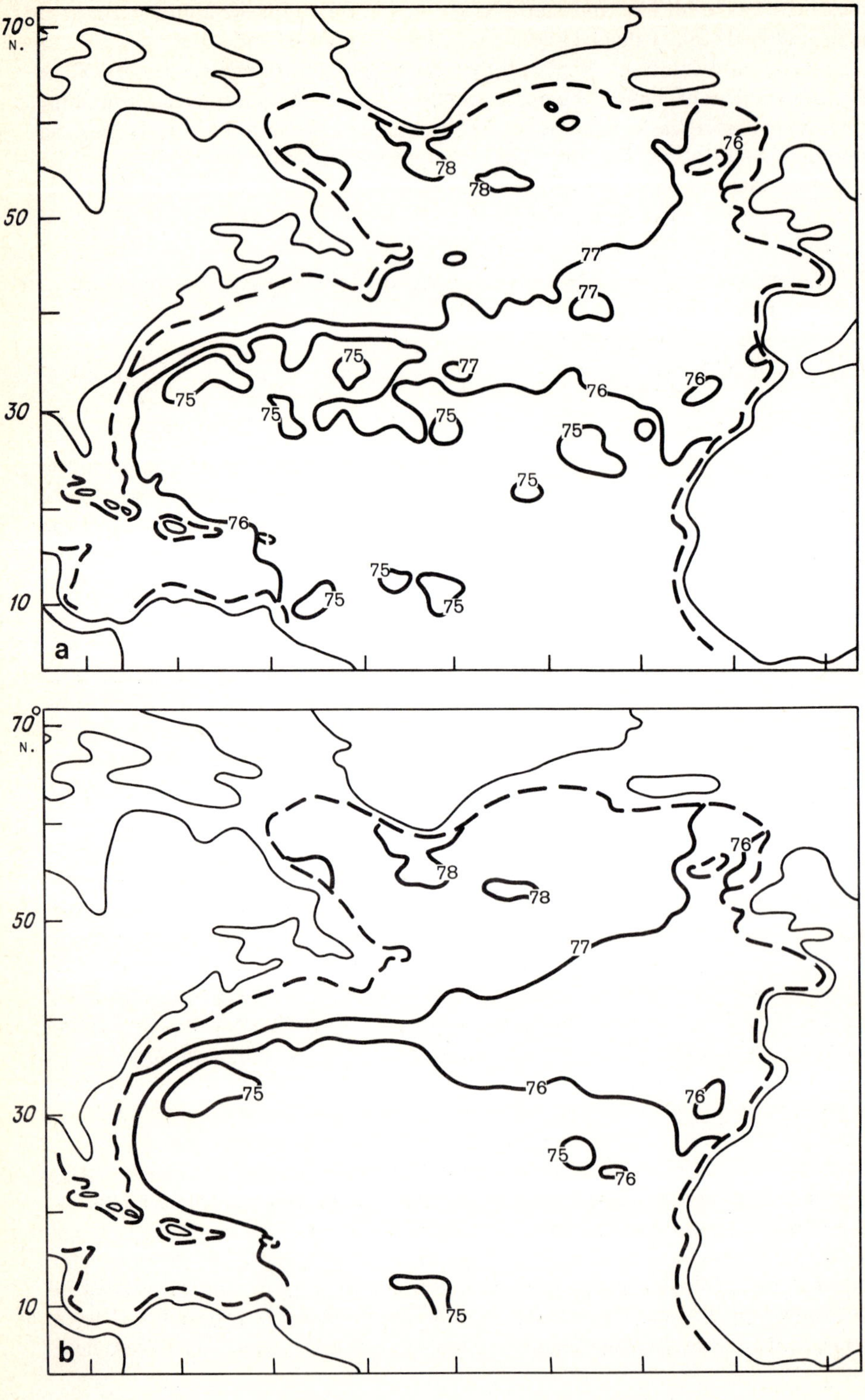
70° N.
50
30
10
a
78
78
76
77
77
75
77
75
76
76
75
75
75
75
75
76
75
75
75
70° N.
50
30
10
b
78
78
76
77
75
76
76
75
76
75

Fig. 42a, b

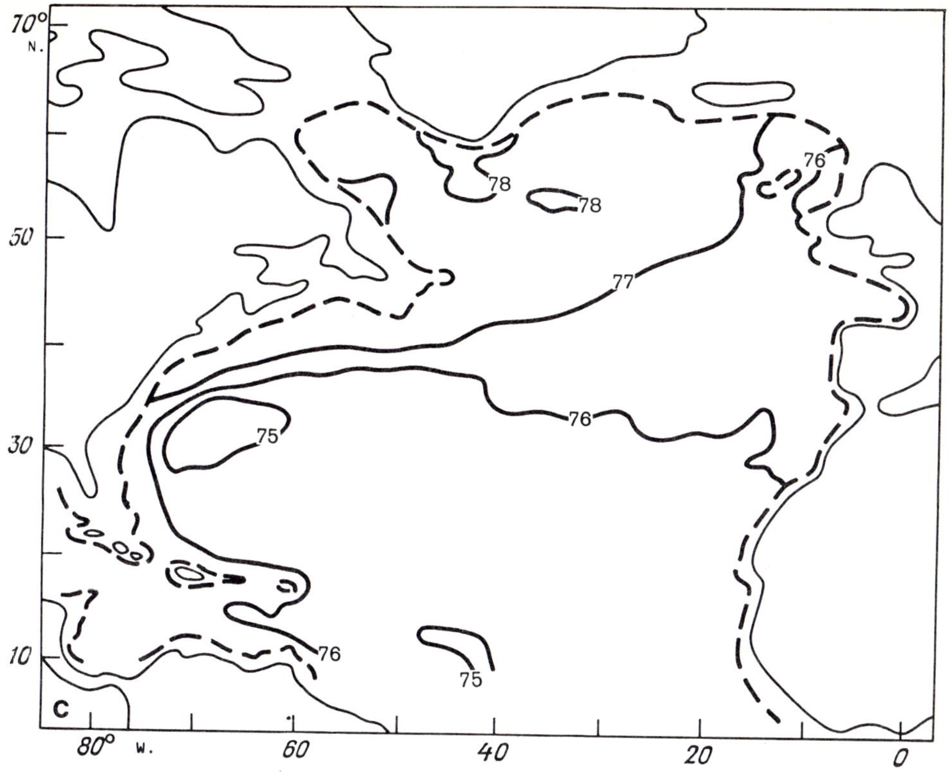

Fig. 42c

Fig. 42a–c. The density of the North Atlantic waters at 1000 m depth: **a** initial density; **b** after 8 days of integration; **c** after 15 days

were adjusted for the summer season. The specified fields are the same as those in the World Ocean current calculations. The equation for the diagnostic model was supplemented by the equation for the density diffusion. The specified density distribution was assigned at all boundaries of the area. The results of the diagnostic calculations were used as boundary conditions and as the initial approximation for the velocity. The equation for the density was approximated by the implicit scheme of directed differences. The time step is assumed to be 8 h. The coefficient of the horizontal turbulent diffusion μ varied from 5×10^6 to 5×10^7 cm s^{-1}. The velocity was calculated in the shifted grid points, i.e. in the centres of 1° "squares" from the known pressure and wind stress values in the four nearest points.

Basic fields variations took place during the first 10 days of the integration (Fig. 42). The adjustment resulted in the isopycnal "straightening", disappearance of small vortexes and meanders. Only large ones, stipulated by the significant bottom relief irregularity, were kept. The qualitative structure of the density and flow fields did not change. Further integration by time results in a progressive smoothing of the solution. Its speed and extent are related to the value of μ. It is important that the temporal variation of the kinetic energy as well as of normalized residuals for the density does not indicate sinuous oscillation even at the first stage of integration. The model does not describe fast wave processes, and only synoptic "noise" is filtered due to dissipative model characteristics, i.e. the solution is smoothed.

In the next numerical experiment two equations were used: the equation of heat transport and that of salt transport (instead of density diffusion). The equation of state is the same as that used earlier in this chapter for calculating density from temperature and salinity. The area, grid set, number of levels, specified fields, nature of boundary conditions, difference scheme and calculation technique are the same as those used in the experiments on density adjustment. The coefficients of vertical turbulent heat and salt diffusion were assumed to be 1 $cm^2\,s^{-1}$ (for the momentum it is, as before, 100 $cm^2\,s^{-1}$), and that of horizontal turbulent diffusion, $10^7\,cm^2\,s^{-1}$.

The calculation was performed for 3 months with the time step equal to 3 h. Qualitatively, the results and nature of the process coincide with those obtained for the density adjustment. The nature of the temporal variation of the kinetic energy and normalized residuals for temperature and salinity did not change and indicates the monotone fall from the specified "diagnostic" state with a gradual moderation in the decrease of energy and residuals.

5.5.4 The Adjustment of the Temperature and Flow Fields by the Non-Linear Model

Let us now consider the results obtained by the non-linear model. The area, grid set and specified fields are the same as those used in Section 5.5.2. The equations for the diagnostic model were supplemented by the heat transport equation with the Dirichlet conditions at all boundaries. The semi-implicit scheme of directed differences with the time step equal to 1 h was used. The velocity values of the boundary are taken from the diagnostic problem solution. The salinity was not calculated but it was taken into account when calculating density from the equation of state. Moreover, the specified climatic distribution of salinity was used. The first calculations, without regard for salinity in the equation of state, indicated that errors (due to the ignored salinity) are rather large, particularly at high latitudes. In contrast to the diagnostic calculations of the World Ocean currents, the preliminary filtration of the fields was not carried out since the synoptic "noise" in the fields was supposed to be removed during the process of adjustment. The coefficients of the turbulent exchange and diffusion are assumed to be $\nu = 10\,cm^2\,s^{-1}$, $\mu = 10^7\,cm^2\,s^{-1}$. The calculation was performed for 105 days. The maximal residuals for velocities, sea surface topography and temperature were 0.03 $cm\,s^{-1}$, 0.04 cm and 0.007°C respectively. The analysis of the solution shows, however, that the adjustment ended in the first 10 days. The basic, fast variations in the temperature and flow fields took place during that time interval, and later on after further integration, the solution changes much slower, progressively smoothing step by step, starting even to be "spoiled". So, after 30 days of integration, the recirculation of the Gulf Stream disappears.

Besides the detailed temporal variation of the kinetic energy at the first stage of integration, i.e. the stage of adjustment, Fig. 43 indicates as well the variation of the kinetic energy with the diurnal averaging for the whole period of integration. Figure 44 illustrates the behaviour of maximal residuals with time for temperature,

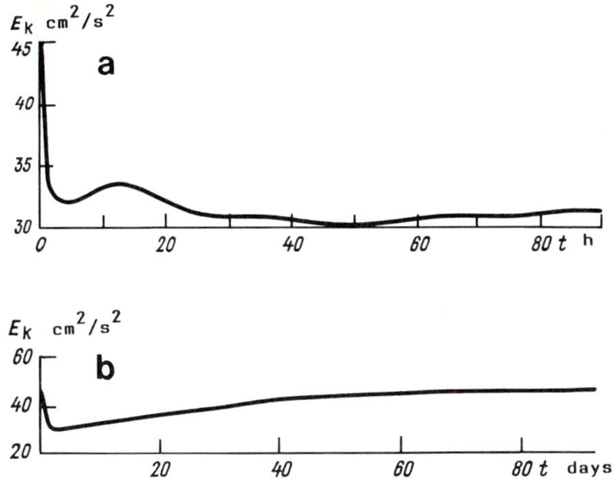

Fig. 43a, b. The time variation of an average kinetic energy: **a** detailed, during the first 90 h; **b** with the daily averaging over 105 days

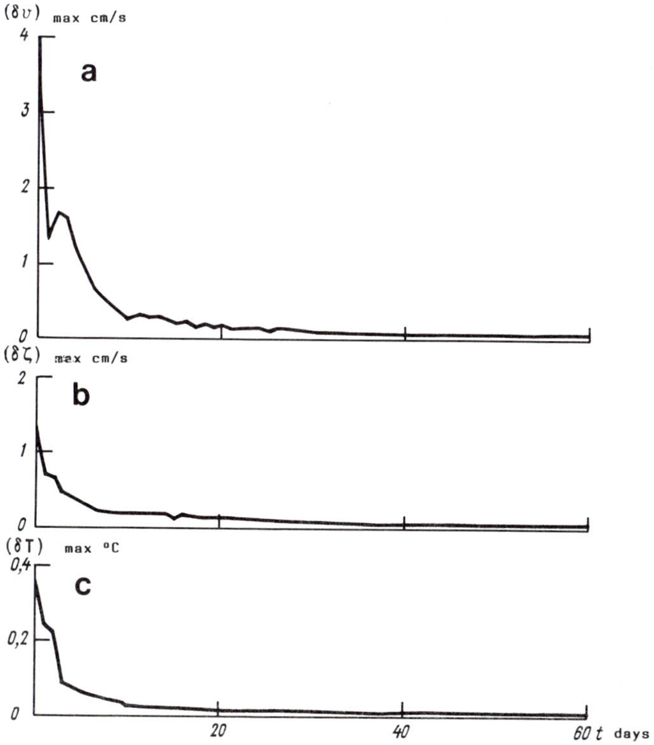

Fig. 44a–c. The time variation of the maximal residuals for the velocity (**a**), the SST (**b**) and temperature (**c**)

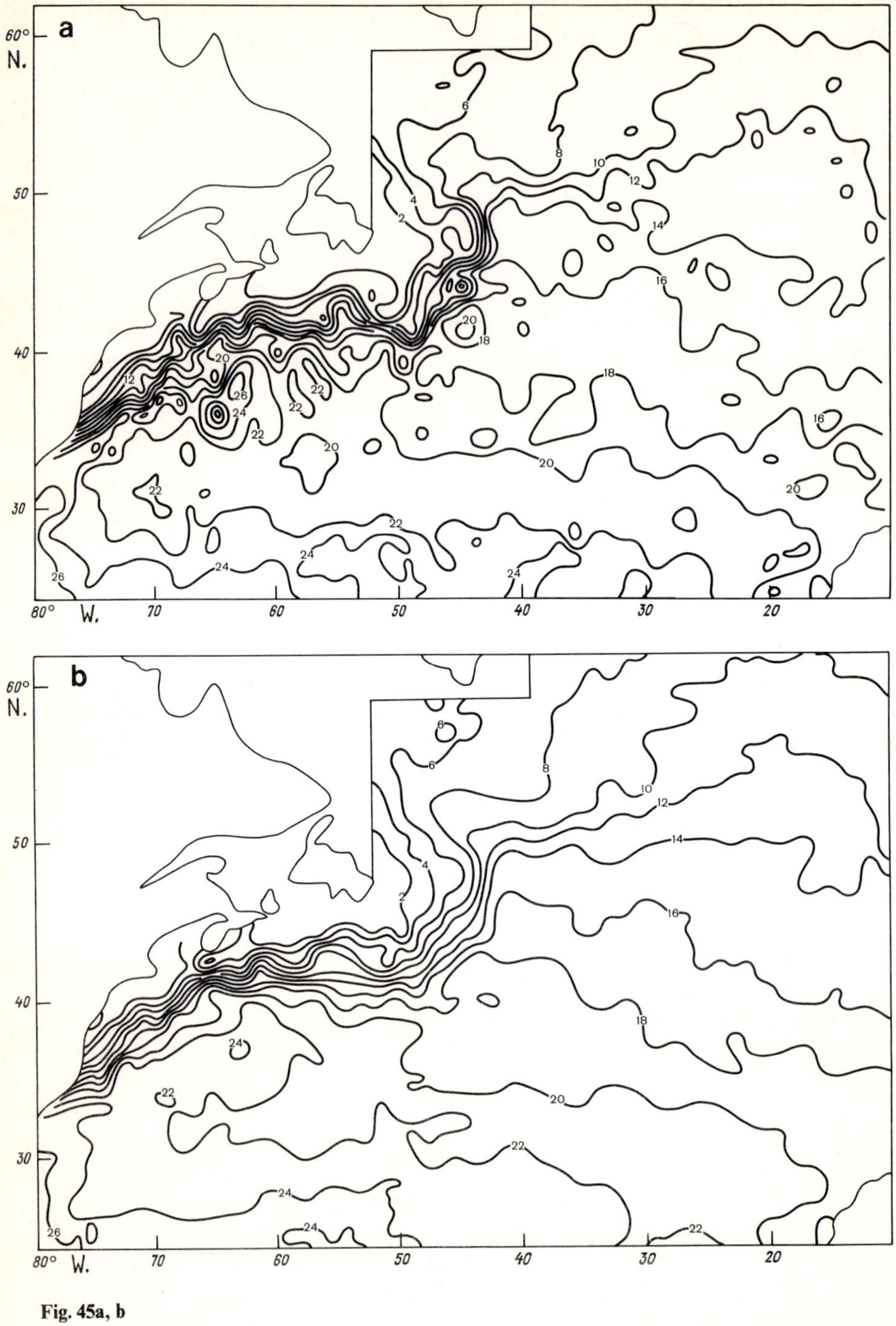
a
60° N.
50
40
30
80° W.
70
60
50
40
30
20
b
60° N.
50
40
30
80° W.
70
60
50
40
30
20

Fig. 45a, b

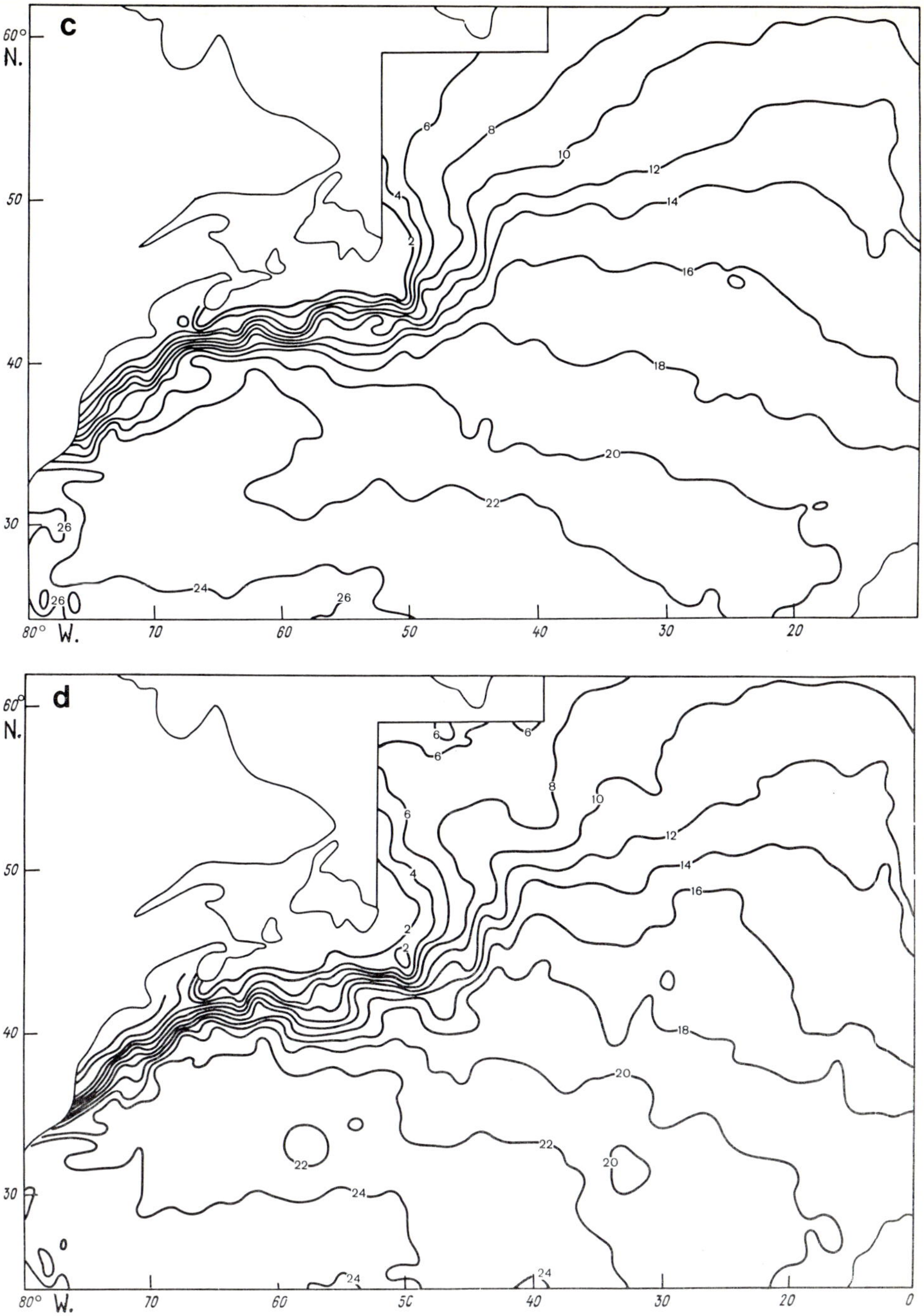

Fig. 45c, d

Fig. 45a–d. The temperature of the North Atlantic waters at 50 m depth: **a** initial temperature; **b** after 10 days of integration; **c** after 25 days; **d** after 65 days

velocity and sea surface, i.e. criteria traditionally used in prognostic models. The tendency for establishing mean kinetic energy at long-term integration as well as the decrease with time of maximal residuals for temperature, velocity and sea surface topography are well pronounced.

First compared with the diagnostic version, a fast, abrupt decrease of mean kinetic energy with sinuous oscillations is observed during the first few days of integration. The energy minimum is reached in 5 days, followed then by a smooth, but gradually retarded increase.

The qualitative structure of the temperature and flow fields is preserved after the adjustment. The most interesting peculiarities are (1) the remarkably increased correlation with the bottom relief, which is observed even in the sea surface topography field—the integral function, and (2) the preservation, both in the velocity and temperature fields, of the well-pronounced Gulf Stream jet with a simultaneous and intensive weakening of small details of circulation (open ocean eddies and jetlike current meanders) related to the synoptic "noise" in the specified fields. The impact of the bottom relief is particularly well pronounced in the area of the Newfoundland bank, and in the area of the highly elevated Mid-Atlantic Ridge.

Compared with the specified temperature field, the adjusted temperature field is much more regular (Fig. 45). It is of interest that the temperature drop in the upper layer across the Gulf Stream jet after 10 days of integration no longer changes (or only slightly). The Gulf Stream velocity after the adjustment decreased and was only 50–55 cm s^{-1} compared with 75–80 cm s^{-1} of the diagnostic version.

As a whole the adjustment period led to a substantial correction of the diagnostic flow fields and specified temperature fields.

The adjustment process of this model is quite different compared with the intensively filtered quasi-geostrophic model used previously. It is important that time derivatives remain in the equation of motion, which makes it possible to describe fast wave processes, in particular, inertial gravitational waves, which are evidently mainly responsible for the field adjustment. The described calculations show neither an increased correlation of the hydrological fields with the bottom relief, nor sinuous variations of the kinetic energy at the adjustment stage when passing to the stage of slow, monotone measurements after further integration by time. The filtered, quasi-geostrophic model, despite the close balance of forces, lacks both effects, i.e. the adjustment is reduced to filtering synoptic "noise" in the specified data due to dissipative model characteristics.

5.6 The Diagnostic Calculations of Flow in the Equatorial Belt of the Ocean

The equatorial belt is one of the active energy zones of the World Ocean which significantly effects the Earth's climate. It is characteristic of the very complicated dynamics of flows with numerous vertical and horizontal irregularities. So, despite a great number of field observations and theoretical studies carried out since the discovery of the equatorial undercurrent (Cromwell et al. 1954; Kolesnikov et al. 1968; Philander and Pacanowski 1980), which made it possible to recognize and explain a number of peculiarities in equatorial dynamics, our knowledge of the phenomenon is far from being sufficient and requires further systematic investigations. The equatorial belt of the Indian Ocean has been studied most poorly, even schematic maps of flows compiled from observation data are lacking. The available maps of the Indian Ocean flows, compiled either from ship-shift data (sea surface flows) or by the dynamic method, are based on a 5° averaging scale, and do not reflect the local structure of the equatorial circulation. Furthermore, the dynamic method is invalid at the equator.

The quasi-geostrophic model is also invalid at the equator, thus producing large errors near it, where non-linear effects should be considered as most important in the force balance. However, the quasi-geostrophic approximation is quite valid for calculating large-scale flows outside a relatively narrow equatorial belt. So, the calculation of flows in the basin covering the equator can be substantially simplified: instead of a universal though rather laborious and cumbersome model carried out at once for the whole basin, a two-stage calculation can be made. At the first stage the flow is calculated in the whole basin by a quasi-geostrophic model, which is easily accomplished with a computer. The grid set is constructed in such way that points nearest to the equator are placed at a distance where the geostrophic balance is a determining factor, i.e. to avoid the equatorial belt. At the second stage the flows of the equatorial belt are calculated by a non-linear model. The velocity at the "liquid" sections of the boundary is taken from the solution obtained by quasi-geostrophic model, and the non-slip conditions are set at the solid west-east boundaries. So, the mutual adjustment of the solution for the specified subareas is carried out via the boundary conditions. It is expedient to construct the internal grid set with some overlapping along the meridian of points by the external grid to minimize the effect of errors in the solution of a quasi-geostrophic problem near the equator.

Thus, the efficient calculation of the basin flows covering the equator requires the application of both the quasi-geostrophic and non-linear model. Moreover, the question of the validity limits of the models is quite natural. In particular, the validity limits of the quasi-geostrophic model should be estimated to determine the necessary reduction of the grid size near the equator when calculating flows of the whole ocean or that part including the equator.

5.6.1 The Equatorial Atlantic Flow Calculations by the Quasi-Geostrophic Model: the Assessment of the Validity Limits of the Models

The modified equation for the sea surface topography, according to Demin and Sarkisyan (1977) and Sarkisyan (1977b), is used in the calculations:

$$\pm\frac{1}{2\alpha}\Delta\zeta+\left(\frac{\beta}{l}H-\frac{\partial H}{\partial y}-\frac{3\beta}{4\alpha l}\right)\frac{\partial\zeta}{\partial x}+\left(\frac{\partial H}{\partial x}\mp\frac{3\beta}{4\alpha l}\right)\frac{\partial\zeta}{\partial y}$$

$$=\frac{\operatorname{rot}\tau}{\rho_0 g}+\frac{\beta\tau_x}{\rho_0 g l}\mp\frac{1}{2\alpha\rho_0}\int_0^H\Delta\rho\,dz+\frac{\beta}{\rho_0 l}\left(\int_0^H z\frac{\partial\rho}{\partial x}dz-H\int_0^H\frac{\partial\rho}{\partial x}dz\right)$$

$$+\frac{1}{\rho_0}\left[\left(\frac{\partial H}{\partial y}+\frac{3\beta}{4\alpha l}\right)\int_0^H\frac{\partial\rho}{\partial x}dz-\left(\frac{\partial H}{\partial x}\mp\frac{3\beta}{4\alpha l}\right)\int_0^H\frac{\partial\rho}{\partial y}dz\right]. \tag{5.14}$$

The summands with the coefficient $3\beta/4\alpha l$ are additional to the bottom stress, which is produced as a result of the allowance for variability of α. The equation also takes the second summand of the right-hand side into account, in addition to the effect of the direct wind impact, which is related to the change of the Coriolis parameter. At mid-latitudes these factors are insignificant. However, their role should increase when approaching the equator.

The horizontal velocity components (excluding the near-bottom boundary layer) are calculated by formula:

$$u = \frac{e^{-\alpha z}}{2\alpha\rho_0\nu}[(\tau_x \pm \tau_y)\cos dz - (\tau_x \mp \tau_y)\sin\alpha z] - \frac{g}{l}\frac{\partial\zeta}{\partial y} - \frac{g}{\rho_0 l}\int_0^z \frac{\partial\rho}{\partial y}d\xi + \tilde{u} \quad (5.15)$$

$$v = \frac{e^{-\alpha z}}{2\alpha\rho_0\nu}[(\tau_y \mp \tau_x)\cos\alpha z - (\tau_y \pm \tau_x)\sin\alpha z] + \frac{g}{l}\frac{\partial\zeta}{\partial x} + \frac{g}{\rho_0 l}\int_0^z \frac{\partial\rho}{\partial x}d\xi + \tilde{v}\,, \quad (5.16)$$

where

$$\tilde{u} = -\left(u\frac{\partial u}{\partial x} + v\frac{\partial v}{\partial y}\right)\Big/ l\,, \qquad \tilde{v} = \left(u\frac{\partial v}{\partial x} + v\frac{\partial v}{\partial y}\right)\Big/ l$$

are the non-linear corrections.

The model assumes the consideration of non-linear corrections in the equation for the sea surface topography, although they are of less importance here. The relations for $\tilde{u}$, $\tilde{v}$ can be readily transformed with the help of the quasi-geostrophic approximation and can be expressed in terms of derivatives with respect to the sea surface topography and density. The approximate expressions are given in Demin and Sarkisyan (1977), and Sarkisyan (1977b).

The calculations were performed for the area from 11°N to 11°S, and from 42°W to 6°E. The technique of calculations is described in Chapter 2. The density fields correspond to the summer season and are taken from the observation data under the Aqualant-II program.

The homogeneous, 2° grid set is used in the first calculations, and the points nearest to the equator were located at 1°N and 1°S. Nine levels are taken vertically from 0 to 1000 m.

The calculations of the right-hand side of the equation for the sea surface topography indicate that baroclinic factors with values one to two orders higher than that of the wind summands contribute considerably. Unlike the mid-latitudes, here the baroclinic β-effect dominates [the underlined term on the right-hand side of Eq. (5.14)]. The wind addition $\beta\tau_x/\rho_0 gl$ has approximately the same value as the vortex of the surface wind stress. At 1°N and 1°S it even exceeds the latter. The characteristic behaviour of the right-hand side with latitude is an abrupt increase near the equator (Fig. 46). The sea surface topography, however, does not indicate any sudden variations near the equator. Furthermore, the sea surface topography is similar to that calculated with a quasi-dynamic method, i.e. from the relation without direct information on the location of the equator. The reasons for these peculiarities will be considered below.

The velocity fields conform with the available concept, however, at 1°N and 1°S the velocity value is overestimated. The use of the non-linear corrections does

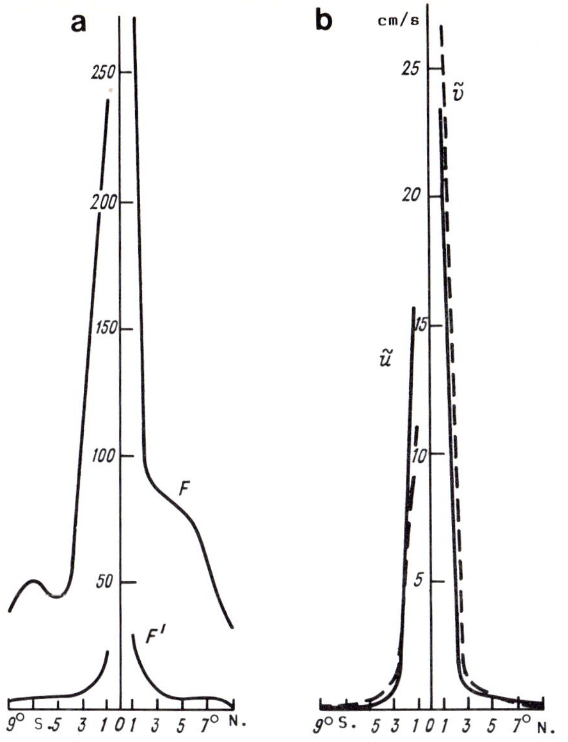

Fig. 46a, b. The variation of the zonal absolute mean values with latitude: **a** the right-hand sides of the equations for ζ and ζ', in arbitrary units (curves F and F' respectively); **b** zonal $\tilde{u}$ and meridional $\tilde{v}$ components of nonlinear additions at the ocean surface

not improve the situation. In some points the velocity value still remains overestimated. The value of non-linear corrections is low, i.e. not more than 1–2 cm s^{-1} over the whole area, excluding 1°N and 1°S, where it increases strongly, when approaching and even at some points exceeding the gradient component value. Figure 46 indicates two basic peculiarities of the field of non-linear corrections: (1) a fast increase when approaching the equator, beginning approximately with 3°N and S, and (2) a fast decrease with depth, particularly in the upper layer. With regard to the basic velocity component, the value of non-linear corrections reaches 100% and more at 1°N and S; at 3°N and S it is within 10%; and at extreme latitudes, it does not exceed 2%. So the validity limits of the model under consideration are somewhere within 1° and 3°N and 1° and 3°S.

In the next cycle of calculation the points nearest to the equator were located at 2°N and S (in the rest, the grid set remains the same).

The sea surface topography and velocity fields differ from those obtained earlier principally only at 2°–3°N and S. The estimations obtained indicate that nonlinearity is of substantial though not of determining importance when being removed 2° from the equator. The use of non-linear corrections this time resulted in a rather noticeable smoothing of the velocity field near the equator, particularly of the meridional component.

So, within the framework of the quasi-geostrophic model the 2° approach to the equator from north and south is similar to the maximum permissible one. The estimates are approximate and to some extent connected with the technique used and the nature of the specified data (for example, with the fact that the density field is specified only up to 1000 m rather than to the bottom of the ocean). If we assume the solution for ζ as a sum of ζ_d and ζ' (Kozlov 1973; Sarkisyan 1977b), where ζ_d is found from Eq. (1.103) and perform the respective calculations, then the right-hand side of the equation for ζ' will appear to be approximately an order lower compared with the right-hand side of the equation for ζ. The amplitude of the field ζ' is equal to 2 cm, while $\zeta_d \approx 33$ cm. The information of the equator location (the Coriolis parameter) is only contained in ζ', which is relatively small. The peculiarities noted above are attributed to this fact.

Let us consider briefly the relative role of some factors of the model. The calculations of ζ performed with and without allowance for normal derivatives in the boundary relations (secondary effects of the bottom stress) resulted in rather insignificant differences: the mean relative error was lower than 5%. The error was localized near the bottom boundary. Its maximum value is observed in the area of the weakest sea surface topography gradients. The ignored wind components in the boundary relations led to a similar insignificant difference: on the average, less than 3%.

The role of the secondary effect of the bottom stress appeared to be negligible: the error was only 0.3%.

The validity limits of the model and the dynamic method should be similar. In this connection it is noteworthy to mention the work of Dubravin (1971), where flow velocities calculated by a dynamic method and observed in the equatorial Atlantic were compared. It was indicated that from 1°30′–2°N to the North and from 1°30′–2°S to the South the difference is insignificant particularly at the subsurface levels. At the same time, near the equator the dynamic method produces strongly overestimated velocities (at the equator, for example, four- to five-fold), which often do not coincide with the observed values or even the direction. This indicates that the geostrophic balance is not met here. To calculate flows in the belt one should account for non-linear terms. The importance of non-linear effects in a 3°–4° belt, including the equator, is mentioned in Holland and Hirschman (1972).

The velocities of the wind–driven gradient flows at the surface are on the average: $u = 0.2\,\text{cm}\,\text{s}^{-1}$, $v = 0.1\,\text{cm}\,\text{s}^{-1}$, and the maximum wind–driven sea surface slope, 0.8 cm. Baroclinicity is a determining factor. In the area under consideration the β-effect is a determining factor, which is attributed to the proximity of the equator.

Qualitatively, the surface flow field quite conforms with the available concept. The South Equatorial current (up to $80\,\text{cm}\,\text{s}^{-1}$), Equatorial countercurrent ($35\,\text{cm}\,\text{s}^{-1}$) and Guinean current ($25\,\text{cm}\,\text{s}^{-1}$) are markedly pronounced. The character of circulation is primarily zonal.

The flow field undergoes significant reconstruction with depth. Under the South Equatorial current along the western boundary a water flow extends from the south to the equatorial zone, and is most pronounced at 100–150 m. At 5°S the flow divides into two parts: one part flows in the north-west direction to the equatorial zone, and the other part turns to the east and flows along 5°S, generating the countercurrent under the South Equatorial current. The water flow to the equatorial zone is clearly observed from the north-west area as well, under the northern branch of the South Equatorial current. A common peculiarity of the vertical structure of the circulation are undercurrents, which are to some extent characteristic of each of the basic flow systems. However, because of the large-scale averaging of the specified density field and model nature, the Lomonosov current is not expressed adequately, and sections of the "broken" jet are observed.

The vertical velocity field in the surface layer has the following peculiarities: abrupt upwelling, particularly south of the equator, in the South Equatorial current; and mainly downwelling in the Equatorial countercurrent, relatively weaker in intensity; alternation of upwelling and downwelling zones of approximately similar velocity values in the Guinean current. The situation is much more complicated in the intermediate waters. The characteristic vertical velocity value is $10^{-3}\,\mathrm{cm\,s^{-1}}$.

5.6.2 The Calculations of Flows of the Equatorial Belts in the Atlantic and Indian Oceans by the Non-Linear Model

The first cycle of calculation is carried out for the equatorial Atlantic (Demin 1975a; Demin and Sarkisyan 1977), based on the same density, wind and bottom relief fields as in the previous section. The problem was solved by a method of reaching steady state with a semi-implicit scheme of directed differences. The boundary conditions are taken from the solution of the quasi-geostrophic problem. The grid sizes were: 2° by x from 1° to 0.5° by y. The optimal value of the time step and the required accuracy of the iteration process, which provides the output of the solution to the steady-state conditions, were determined by experimental tests and were assumed as follows: $\Delta t = 5 \times 10^3\,\mathrm{s}$, $\max|(u^{n+1} - u^n), (v^{n+1} - v^n)| < 0.1\,\mathrm{cm\,s^{-1}}$ (the convergence process was, furthermore, controlled by the kinetic energy, temporal variation). At the given accuracy the results do not depend on the specified approximation. The time of reaching steady-state conditions is 1–2 months depending on the input parameters (particularly on value μ) and specified approximation.

The first calculations were performed for the grid set with a size 2° by x and 1° by y; parameters: $\nu = 10\,\mathrm{cm^2\,s^{-1}}$; μ varied from 10^{10} to $10^6\,\mathrm{cm^2\,s^{-1}}$. Figure 47 illustrates the dependence of the solution on the value μ. In the case of high μ of the order of $10^9\,\mathrm{cm^2\,s^{-1}}$ and higher, the solution is considerably smoothed and there is no intensification of equatorial flow. The calculation of summand values in the equation of motion indicates that the "weight" of the horizontal exchange exceeds the "weight" of non-linear summands by one order, i.e. the importance of the latter is relatively low, and the non-linear problem in this case is, in principle, a disguised linear problem. Furthermore, the solution at high μ is closely related to the boundary conditions for u, v. With a μ decrease to $10^8\,\mathrm{cm^2\,s^{-1}}$ and less, a markedly pronounced intensity peak of the zonal flow component is observed at the equator and at subsurface levels and a minimum of the meridional component is observed. The role of the non-linear effects becomes very significant at $\mu = 10^7\,\mathrm{cm^2\,s^{-1}}$; their "weight" on average in the upper layer is an order higher than that of the horizontal exchange. Almost all non-linear summands demonstrate an increase in "weight" when approaching the equator (at high μ the equatorial "intensification" of non-linear summands was lacking). The dependence of the solution on the boundary conditions weakens considerably. At $\mu = 10^6\,\mathrm{cm^2\,s^{-1}}$ the results change already insignificantly, i.e. the relative role of the numerical viscosity becomes higher.

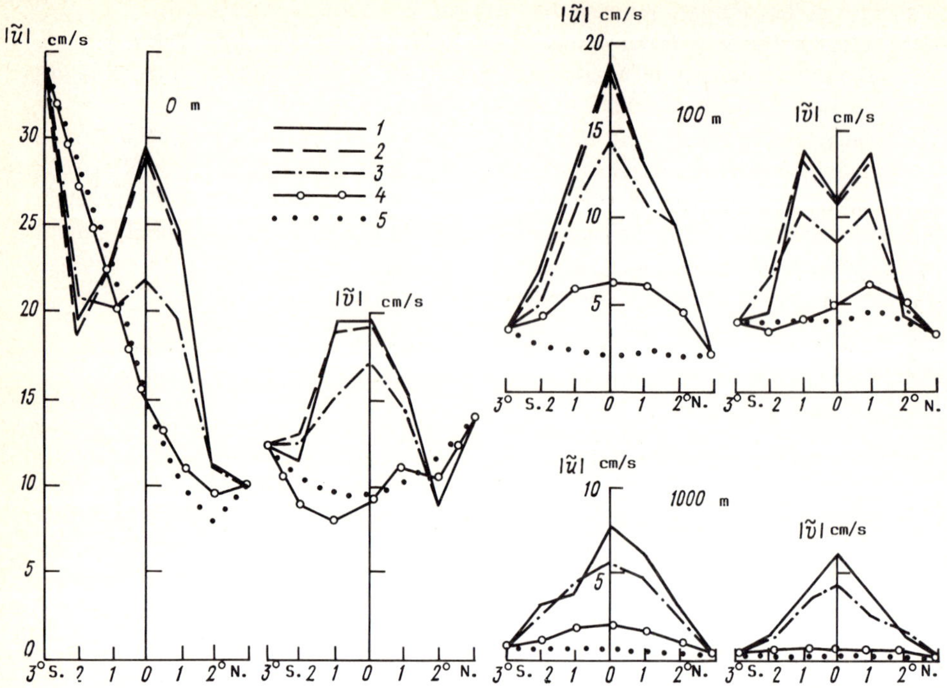

Fig. 47. The intensity of zonal mean currents with different values of μ at 0, 100 and 1000 m as a function of latitude (*1* $\mu = 10^6\,cm^2\,s^{-1}$, *2* $10^7\,cm^2\,s^{-1}$, *3* $10 - cm^2\,s^{-1}$, *4* $10^9\,cm^2\,s^{-1}$ *5* $10^{10}\,cm^2\,s^{-1}$)

A number of calculations by the linear model (non-linear summands were neglected in both the equation of motion and sea surface topography) were carried out for comparison with the solution of a completely non-linear problem. At $\mu \geqslant 10^9\,cm\,s^{-1}$ the results practically coincide. The differences increase with a decrease in μ, particularly at the equator, and at $\mu = 5 \times 10^7\,cm^2\,s^{-1}$ they are very significant (Fig. 48). In the linear version the intensity of the meridional component exceeds the intensity of the zonal component at the surface, which contradicts the concept on the zonal nature of the equatorial flows. Instead of a minimum of meridional component at the equator, its maximum is observed. The differences are more marked in the eastern area, where the density field is less smoothed. As a whole, they are the largest at the equator. The importance of non-linearity on the average decreases with the distance from the equator and the increase of depth. The difference between the linear and non-linear version decreases significantly as well. In the upper ocean layer the maximum "weights" along non-linear summands have $v\,\partial v/\partial y$ and $v\,\partial u/\partial y$. $w\,\partial u/\partial z$ and $w\,\partial v/\partial z$ gain in importance with increased depth.

Briefly, some other factors of the model will be mentioned. The sea surface wind stress plays a significant role in the surface currents. When summands with τ_x, τ_y are neglected in the model equation, the importance of the meridional velocity component in the surface currents increases significantly, and the current divergence at the equator deteriorates markedly. The importance of the direct wind effect decreases rapidly with depth and at 100 m, it is insignificant. The importance of the

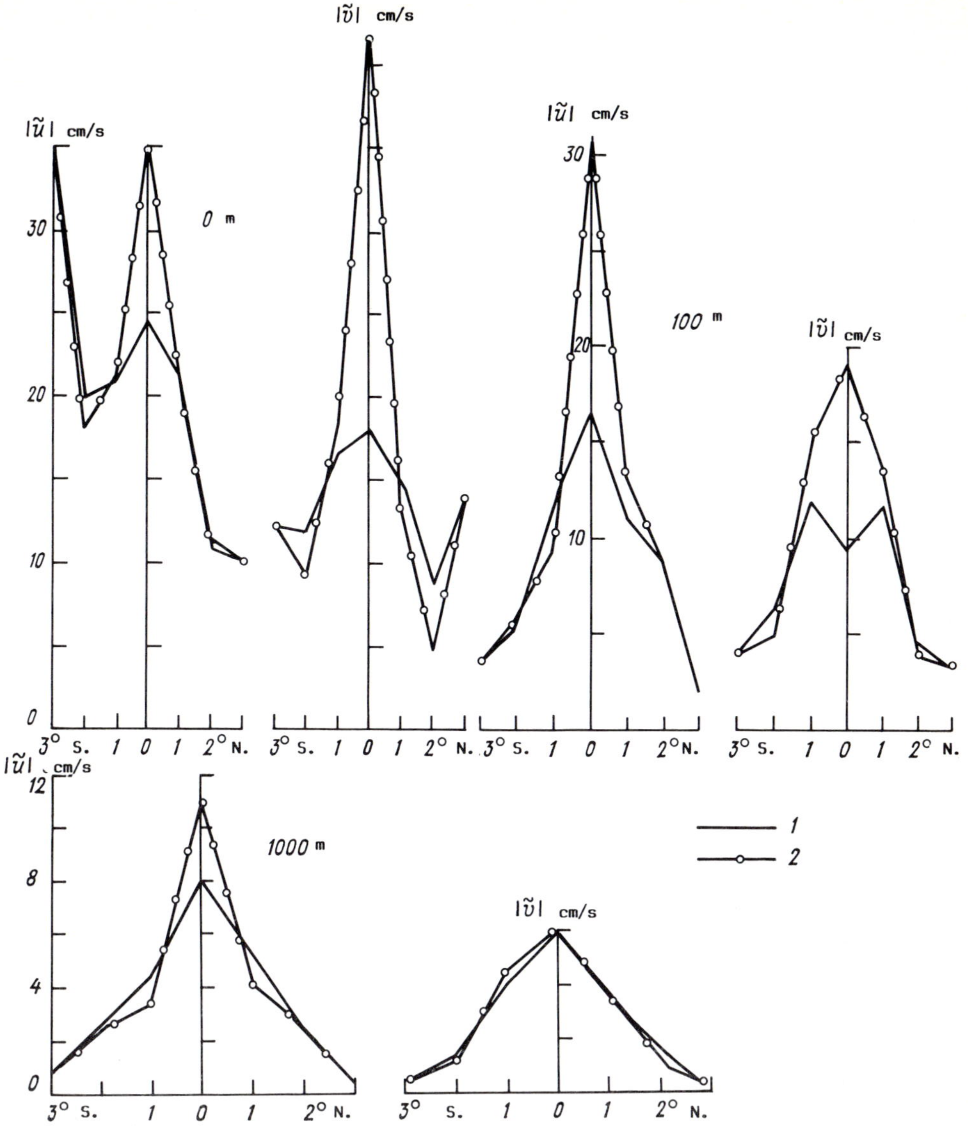

Fig. 48. The intensity of zonal mean currents in nonlinear (*continuous lines*) and linear (*dash-dotted lines*) versions at 0, 100 and 1000 m with $\mu = 5 \times 10^7\, cm^2\, s^{-1}$)

Coriolis force, only 1° from the equator, appeared to be so great, that it cannot be neglected even when calculating currents in a narrow equatorial zone.

Let us now consider the force balance in Eq. (1.130) for ζ. The averaged absolute values of the domain according to the right-hand side summands of the equations are (in CGS units): direct wind effect (VI), 2×10^{-17}; baroclinicity effect, 3.4×10^{-15}. The fraction of the basic summand (II) is 3.2×10^{-15}; the fraction of the second summand (I), related to the bottom relief effect, is 3.1×10^{-16}; the iterative part (all the rest summands) is 3×10^{-16}. The contribution of summands VII and VIII is negligible. If in the first approximation only basic baroclinic summands are kept on the right-hand side, the resultant, simplified equation has a known solution (1.103). So, even in the equatorial zone the solution for ζ is determined in the same way as at mid-latitudes, primarily by this integral. However, one should bear in mind that even relatively insignificant variations in the sea surface topography can result in rather substantial variations in the velocity field near the equator.

Briefly, the peculiarities of the sea surface topography and flow fields should be noted. The general sea surface slope from west to east is well pronounced, although its slope is non-monotone and disturbed by local maxima. The surface flows are mainly generated under the trade-wind effect. The velocity value is up to $75\,\mathrm{cm\,s^{-1}}$. On average, it is $30\,\mathrm{cm\,s^{-1}}$. The equatorial divergence is well pronounced in the western area. As a result of the non-monotonicities of the west-eastern sea surface slope, the equatorial subsurface countercurrent, as in the quasi-geostrophic model, has sections of "breaks". The non-monotonicity of the slope is first and foremost connected with the peculiarities of the specified density field, which reflects inadequately the narrow jet of the countercurrent due to the large (2°) averaging scale. To demonstrate that a wavy sea surface slope along the equator is not generated artificially by the model, the following numerical experiment was carried out. Keeping the same boundary conditions of u, v, ζ, i.e. those obtained from the baroclinic problem, we neglect the density anomaly inside the area. In this case the information on the sea surface slope is contained in the boundary conditions, and there are no disturbing sources inside the area. In this case the sea surface is monotonously inclined from west to east along the equator, and as a consequence, a continuous eastward jet is clearly pronounced under the surface trade-wind flow (Fig. 49). The countercurrent has a convergence nature, therefore, a narrow area

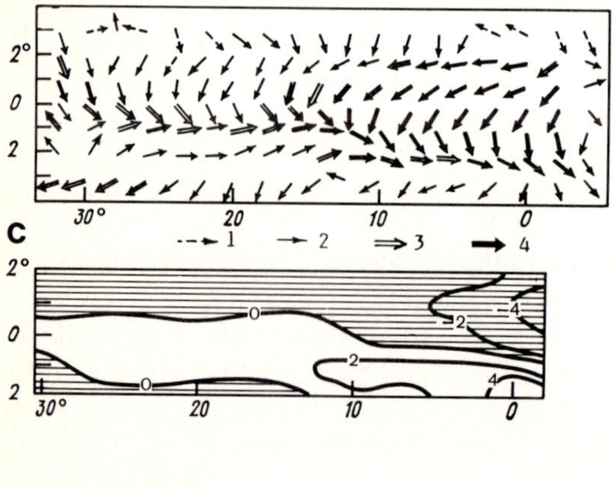

Fig. 49a–c. The currents schemes in a stylized version (inside the domain $\rho = 0$, boundary conditions are baroclinic): **a** the SST, in cm; **b** the currents at 75 m *1*, $5\,\mathrm{cm\,s^{-1}}$; *2* 6–15; *3*, 16–25; *4*, 26–50); **c** the field of the vertical velocity at 200 m (the upwelling areas are *shaded*, the isotachs are drawn at $2 \times 10^{-3}\,\mathrm{cm\,s^{-1}}$)

extending along the equator with a downwelling zone in the vertical velocity field bordered from north and south by upwelling zones corresponds to the area under the countercurrent. The major part of the area in the upper ocean layer is under the upwelling effect, which is related to the divergence of the equatorial trade-wind flows. This version certainly cannot be of oceanographic importance, it is only a qualitative model test. Let us emphasize that the results obtained pertain to the solution of the problem for some mixed version rather than for a homogeneous ocean. Indeed, the boundary conditions for the sea surface, which is responsible for the very existence of the countercurrent, should have been taken from the homogeneous model as well. However, in the homogeneous model the sea surface slope for the whole area from 11°N to 11°S, as mentioned above, is only 0.8 cm. The intensity of the countercurrent in this case would be actually low.

The drawback of the calculations performed is a rather crude resolution of the grid set, in particular meridionally, across the Lomonosov current jet, and vertically in the upper layer. Therefore, two levels, 25 and 50 m, were added to the next calculations and the grid size by "y" near the equator was one-half as much. The required arrays were obtained by interpolation.

The surface flows changed rather significantly. The intensity increased drastically and the trade-wind nature of the flow became clearly defined. The local minimum of the meridional component is observed here even at the surface though shifted 0.5° from the equator. The difference decreases rapidly with depth and at 100 m it is insignificant. Intense changes in the surface flows are mainly related to the belitling the vertical grid mesh. This influences substantially the vertical viscosity which is of great significance for the surface flows. It is important that the nature of the countercurrent has not changed qualitatively: the jet still has "breaks" in the same sections of the area. With ν increased to $100\,\mathrm{cm}^2\,\mathrm{s}^{-1}$, the surface flows smoothed considerably, and the intensity dropped abruptly. However, in this version as well the variations are significant only within the limits of the surface layer and diminish rapidly with depth. The nature of the countercurrent remains practically unchanged.

Thus, a simple grid division with the interpolation of the specified data did not result in a significant change of results outside the narrow surface layer and did not improve the nature of the countercurrent qualitatively. The used density field, as already mentioned, corresponds to a 2° grid mesh and does not reflect narrow, jetlike currents such as the Lomonosov current with sufficient adequacy. The calculations of the equatorial circulation should be based on more detailed density fields with less spacial averaging.

Below are examples of such calculations of the equatorial zones for the Indian and Atlantic Ocean from 7.5°N to 7.5°S (Demin and Ibraev 1985; Sarkisyan et al. 1984a, b). The calculations are performed by a non-linear model for 15 levels by height (from 0 to 5000 m) on the grid set with a size 1° by "x", and from 1° to 15′ by "y" (near the equator). The specified multi-year mean seasonal density and wind fields as well as the ocean bottom relief field are the same as in Section 5.2. The model, numerical method and calculation technique are in agreement with those applied above. Furthermore, for the equatorial Atlantic a supplementary numerical experiment was carried out. (Demin and Ibraev 1985), where for the sea surface

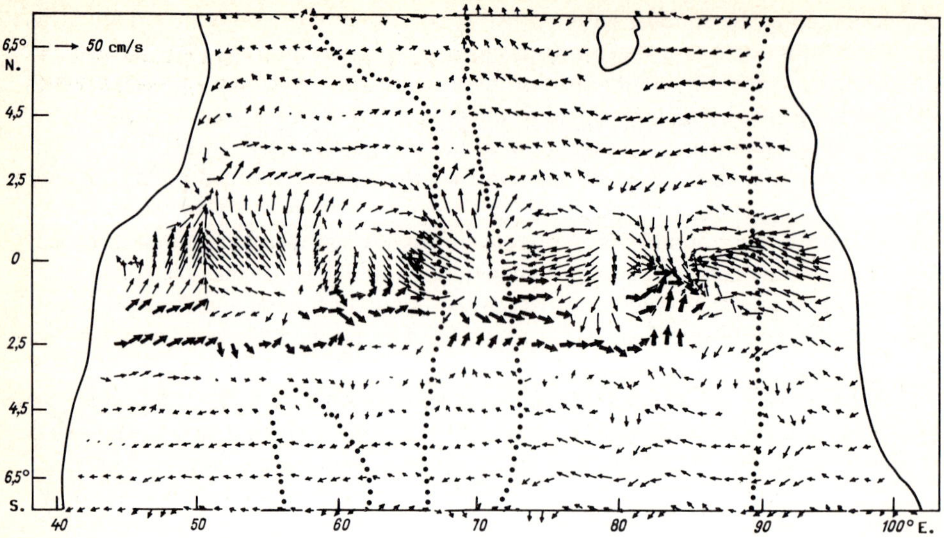

Fig. 50. The currents of the Indian Ocean equatorial zone at 100 m depth in the winter season. *Dotted lines* denote the main underwater ridges, *thick arrows*, the equatorial undercurrent

topography the Neuman problem was solved instead of the Dirichlet problem (cf. Sect. 2.2.2). The boundary velocity values at liquid (north and south) boundaries are obtained by the quasi-geostrophic model; at solid boundaries (west and east), non-slip conditions are constructed. The values of the horizontal and vertical turbulent exchange coefficients are, as before, assumed to be equal to the generally accepted values for the equatorial zone: 10^7 and $10^2\ \mathrm{cm}^2\,\mathrm{s}^{-1}$ respectively (O'Brien 1979; Semtner and Holland 1980). The time step is specified as 1 h. The process is adjusted in 1–2 months, depending on the input parameters.

The most interesting peculiarities of the equatorial Indian Ocean circulation is the system of alternate cells of divergence and convergence located along the equator in a narrow belt from 5°N to 1.5°S, which are clearly observed already in the subsurface layers (Fig. 50). These cells are preserved in deep-sea water, being somewhat transformed and shifted in space, i.e. they are of a quasi-geostrophic nature. In the vertical velocity field they conform with zones of intense upwelling and downwelling (Fig. 51). North and south of the 3° belt, the obtained flow schemes agree qualitatively with modern concepts, demonstrating all three basic elements of circulation: South Equatorial, monsoon and Somalian currents. It is of interest that despite the clearly pronounced seasonal variations of the atmospheric circulation of the Indian Ocean (summer and winter monsoons), which are responsible for the main reconstruction of the general system of flows, the equatorial cell system, except for a small western area, is maintained without qualitative changes. When interacting with adjacent quasi-zonal flows, it affects them significantly. So, the subsurface countercurrent (an analogue of the Tareev current) appeared to be substantially drifted to the south, and depending on the location and nature of cells, it either approaches the equator or moves away from it, passing by as a flow of 100–300 km width in the belt from 0° to 2.5°S, i.e. the line of the jet mid-stream is of a clearly

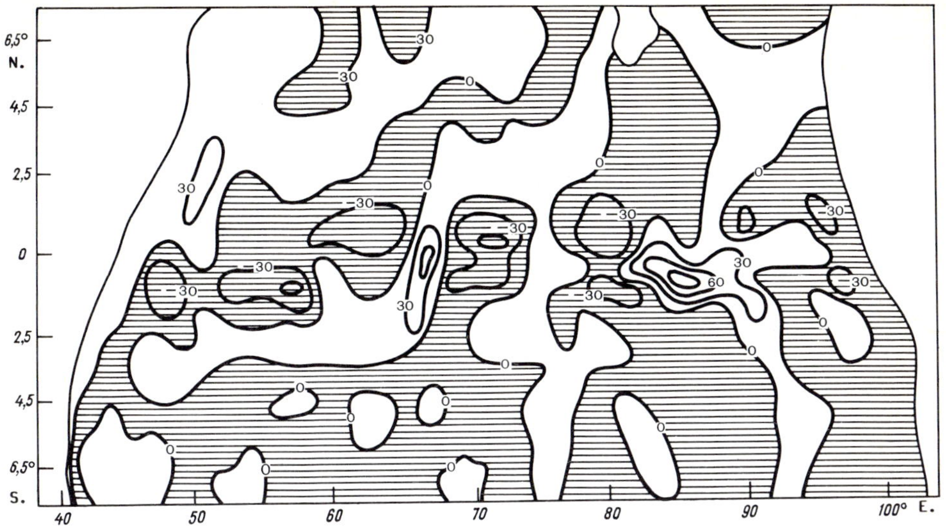

Fig. 51. The field of the vertical velocity in the Indian Ocean equatorial zone at 100 m depth in the winter season, at 10^{-5} cm s^{-1} (the upwelling areas are shaded)

pronounced wave nature. The density field analysis showed that the horizontal density distribution from west to east along the equator is wavelike at all depths and has some peculiarities, i.e. maxima and minima (cells of heavy and light water), which correspond to cells of divergences and convergences (upwelling and downwelling). Cells in the density and circulation fields correlate with the large-scale peculiarities in the bottom relief: ridges and abyssal plains. So, the most powerful zone of divergence and upwelling is located (in summer and winter seasons) at 68°–70° E, where the equator crosses the mid-ridge. The impact of the bottom relief is observed outside the 3° belt also. However, its effect near the equator has some peculiarities, which are mainly attributed to the quite different force balance there. So, calculations performed by a quasi-geostrophic model were not indicative of the existence of the equatorial cells, though outside the equatorial belt the circulation pattern was qualitatively consistent with that obtained here and demonstrated the bottom relief effect as well.

To check the model response to the changes in the input parameters, a number of additional, numerical experiments were carried out. First, the turbulent exchange coefficients were varied. It turned out that even in the case of strongly increased coefficients of the horizontal exchange, i.e. up to 5×10^8 cm^2 s^{-1}, the cellular structure of the circulation at the equator is maintained, though the intensity and number of cells decreased, i.e. the weak cells "are diffused". The change of the vertical exchange coefficient does not result in the destruction of the cell system, affecting, principally, only the surface layer flow. Second, an experiment was carried out, in which the density field within 1.5°N–1.5°S was "corrected" in such a way that isopycnals in the upper layer had a regular slope from east to west. As a result, a monotone sea surface slope from west to east was obtained and consequently, a continuous eastward jet at the equator. The system of the equatorial cells disappeared, i.e. it is determined by the specificity of the density field rather than generated by the model. This has been confirmed by the calculations considered below of the equatorial Atlantic currents, which did not indicate such cells. Third, experiments with the filtration of the specified fields were carried out, whereby the cellular flow structure is kept at the equator. So, the maps given here were compiled on the basis of calculations performed on the double-smoothed (cosine filter) specified fields.

As mentioned previously, for the equatorial Atlantic two numerical experiments were carried out. In one of the experiments the Dirichlet problem was solved for the sea surface, and in the other experiment, the Neuman problem was solved. A number of results obtained from the second experiment are shown in Figs. 52 and 53. To avoid overburdening of the map, the velocity vector arrows in Fig. 52 are shown throughout with a 1° step. The sea surface topography indicates a system of three anti-cyclonic gyres north of the equator, a system of at least three cyclonic gyres south of the equator and a west-east slope of the sea surface in the central section, in the vicinity of the equator. A system of three equatorial countercurrents is clearly observed in the subsurface circulation field: a northern, southern and the Lomonosov current proper. The latter looks like a continuous, intense, jetlike current (up to 100 cm s^{-1}) with marked differences in velocities compared with the abjacent areas. Other peculiarities should be mentioned: markedly pronounced water flow to the equator, both from the north and south near the South American coast; a marked impact of the bottom relief (even at a depth of 100 m); markedly pronounced flow convergence, almost over the whole jet of the Lomonosov current (except for a small area between 10° and 5° W, where a local divergence zone is located, and the jet considerably slackens).

Compared with the results obtained in the experiment with the Dirichlet problem for the sea surface, the solution is adequately improved. The variations are particularly significant in the southern area, which is characteristic of the weakest density gradients. The west-east drop of the sea surface along the equator in the Dirichlet problem is disturbed by two local rises, the most marked of which is located between 12° and 7° W. The total sea surface difference in the Dirichlet problem is much less: 30 cm compared with 45 cm in the Neuman problem. The general structure of the sea surface topography north of the equator is the same: both schemes indicate a system of three anti-cyclonic gyres. The Dirichlet problem in the subsurface current field lacks the southern, and weakest branch of the equatorial countercurrent (the current flows in a westerly direction) and the northern branch of the countercurrent is much less pronounced. The system of water flow to the equator in the western part is maintained as well as the location of the Lomonosov current. As in Fig. 52 it is shifted south of the equator, which is attributed to the nature of the specified density field (which does not change in the diagnostic problem). However, the countercurrent velocity is much lower, and it does not look like an intense jet. Furthermore, in the area between 12° and 7° W (in the zone of the local elevation of the sea surface) the jet is "broken". So, the modified model version, which includes the Neuman problem for the sea surface, has led to significantly more realistic results. At low latitudes the accuracy of the sea surface calculations is of particular importance. Due to the smallness of the Coriolis param-

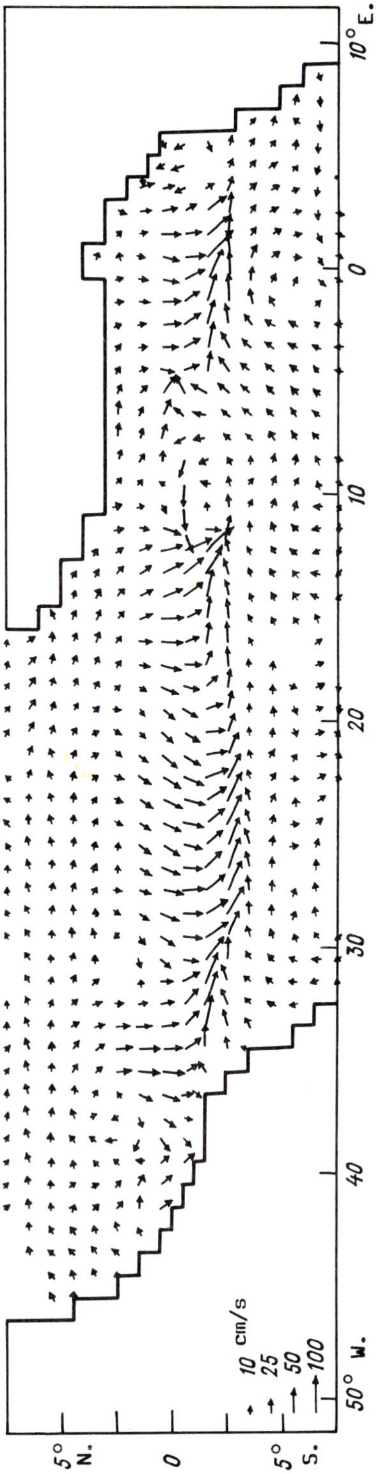

Fig. 52. The currents of the equatorial Atlantic at 100 m depth in the summer season

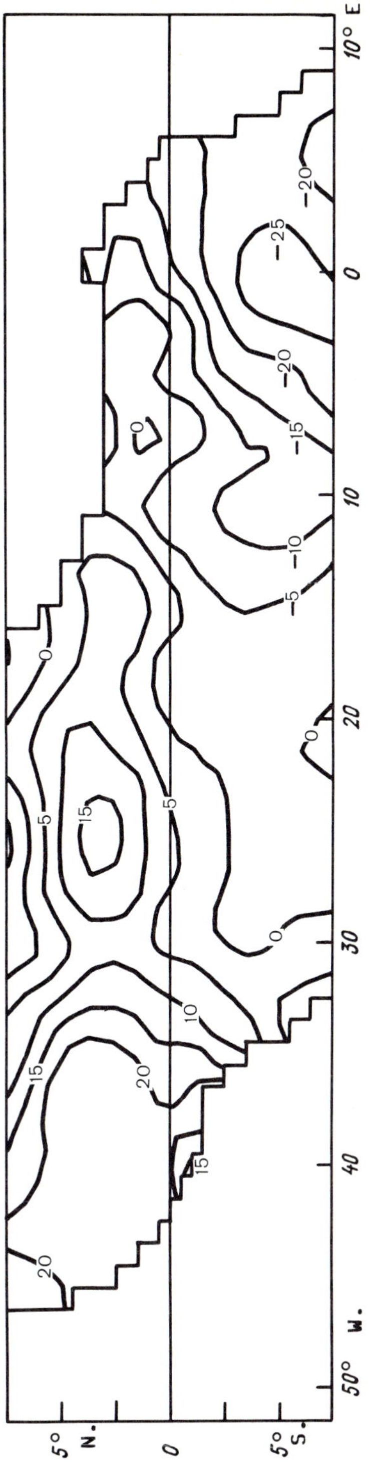

Fig. 53. The topography of the transformed level of the equatorial Atlantic in the summer season (cm)

eter, the relatively small pressure factors are responsible here for much stronger flows than at mid-latitudes.

As a whole, the nature of the equatorial circulation of the Atlantic and Indian Oceans differs significantly. The general sea surface slope in the equatorial Atlantic (Fig. 53) occurs in both seasons from west to east, whereas in the Indian Ocean it is, in contrast, from east to west (the slope is 20–30 cm). Instead of a system of equatorial cells of divergences and convergences of the Indian Ocean, in the Atlantic a system of anti-cyclonic gyres north of the equator and a system of cyclonic gyres south of the equator take place. The cellular structure formation of the Indian Ocean is certainly associated not only with the peculiarities of the bottom relief (its effect is as strong as that in the Atlantic), but evidently also with the peculiarities of the monsoon circulation of the atmosphere. The issue of similarity of the Tareev current with the Lomonosov and Cromwell currents is not an ordinary one. If the last two currents are easily attributed to a general west-east slope of the sea surface (and to the respective isopycnal slope), produced by relatively stable year-round trade winds, the Tareev current is evidently generated by the dynamics of local areas of the required west-east sea surface slope (more developed in the winter season) as well as by the dynamics of the equatorial cell systems of divergences and convergences.

The existence of dynamic cells in the equatorial ocean zones, i.e. zones of divergences and convergences in the Indian Ocean and anti-cyclonic and cyclonic gyres in the Atlantic, agrees well with the cellular structure of carbon distribution in the upper precipitation layer (Romankevitch 1977). The map of currents compiled from ship-drift data and given in "The Atlas of Oceans" (1977) provides some indication of the possible availability of a cell system in the Indian Ocean. Despite a large, 5° averaging scale, which considerably suppresses the local peculiarities, the map indicates a system of alternate vectors of zonal and meridional directions in the equatorial region, unlike quasi-zonal currents south and north of it.

To check the obtained peculiarities of the equatorial circulation in the Indian Ocean, further investigations are certainly required, which would include both instrumental measurements and numerical modelling, which should first and foremost be based on the more accurate data of hydrological observations. A 1° averaging scale of the density field is still too large for the equatorial zone, and is comparable, for example, with jet widths of countercurrents. Furthermore, the quality of the density fields here should be much higher than at mid-latitudes, since density gradients here are much smaller. The increased demands upon the quality of the climatic density fields are stipulated by the clearly expressed wave dynamics of the equatorial ocean zone.

5.7 The Calculation of Flows in the Black Sea Offshore Zone

The majority of calculations of the Black Sea flows (Gamsakhurdia and Sarkisyan 1975; Phylippov 1968) are based on large grid sizes of 0.5° and larger. They reflect therefore only the basic characteristics of the sea circulation. In particular, all known schemes of the western Black Sea flows, based on numerical calculation results, show only the main Black Sea flow (MBF); the local details of the circulation, for example, those in Zenkovitch (1974), are lacking. This section deals with some results of diagnostic and adjustment calculations of flows in the western Black Sea offshore region on high resolution grid sets (Demin and Trukhchev 1984a, b, 1985).

5.7.1 The Diagnostic Calculations of Flows for the Summer and Fall Seasons

The calculations are based on two hydrological surveys, carried out near the Bulgarian coasts in the fall and summer seasons. In the first case the area is considered between the Bulgarian coast and 29°30′E, 43°45′N and 42°N. The grid size is 5 miles, 12 levels are taken vertically: 0, 10, 20, 40, 60, 80, 100, 200, 500, 1000, 1500 and 1800. The density field is constructed on the basis of the hydrological survey, performed on board the research vessel "Orbeli" in October 1976. The sea surface wind stress is calculated by the average for the period of the atmospheric pressure field survey by the Akerblom formula.

The first cycle of calculations is carried out by the quasi-geostrophic model. The equation of the sea surface topography was approximated in two ways: by the scheme of directed differences and the Ilyin scheme (1969). The solution of the difference problems was carried out according to the Gauss-Zeidel method. The sea surface at the basin boundary was found by the method of "sweep" dealt with in Chapters 2–4. The water flux at liquid sections of the boundary was specified by the quasi-dynamic method. The vertical turbulent exchange coefficient was assumed to be $10\,\mathrm{cm}^2\,\mathrm{s}^{-1}$.

The sea surface topography calculation results are given in Fig. 54. It can be seen from Fig. 54 that the SST calculated by a quasi-geostrophic model and quasi-dynamic method differ considerably only in the near-shore shallow-water zone. The differences are even less in solutions obtained with schemes of directed differences and the Ilyin scheme. Both results, despite the sharp differences in the grid scale, coincide with the earlier obtained results of diagnostic calculations of large-scale ocean currents (Sarkisyan 1977a).

The main characteristics of the sea surface topography are as follows: the anti-cyclonic circulation system in the south, the cyclonic system in the north, the MBF jet near the open boundary and a weak gradient field in the offshore zone. The sea surface topography adjusts well with the surface flow structure, particularly in the shallow-water part of the area. This indicates that the direct wind effect is relatively low, and even at the surface the flows are principally of a gradient nature. Indeed, the velocity component of the wind drift does not exceed $8\,\mathrm{cm\,s}^{-1}$, when the characteristic value of the gradient component in the offshore area is 40–$50\,\mathrm{cm\,s}^{-1}$.

The main characteristics of the sea surface flow field are: an intense jet (up to $100\,\mathrm{cm\,s}^{-1}$), which turns south-eastward, south of the Kaliakra Cape, generating here a cyclonic meander and an anti-cyclonic gyre in the central basin with a centre opposite the Burgas Bay. The anti-cyclonic circulation system is also displayed in the flow pattern of Zenkovitch (1974). The gyre is evidently generated by the MBF jet under the impact of local conditions mainly the effect of the bottom relief. The nature of circulation changes insignificantly. The reconstruction of circulation

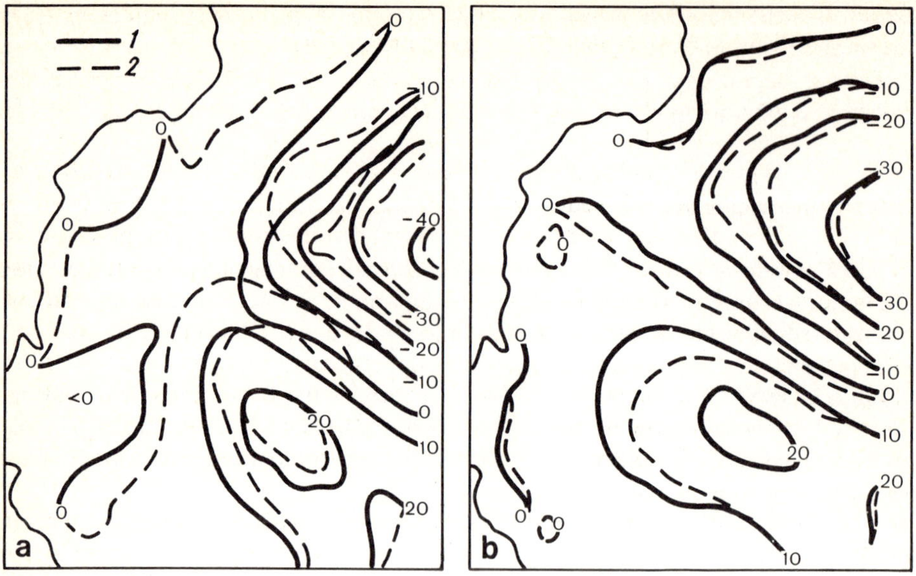

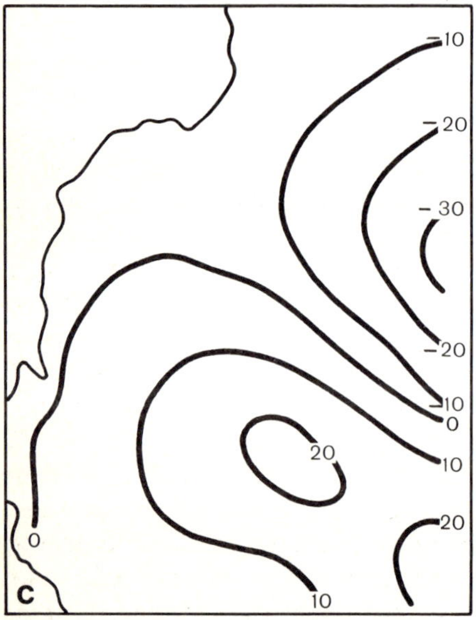

Fig. 54a–c. The topography of the transformed level of the western part of the Black Sea in the autumn season in diagnostic models, in cm: **a** the quasi-geostrophic model (*1*) and the quasi-dynamic method (*2*); **b** a non-linear model, calculated by the Ilyin scheme (*1*) and the scheme of directed differences (*2*). The SST at the boundary is calculated by the quasi-dynamic method, $\mu = 7 \times 10^5\,\mathrm{cm^2\,s^{-1}}$, $\nu = 10\,\mathrm{cm^2\,s^{-1}}$; **c** a non-linear model, calculated by the Ilyin scheme. The level at the boundary is calculated from Eqs. (1.90), (1.91) of Chapter 1; μ and ν from Eqs. (7.1) and (5.2) of Chapter 5

is only observed in the near-bottom layers, which is attributed to a new flow generation.

Let us now consider the results obtained by the non-linear model. The modified version, with allowance for variability of the vertical and horizontal turbulent exchange coefficients ν and μ, is used in the calculations. The first was determined from Eq. (5.12) by the technique of Marchuk et al. (1976a). The thickness of the upper, quasi-homogeneous layer was limited from the top by meeting the conditions $\nu \leqslant 1\,\mathrm{cm^2\,s^{-1}}$, and from the bottom by the second level height (10 m). Furthermore, the upper limit (equal to $100\,\mathrm{cm^2\,s^{-1}}$) was used for ν to avoid abrupt "bursts" in the solution (including value v), which are clearly observed at the first stage of integration. Coefficient μ was calculated by the known formula:

$$\mu = (Ch)^2 \left| \frac{\partial u}{\partial y} + \frac{\partial v}{\partial x} \right|, \tag{5.17}$$

where h is the grid size, and coefficient C is ten times smaller than for the atmosphere. The formula (5.17) was selected for the sea by numerical experiments in such a way that the mean value of μ equals $7 \times 10^5\ \mathrm{cm^2\,s^{-1}}$, resulting from the Joseph-Sender's (1958) formula. The value C is 0.5, and the range of the μ spacial variation is from 10^4 to $10^7\ \mathrm{cm^2\,s^{-1}}$. The velocity values at liquid boundary sections are obtained by the quasi-geostrophic model; at solid sections, non-slip conditions are taken.

The solution was sought by the method of reaching the steady state by semi-implicit schemes, constructed with the help of the method of directed differences, the Ilyin method and the modified method of directed differences (Roache 1976), differing from the conventional method in the introduced weight coefficients, and consequently, a higher approximation order. The differential problems were solved by the Gauss-Zeidel method. The schemes make calculations possible with small enough μ values (at 5 mile size, up to $10^4\ \mathrm{cm^2\,s^{-1}}$), though they limit rather strictly the time step: at $\mu = 7 \times 10^5\ \mathrm{cm^2\,s^{-1}}$ for the scheme of directed differences, the optimum step is about 15 min, for the Ilyin scheme it is around 10 min. The time of reaching the steady state is about several days, depending on the scheme, input parameters, specified approximation and techniques of parameterization of the vertical and horizontal turbulent exchange. In the deep water part of the basin, the solutions of all three schemes are similar (Fig. 54) and adjust well with that from the quasi-geostrophic model, i.e. the force balance is similar to the geostrophic one. This is revealed by the current fields comparison.

The numerical experiments carried out when varying μ and v, which were assumed both constant and variable, indicated that in the deep water part of the basin the variations of the solutions are insignificant.

Appreciable variations are observed only in the near-shore zone and somewhat less variations are observed near the open boundaries. The latter results in some improved adjustment of the solution with the boundary conditions. Such complicated calculations might be not reasonable in diagnostic problems, at least for the deep sea. It is of importance only that the coefficients were not strongly overestimated.

Briefly, the force balance in the equation of motion should be mentioned. The Coriolis force, vertical turbulent exchange and horizontal, non-linear summands are significant in the upper sea layer. In the intermediate and abyssal water layer the balance is close to the geostrophic. The role of the non-linear summands is of particular importance in the near-shore and bottom slope areas. For example, at a depth of 80 m, where the shelf-slope boundary is approximately located, the mean "weight" of $v\,\partial v/\partial y$ almost coincides with the "weight" lu and lv.

The numerical experiments, when varying boundary conditions for the velocity, indicated that the anti-cyclonic gyre and the MBF jet were maintained in all versions, i.e. these are rather stable elements of circulation in the area under study.

Furthermore, a numerical experiment was carried out when sea surface topography was specified at the boundary by the quasi-dynamic method. It is of interest

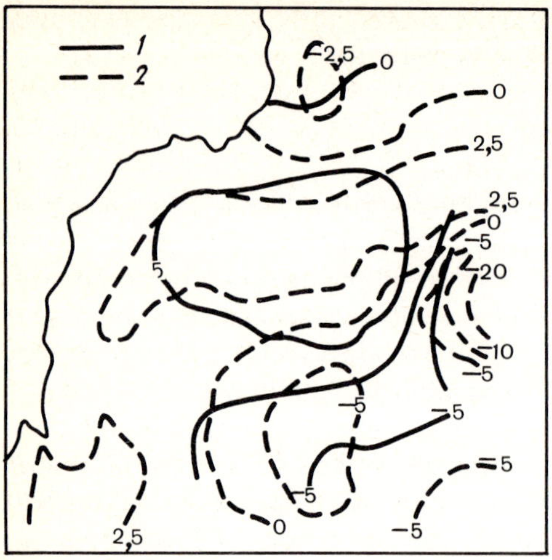

Fig. 55. The same as in Fig. 54, but for the summer season in diagnostic models, in cm (*1*, a non-linear model; *2*, is a quasi-geostrophic model)

that despite some disadjustment of the boundary conditions for velocity and sea surface, which was observed in the experiment, the nature and time of reaching steady state did not change significantly. There were less variations in the solution compared with velocity variations at the boundary. This results from the fact that the quasi-dynamic method provides the major solution of the equation.

The flow fields calculated by the non-linear model (Fig. 57) adjust well with results obtained for the quasi-geostrophic model at the deep water part of the area and differ qualitatively from them in the shelf zone. The ageostrophic effects are of significant importance in the shelf zone. So, the quasi-geostrophic model as well as dynamic method are seldom applicable here. The deep water currents in the quasi-geostrophic non-linear models are similar enough. The balance in the abyssal layers is already close to the geostrophic.

The Ilyin scheme and that of directed differences provide close enough results for both the sea surface topography and flow fields. In the first case the flow field is somewhat less smooth. The velocity in the MBF jet increases by 10–15%.

The second run of calculations is carried out for the area located between the Bulgarian shore and 29°17′E, 43°35′N and 42°35′N. The grid size is 5 miles. Nineteen levels are taken vertically: 0, 5, 10, 20, 30, 40, 60, 80, 100, 150, 200, 300, 500, 750, 1000, 1250, 1500, 1750, 2000. The density field is constructed on the basis of data from the Bulgarian expedition on board the research vessel "Researcher" in August 1981. The sea surface wind stress is calculated by the Akerblom formula from the atmospheric pressure field mean for the period of survey.

At the first stage, as before, the calculations were performed by the quasi-geostrophic model. The equation for the SST was approximated by the Ilyin scheme (1969). The differential problem was solved according to the Gauss-Zeidel method. The vertical turbulent momentum exchange is assumed to be $10\,cm^2\,s^{-1}$. The sea surface topography at the basin boundary was determined from the boundary equations of the first order, based on the tangential derivatives of the sea surface.

At the second stage, a non-linear model is used. The problem was solved by the method of reaching steady state by the semi-implicit scheme of directed differences.

The technique of calculations coincides primarily with that used for the fall season. In contrast to those calculations, the first derivatives in the equation of the sea surface topography were, however, approximated by the central differences, i.e. with second order of accuracy (in the equation of motion, as before, with directed differences). Equation (1.130) is not an equation with a small parameter at the highest derivative (unlike the equation of the sea surface topography in the quasi-geostrophic model, i.e. the equation of vorticity); on the contrary, the main summand on the left-hand side is the one with the second derivative of the sea surface. This indicates the effect of the barotropic part of the pressure gradient. The time step is assumed (as before) to be 15 min. The time for reaching steady state was several days. The velocities in the open boundary sections were specified from the solution obtained with the quasi-geostrophic model. The technique of calculating μ and v and the boundary values of the sea surface is the same as before. The vertical velocity was determined, as in the quasi-geostrophic model, from the continuity equation differentiated by z (making it possible to use directly both boundary conditions for w). The solution of the equation coincides with the solution of the continuity equation when observing the integral condition of continuity $\operatorname{div} \mathbf{s} = 0$ ($\mathbf{s}$, total flow).

Let us now consider the calculation results. The basic peculiarities of the surface circulation shown both in the sea surface topography schemes (Fig. 55) and flow maps (Figs. 57, 58) are: an anti-cyclonic gyre in the central area, a cyclonic system in the northern area, the periphery of the MBF jet near the open boundary, a quasi-zonal east flow near the Kaliakra Cape and a relatively weak flow primarily of a southern direction near the shore. All these characteristics are revealed by both models, though differences are significant. The non-linear model indicates, north of Kamchia, the northern offshore flow, i.e. the eastern periphery of the anti-cyclonic gyre. The latter is markedly pronounced, while the southern cyclonic system became

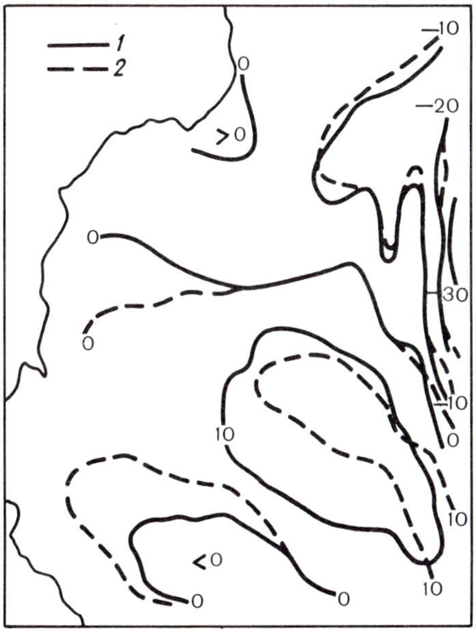

Fig. 56. The same as in Fig. 54, but for the autumn season after the adaptation, in cm. Version A (*1*, the Fiadeiro-Veronis scheme; *2*, the scheme of directed differences)

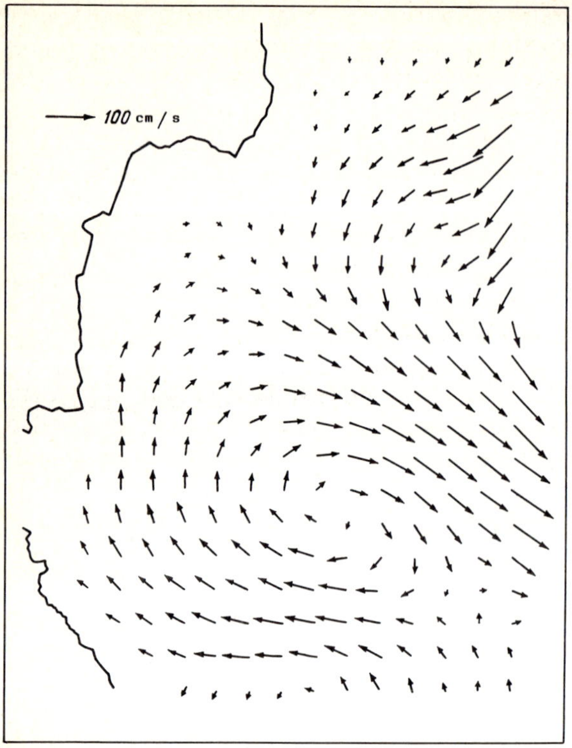

Fig. 57. The sea-surface currents of the western part of the Black Sea in the autumn season, calculated by the non-linear diagnostic model (the parameters of the model are the same as for the level in Fig. 54c)

weaker and decreased in the meridional direction. The obtained schemes on meridional flows differ from a similar one for the fall season. Instead of the anti-cyclonic gyre located against the Burgas Bay in the fall season, a cyclonic system was generated. The MBF jet shifted to the east and north of the cyclonic gyre, the anti-cyclonic one was generated, etc. All this indicates that the temporal variation of flows near the Bulgarian shore can be of a qualitative nature.

The current reconstruction with depth, in contrast to the fall season, is rather significant. At a depth of 100 m the circulation is expressed by the MBF jet and two gyres located along the bottom slope: the anti-cyclonic gyre in the north and cyclonic gyre in the south with the interface at approximately 43°15′N. At a depth of 500 m the northern flow is clearly observed. West of it a cyclonic gyre is located, extending meridionally along the bottom slope. Both models provide closer results, though velocity values in the geostrophic model are significantly higher. At a depth of 1000 m the circulation is expressed by one anti-cyclonic gyre. Its western periphery is a flow in a northern direction. As a whole, the circulation in the abyssal layer is close to that obtained for the fall season, which also indicates the northern jet as a main characteristic.

The vertical velocity fields present a rather complicated situation due to a number of available zones of convergences, divergences and unequally directed gyres. The main peculiarity of the upper layer is the upwelling in the southern part and less intense downwelling in the northern.

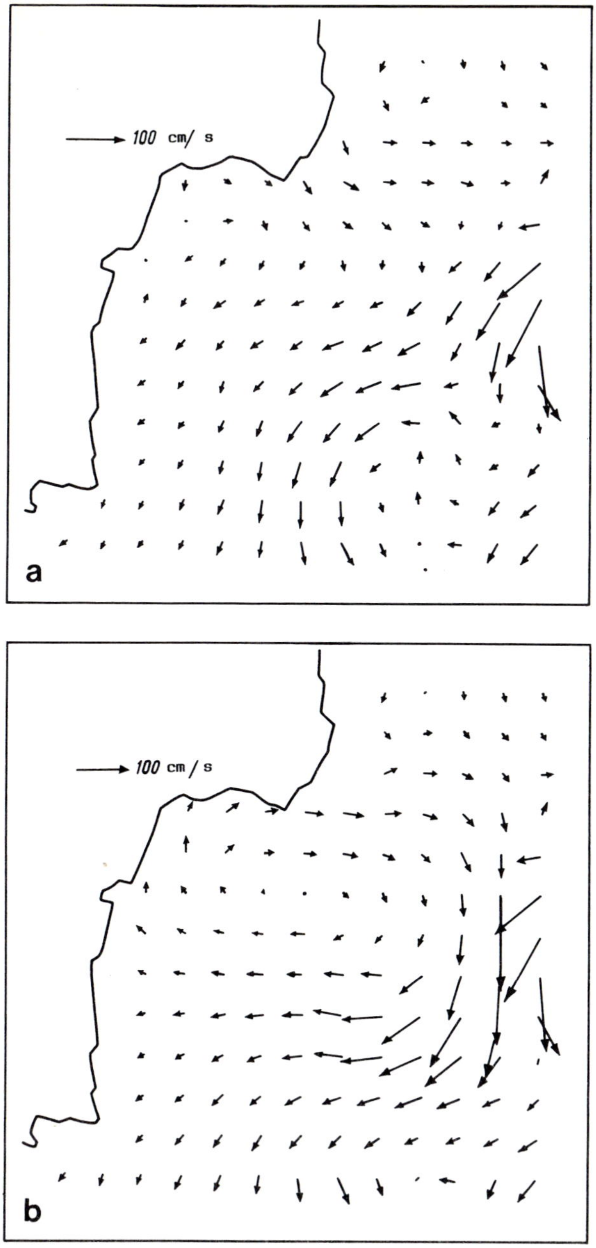

Fig. 58a, b. The same as in Fig. 57, but for the summer season in the diagnostic, quasi-geostrophic (**a**) and non-linear (**b**) models

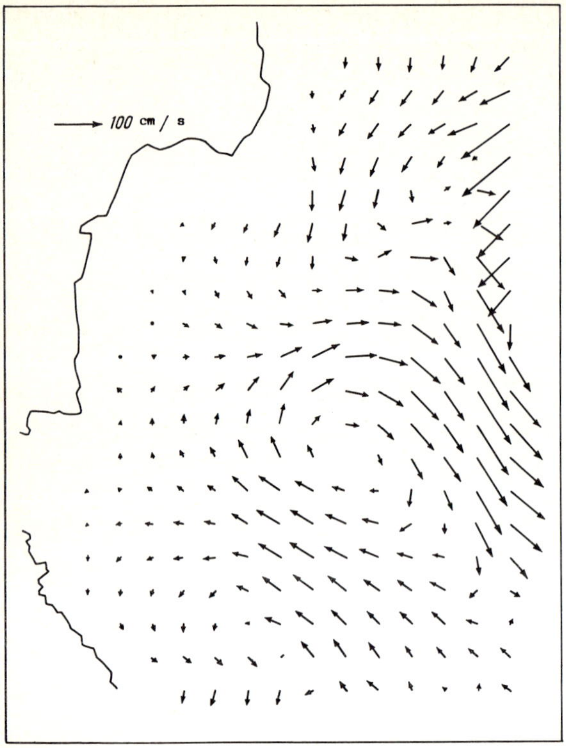

Fig. 59. The sea-surface currents of the western part of the Black Sea in the autumn season after the adaptation

The dominating force balance in the quasi-geostrophic model will now be considered. The main summands in the equation of forces are the Coriolis force and the pressure gradient. The vertical viscosity and non-linear summands contribute considerably to the upper layer. The differences in results from the two models are attributed to the force balance and the existence of a feedback effect of the model based on the primitive equations of motion as well as to filtering characteristics of the quasi-geostrophic model used. They are, in part, certainly connected with the approximation, in particular with the computational viscosity. All these differences significantly effect the relatively weak flows in the shelf zone.

5.7.2 Numerical Experiments on the Calculation of the Vertical Velocity Component

Maximum inaccuracy is observed in the numerical models of the sea flows when calculating vertical velocity. The latter is usually determined from the vertically integrated equation of continuity through the divergence from the horizontal velocity. Sometimes it is calculated from the vertically differentiated (preliminary) equation of continuity, which results in the equation of the second order and makes it possible to use directly both boundary conditions, i.e. at the surface and at the sea bottom. In some models it is determined from the formulas obtained when substituting into the continuity equation, the expressions of the horizontal flow velocity components, obtained from the quasi-geostrophic model (Sarkisyan 1977a). Occasionally, in order to calculate the vertical velocity, the vorticity equation is used (Seidov 1977). More complicated methods are known (Marchuk 1967a) based on the application of special equations for the vertical velocity, which were obtained from the specified system of the model equations. To determine w one should solve three-dimensional, boundary-value problems. The instrumental measurements of the vertical velocity are very complicated and principally all available information is based on indirect methods.

The errors in the vertical velocity measurements can be stipulated by: (1) the roughness of the difference approximation of the model equation (due to peculiarities of the force balance in the continuity equation, i.e. small difference of large values, relatively small errors in the horizontal velocities can result in substantial errors for the vertical velocity); (2) insufficient adjustment of the model equations which may result from the inadequacy of assumptions and simplifications specified in each separate model equation; (3) incompleteness of the hydrodynamic model. In the diagnostic models insufficient adjustment of the specified density, wind and bottom relief fields can represent an additional source of errors.

The extent of inaccuracy due to these factors differs according to the various hydrodynamic characteristics. So, the sea surface topography (the integral function), as a rule, adjusts qualitatively well in many models. The horizontal velocity differs to a wider extent, particularly in the abyssal layers. The vertical velocity may differ so much that it is often difficult to compare it even qualitatively.

A number of numerical experiments on the calculation of the vertical velocity are carried out in Sarkisyan et al. (1983a, b) based on two diagnostic models, quasi-geostrophic and non-linear. The calculations are performed for the previously mentioned Black Sea area based on the data of the hydrological survey in the fall season. The technique of calculations and the parameters of the quasi-geostrophic model are the same as before. The non-linear problem was solved by the method of reaching the steady state by the semi-implicit scheme of the directed differences. The vertical turbulent exchange coefficient is assumed to be $10\,\mathrm{cm}^2\,\mathrm{s}^{-1}$ as in the quasi-geostrophic model, that of the horizontal $7 \times 10^5\,\mathrm{cm}^2\,\mathrm{s}^{-1}$. The remaining parameters of the model and numerical algorithm are the same as before. In both models identical boundary conditions for w are specified: "rigid lid" at the surface, and non-slip at the bottom.

For calculating w the following ways are used in both models: 1. Calculation from the continuity equation, integrated by z; 2. Calculation directly from the continuity equation. Three-point difference approximation of $\partial w/\partial z$ is used on the irregular grid net, which results in a differential equation of the second order, and makes it possible to specify two boundary conditions. The difference problem is solved by the sweep method or by the iteration method. In general, however, the vertical grid size can be at some depths regular (as in this problem) and the diagonal prevailing of the matrix will be disturbed. To avoid this and to solve the difference equation, a method of non-monotonic sweep (Samarsky and Andreyev 1976; Samarsky and Nikolaev 1978) was used (certain restrictions are imposed on a number of equidistant points located one after another); 3. Calculation from the continuity equation preliminarily differentiated by z, making it possible to meet both boundary conditions even on the regular grid set. The same method of non-monotonic sweep is used. The solution of the differentiated equation at assumed boundary conditions is as follows:

$$w = -\int_0^z \left(\frac{\partial u}{\partial x} + \frac{\partial v}{\partial y}\right) d\xi + \frac{\int_0^H \left(\frac{\partial u}{\partial x} + \frac{\partial v}{\partial y}\right) d\xi}{H} z \,. \tag{5.18}$$

So, unlike the continuity equation, a straight line is introduced artificially to "correct" the solution if, due to some reason the relation $\operatorname{div} \mathbf{s} = \int_0^H \left(\frac{\partial u}{\partial x} + \frac{\partial v}{\partial y}\right) d\xi = 0$ is not met. It follows, in particular, that differences in the solutions should, in general, increase with a depth, if $\operatorname{div} \mathbf{s} \neq 0$.

Let us introduce the following symbols: $w^{(0)}$ the vertical velocity field obtained from the integrated continuity equation; $w^{(1)}$ obtained from the continuity equation directly; $w^{(2)}$ obtained from the differentiated continuity equation.

Let us first consider results of calculation for w by the quasi-geostrophic model (Fig. 60). Consideration is given to an adequate qualitative relationship of fields $w^{(0)}$ and $w^{(1)}$ even in the lower sea layers. Velocity values are similar as well, i.e. the method of solving the continuity equation did not substantially affect the results. The field $w^{(2)}$ differs from them very much, even in the upper 100-m layer. The vertical profiles of $w^{(2)}$, as a rule, are much smoother because of the existence of the source in Eq. (5.18), which increases linearly with depth and has a sign opposite to the one in the solution of the continuity equation. Consequently, velocity values in $w^{(2)}$ decrease substantially compared with $w^{(0)}$ and $w^{(1)}$, particularly in the abyssal and bottom layers, where they are generally the largest. The maximum differences of $w^{(2)}$ from $w^{(0)}$ and $w^{(1)}$ are therefore observed in the abyssal and bottom layers of the open sea as well as in the shallow-water offshore areas (i.e. when the calculation level is close to the sea bottom). The significant difference of fields $w^{(0)}$ and $w^{(1)}$ from $w^{(2)}$ is indicative of the inadequate fulfillment of the integral continuity conditions. The errors in fields u, v are strongly reflected on w, though they can be relatively low and hardly noticeable on the maps of horizontal flows.

Let us consider now the calculation results from the non-linear model. First, fields $w^{(0)}$ and $w^{(1)}$ are still closer to each other (Fig. 61) than in the quasi-geostrophic model; and second, in contrast to the latter, they are also close qualitatively, at least in the upper sea layer, to $w^{(2)}$. This indicates that differences between the results from different calculation methods decrease in the more complete and adjusted model with a feedback, and the integral condition of continuity is better fulfilled. All three versions, as in the quasi-geostrophic model, are best adjusted in the upper part of the deep-water area. The differences increase in the offshore, shallow-water area with depth, however, much less compared with a quasi-geostrophic model.

The way in which w is calculated practically does not effect the field ζ. There are some differences in the horizontal velocity fields obtained by the non-linear model, however, they are usually only in the abyssal water layers and are only of a quantitative nature (the feedback is missing in the quasi-geostrophic model). The vertical velocity calculated within the non-linear model adjusts better with the

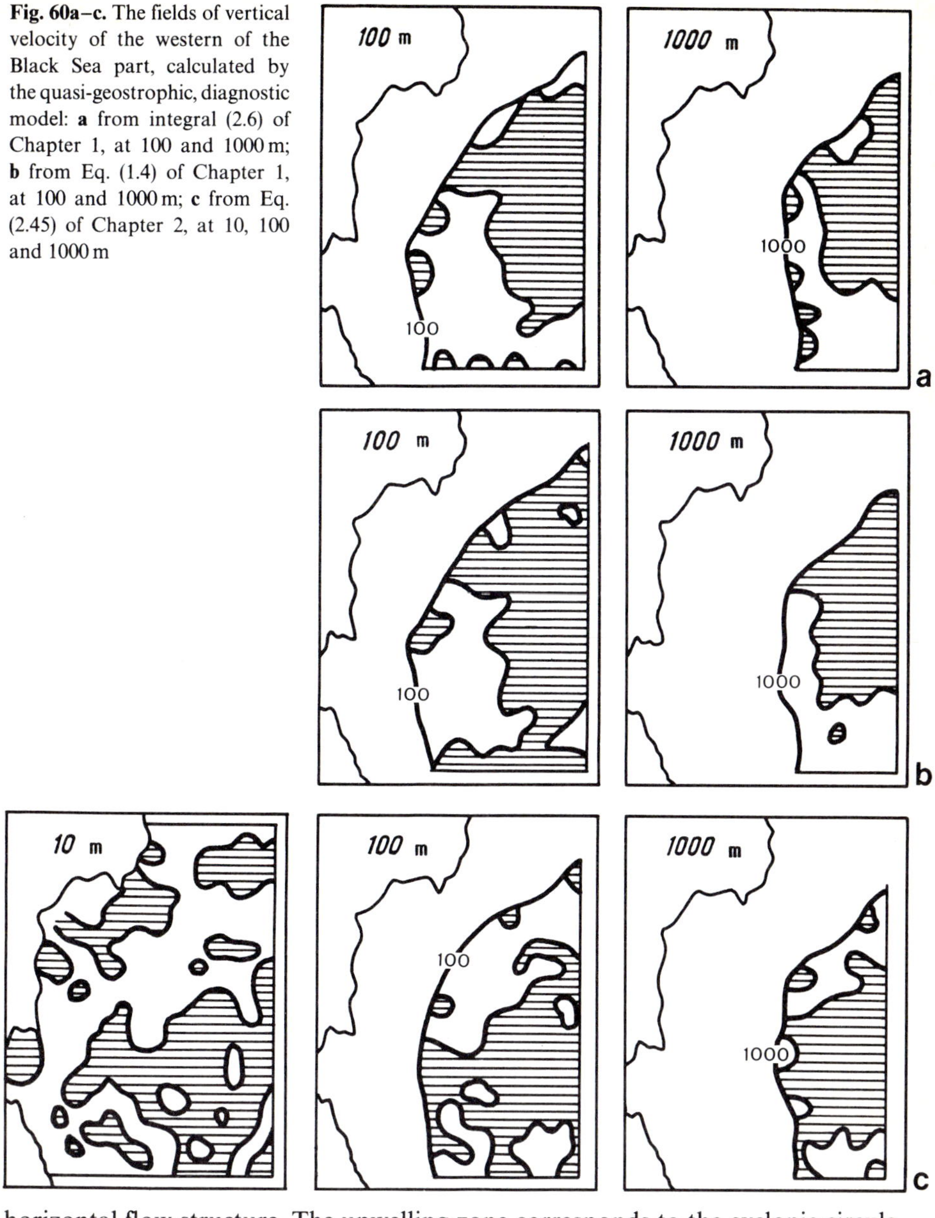

Fig. 60a–c. The fields of vertical velocity of the western of the Black Sea part, calculated by the quasi-geostrophic, diagnostic model: **a** from integral (2.6) of Chapter 1, at 100 and 1000 m; **b** from Eq. (1.4) of Chapter 1, at 100 and 1000 m; **c** from Eq. (2.45) of Chapter 2, at 10, 100 and 1000 m

horizontal flow structure. The upwelling zone corresponds to the cyclonic circulation in the northern area; the downwelling zone corresponds to the anti-cyclonic gyre in the central part; and once again the upwelling zone corresponds to the flow divergence in the southern part. The correspondence, however, decreases with depth, and at 1000 m, for example, it is not observed.

As a whole, fields w obtained by these two models differ significantly, and the differences are not only of a quantitative but also of a qualitative nature. Let us emphasize again that despite significant differences in fields w, the horizontal flow

Fig. 61a–c. The fields of vertical velocity, calculated by the non-linear diagnostic model. The denotations are the same as in Fig. 60

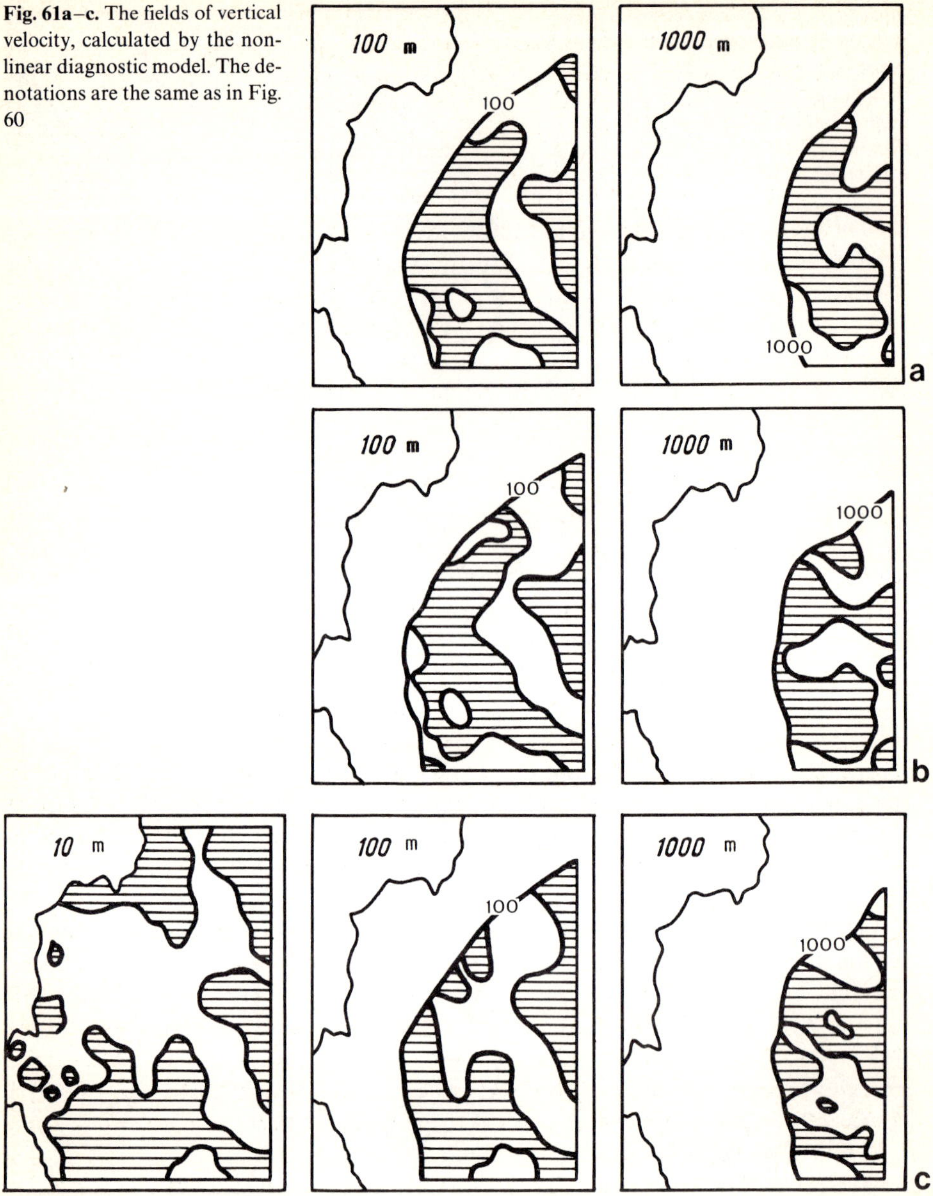

fields differ strongly only in the offshore, shallow-water zone, while in the abyssal water zone they are very close.

The use of the other difference scheme, i.e. the Ilyin scheme (1969) (often used in the diagnostic calculations) in place of the scheme of directed differences, did not result in a qualitative variation of w either in the quasi-geostrophic or non-linear model. The w differences in these models as well as $w^{(0)}$, and $w^{(1)}$ fields with $w^{(2)}$ within the framework of either model are much more significant than differences obtained when using these two difference schemes.

Thus, the extent of the hydrodynamic completeness and adjustment of the model equations is of great importance for calculating w, much greater than for the horizontal velocities. The quasi-geostrophic approximation, in most cases quite applicable to the horizontal flows, is rather rough for the vertical velocity. The calculation of w from the differentiated continuity equation seems to be the most reasonable among all tested methods.

In conclusion, it is worth noting that the vertical velocity field is one of the finest criteria with respect to the quality of the numerical model. The adjustment calculations can provide the most reliable values of w. The fields adjustment in the diagnostic models is not as adequate. To provide a rather accurate observance of the integral continuity condition, one should apply conservative difference schemes, which meet the concentration law on the grid set. In this case one can hope to obtain practically similar solutions according to all three considered methods.

5.7.3 The Adjustment of the Density and Flow Fields

The adjustment calculations are performed for the same two areas of the western Black Sea and on the same grid net based on the same data from two hydrological surveys made in the summer and fall seasons as diagnostic calculations described at the beginning of this section.

The system of equations of the non-linear diagnostic model, modified with allowance for μ and v variations, is supplemented by the equation of the density diffusion (2.48). The boundary conditions for the velocity, time step, calculation technique for μ and v and boundary values ζ are the same as those used earlier in the diagnostic calculations. The adjustment process was governed by the nature of the temporal variations of the kinetic energy, by the adjustment criterion, Eq. (5.10) of Sect. 5.4 and the visualization of the solution. The adjustment time was several days.

The main cycle of calculations is carried out with the help of a semi-implicit scheme of directed differences. Furthermore, numerical experiments were carried out with the Ilyin scheme (1969) as well as the Fiadeiro-Veronis weighted-mean scheme (to approximate: the equation of the density diffusion). The solutions of the difference problems were carried out with the Zeidel iteration process. The calculation results from all three schemes appeared to be close (Fig. 56). The diffusion coefficients μ_1 and v_1 are assumed equal to the corresponding average values μ and v (7×10^5 and $10\,\mathrm{cm}^2\,\mathrm{s}^{-1}$).

The specified density fields assumed from the observations and calculated by the diagnostic version of the model flow field are specified as initial approximations.

Two basic experiments were carried out on the fall-season survey data. The nature of the boundary conditions was different: A, the specified density distribution is assigned at all basin boundaries; B, similar to the version A, without wind stress ($\tau_x \equiv \tau_y \equiv 0$).

The results obtained in experiments A and B, as a whole, are close (the differences will be considered in detail below). In general, they adjust with the diagnostic version as well: two major dynamic peculiarities of the basin are reflected in all cases, i.e. the MBF jet with a cyclonic "loop" in the northern area and the anti-cyclonic gyre in the central part (Figs. 56 and 59). However, this stage of adjustment resulted in a number of substantial variations, particularly in the shelf zone, where the results differ qualitatively: the diagnostic model provides the northern offshore flow, whereas after the adjustment three offshore flows appear, forming the western peripheries of three gyres, the southern one in the northern and southern basin, and a northern one in the central part. Other significant differences are observed: a cyclonic gyre is generated in the southern part of the area (in the diagnostic version it was hardly pronounced); due to abrupt attenuation of the intense offshore flows at the shelf, the basic circulation systems enforced relatively the MBF and the anti-cyclonic gyre (where velocity dropped, though much less); the correlation with the bottom relief was enhanced; the peculiarity of the latter is responsible for both the generation of all three gyres, and the velocity gradient in the bottom slope zone. So, the adjustment stage led to a rather significant correction of the results of the diagnostic calculations.

It is noteworthy that the adjustment attempt within the framework of the quasi-geostrophic model resulted only in smoothing of the specified density and flow fields. It is of interest that prognostic calculations (Trukhchev and Stanev 1983), performed by the quasi-geostrophic model for the same basin, provide the same result.

The wavelike curve of isolines ζ in the MBF zone (Fig. 56) is observed in both the turbulent exchange variable and constant coefficients. It is also kept when using various techniques in calculating ζ at the boundary: from both the complete boundary relations and calculations performed by the quasi-dynamic method. It is reflected as well in all three solutions, obtained from various difference schemes, i.e. the meander in the MBF is not stipulated by the peculiarities of the calculation technique. It might be a result of the inadjustment of boundary conditions with the problem solution. The density field in the area, due to poor hydrological observations, is mainly obtained due to inter- and extrapolation and is therefore particularly inaccurate. In particular, it might be far from reflecting adequately the effect of such an important factor as the bottom relief, which is here rather irregular. The diagnostic models did not indicate any peculiarities in the solution here, i.e. the meander is not a result of errors when calculating the boundary velocity values. These experiments emphasize once more the difficulties in specifying boundary conditions, especially for the density in the open boundary sections.

The role that the wind plays in the models of adaptation is marked even in the sea surface topography field. It is much more important than in the diagnostic models. Putting the wind "in" or "out", under otherwise equal conditions, results, correspondingly, in the intensification or weakening of the anti-cyclonic gyre, and in weakening or intensification of the cyclonic gyre in the south-western basin. Nevertheless, qualitatively all characteristics of the solution are maintained.

One of the important differences from the specified density field is the pycnocline "compression" after the adjustment. In the quasi-geostrophic model (Trukhchev and Stanev 1983) this peculiarity is not observed. As a whole, the differences are substantial enough, though the general nature of the isopycnal distribution remains.

Let us consider now the results for the summer season. The numerical model, parameters and technique of adjustment are the same as in the calculations for the

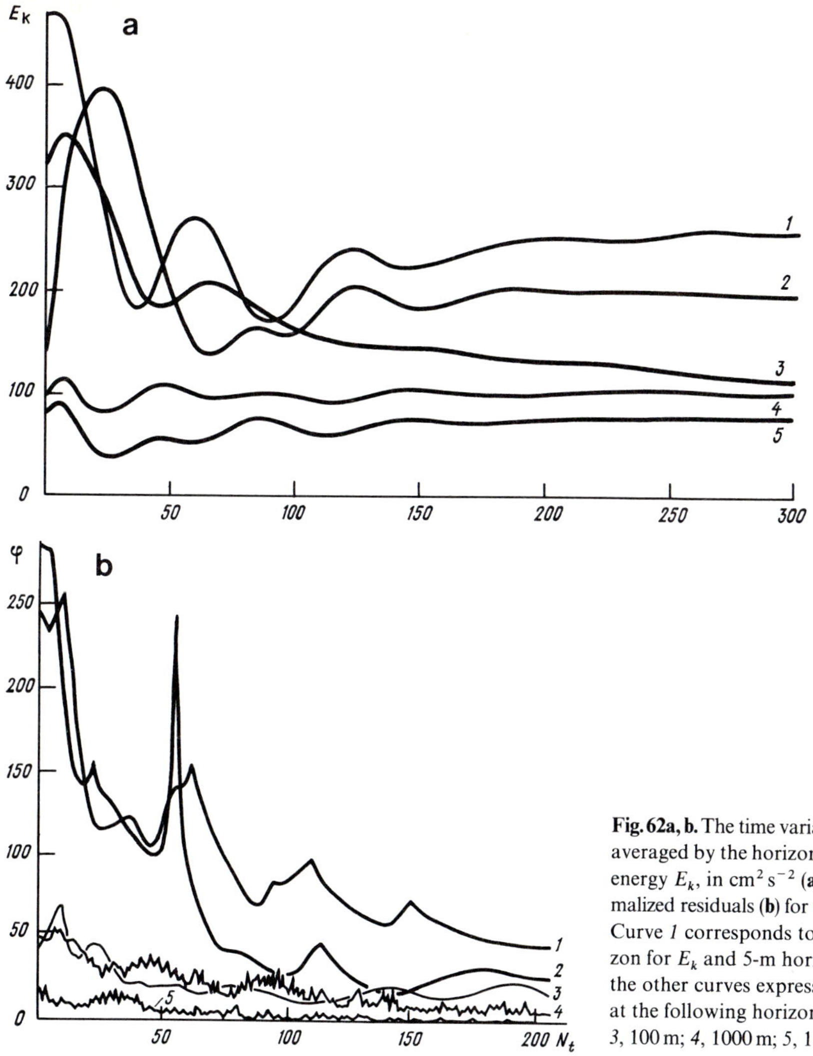

Fig. 62a, b. The time variation of the averaged by the horizontal kinetic energy E_k, in $cm^2\,s^{-2}$ (**a**) and normalized residuals (**b**) for the density Curve *1* corresponds to 0-m horizon for E_k and 5-m horizon for φ, the other curves express E_k and φ at the following horizons: *2*, 30 m; *3*, 100 m; *4*, 1000 m; *5*, 1500 m

fall season. The adjustment process was governed by the same criteria. In the temporal variation of the kinetic energy and normalized residuals for the density:

$$\varphi = \frac{\rho^{n+1} - \rho^n}{\sqrt{u^2 + v^2}\sqrt{\left(\frac{\partial \rho}{\partial x}\right)^2 + \left(\frac{\partial \rho}{\partial y}\right)^2}},$$

the time of completion of the adjustment process agrees with the time of completion of fast and intense wavelike oscillations of these characteristics, stipulated, apparently, by long internal waves, and the time of the beginning of the stage of slow, regular variations (Fig. 62). The field adjustment ends after approximately 150–175 steps by time, i.e. about 2 days of model time in the models of constant exchange coefficients, and in 1.5 days in the model with the variable coefficients. It is necessary in the prognostic version to make 700 steps by time, in this case the solution is not related to the initial approximation, and is determined only by the boundary conditions.

The qualitative structure of the circulation remains after the field adjustment. The most substantial variations took place in the offshore zone and along the bottom slope. Both gyres became much more distinct and were pressed to the bottom slope from one side, and to the MBF jet, from the other. The anti-cyclonic gyre, being narrowed in width, extended meridionally along the bottom slope, covering the whole northern area, with a much pronounced centre near the cyclonic gyre. The weak flows of the shallow-water shelf zone became still weaker, so that strong flows of the abyssal water part became relatively (not absolutely) more intense. The interface of these two dynamic systems extends along the bottom slope. This peculiarity is reflected both in the velocity field and in the sea surface topography field, which is a very weak gradient in this shelf zone, and with a markedly pronounced relief in the abyssal water part. The reconstruction of the gyres and the boundary of the flow systems are evidently related to the bottom relief effect, which becomes, as a rule, more pronounced after the field adjustment. A more distinct development of the correlation of hydrological fields with the bottom relief after the adjustment might be connected not only with the pure dynamic adjustment of input fields, but also with the filtration of high-frequency "noise", which to some extent can conceal the bottom relief effect in the specified fields.

The obtained flow scheme adjusts well with data from instrumental measurements, carried out in the offshore area of Varna within the framework of the international program "Kamchia", as well as with the modern concept on the existence of the quasi-zonal, eastward transport near the Kaliakra Cape in the summer season. At depths of 100, 500 and 1000 m (considered previously) the qualitative circulation structure is the same as in the diagnostic model.

As in calculations for the fall season, some inadjustment is observed in the boundary conditions with the problem solution. It is of interest that in both cases it is pronounced in the northern, relatively shallow-water section of the boundary. The adjustment is adequate in the southern deep-water area.

The numerical experiment with constant coefficients μ and ν equal to variable averages (7×10^5 and $10\,\mathrm{cm^2\,s^{-1}}$) resulted in qualitatively similar fields. The boundary between the flow systems near the area of the bottom slope become less distinct, and gyres become less smooth, with some "inflow" to the shelf. The solutions differ appreciably only in the surface and near-bottom layers, where the intensity of flows somewhat increased. The differences are first of all stipulated by the importance of the vertical viscosity, particularly in the upper layer. At constant coefficients the "weight" of the vertical viscosity in the upper layer is two to three times as greater compared to the variable ones. It is interesting that despite the decrease of the mean vertical viscosity, the field adjusts much faster with variable coefficients. As a whole, the force balance of the equation of motion for the adaptation model is the same as in the diagnostic model.

So, for both seasons the adjustment stage resulted in a sufficient correction of the diagnostic flow fields and specified density fields. Summarizing the results of the calculations, two general characteristics of the adjustment can be noted: the increased role of the bottom relief in terms of the more distinct correlation with both the flows and density fields (reconstruction of the circulation at the shelf and near-bottom slope, a more distinct boundary between the flow systems along the bottom slope, a more distinct imitation of some eddies, hardly noticeable in the specified density field); and relative (not absolute) intensification of the strong flows along with the weakening of weak circulation systems.

References

Allakhverdova TS, Demin YuL (1985) Calculation of the North Atlantic waters climatic circulation, by a non-linear model. Tr GMTS SSSR 62:46–73 (in Russian)

Arakawa A (1966) Computational design for a long-term numerical integration of the equations of fluid motion. Part 1. Two-dimensional incompressible flow. J Comput Phys 1, 17:119–143

Arsenyev VS, Galerkin LI (1976) Dynamical topography of the Pacific ocean relative to the 4000 dbar reference level. Tr VNIIGMI-MTSD 33:67–76 (in Russian)

Brekhovskikh AL (1980) Diagnostic calculation of the world ocean surface currents. Okeanologiya 20, 4:607–612 (in Russian)

Brekhovskikh AL, Demin YuL, Shakhanova TV (1983) Climatic structure of the sea surface topography and the world ocean surface gradient currents in summer and winter seasons. Okeanologiya 23, 6:917–927 (in Russian)

Broshe P, Sündermann J (1985) The Antarctic circumpolar current and its influence on the Earth's rotation. IAMAP/IAPSO joint assembly abstracts, August 5–16, 1985, Honolulu, Hawaii

Bryan K (1969a) A numerical method for the study of the world ocean circulation. J Comput Phys 4, 3:347–376

Bryan K (1969b) Climate and ocean circulation. Part 3. The ocean model. Month Weath Rev 97, 11:806–827

Bryan K, Cox M (1972) An approximate equation of state for numerical models of ocean circulation. J Phys Oceanogr 2, 4:510–514

Bulatov RP (1971) Circulation of the Atlantic ocean waters in various time-space scales. In: Okeanologicheskiye issledovaniya, No 22. M., Nauka, pp. 7–93 (in Russian)

Bulatov RP, Pojarkov SG (1975) Topography of the Atlantic ocean surface. Okeanologiya 15, 6:995–1001 (in Russian)

Bulatov RP, Demin YuL, Pojarkov SG (1975) Vertical circulation of the Atlantic ocean waters. In: Okeanologicheskiye issledovaniya, No 28, M, Nauka, pp 94–104 (in Russian)

Bulatov RP, Demin YuL, Pojarkov SG (1980) The main features of three-dimensional fields of the Atlantic ocean currents (climatic circulation). In: Okeanologicheskiye issledovaniya, No 31, M, Sov Radio, pp 91–105 (in Russian)

Burkov VA (1968) Circulation of waters. In: Gidrologiya Tikhogo okeana. M, Nauka, pp 206–289 (in Russian)

Burkov VA (1980) General circulation of the world ocean. L, Gidrometeoizdat, p 252 (in Russian)

Carrier GF, Robinson AR (1962) On the theory of the wind-driven ocean circulation. J Fluid Mech 12, 1:49–80

Charney J, Fjortoft R, Neumann J (1950) Numerical integration of the barotropic vorticity equation. Tellus 2:237–254

Cox MD (1975) A baroclinic numerical model of the world ocean: preliminary results. In: Numerical models of ocean circulation, NAS Washington, DC, pp 107–120

Cromwell T, Montgomery RB, Stroup ED (1954) Equatorial undercurrent in the Pacific ocean, revealed by new methods. Science 119:648–649

Defant A (1941) Die absolute Topographie des physikalischen Meeresniveaus und der Druckflächen sowie die Wasserbewegungen im Atlantischen Ozean. Deutsche Atlanische Exped. "Meteor", 1925–1927. Wiss Ergeb Bd 6, Teil 2, IFR 5, p 260

Demin YuL (1975a) A non-linear baroclinic model and calculation of the equatorial Atlantic currents. Izv Akad Nauk SSSR. Fizika atmosfery i okeana 2, 5:534–537 (in Russian)

Demin YuL (1975b) Diagnostic calculation of the tropical Atlantic currents. Meteorol Gidrol 1:48–57 (in Russian)

Demin YuL, Brekhovskikh AL (1985) About the diagnostic and semi-diagnostic calculation of the world ocean currents. In: Komplexnii globalnii monitoring Mirovogo okeana, v 3, L, Gidrometeoizdat, pp 100–113 (in Russian)

Demin YuL, Ibraev RA (1985) About the boundary-value problem for the sea-surface topography in sea currents models. Izv Akad Nauk SSSR. Fizika atmosfery i okeana vol 21, No 12 (in Russian)

Demin YuL, Mikhailichenko JG (1981) About the abyssal waters circulation of the Atlantic ocean. Okeanologiya 21, 3:425–432 (in Russian)

Demin YuL, Sarkisyan AS (1977) Calculation of equatorial currents. J Mar Res 35, 2:339–356

Demin YuL, Trukhchev DI (1984a) Numerical modelling of currents near the western coast of the Black Sea. Meteorol Gidrol 2:54–61 (in Russian)

Demin YuL, Truhkchev DI (1984b) The non-linear model of the adaptation of fields of density and currents in the sea. Izv Akad Nauk SSSR. Fizika atmosfery i okeana 20, 12:1171–1182 (in Russian)

Demin YuL, Trukhchev DI (1985) Semi-diagnostic calculation of currents and density field in the coastal zone of the Black Sea. Morskoi gidrofizicheskii jurnal, No. 1 (in Russian)

Demin YuL, Usychenko IG (1984) Numerical investigation of seasonal climatic circulation of waters in the region of the Mozambique Current. In: Okeanologicheskiye issledovaniya, No 38, M, Radio i Svjaz, pp 36–48 (in Russian)

Demin YuL, Hagen E, Gurina AM (1983) About the mesoscale variability of coastal upwelling near north-western Africa. Izv Akad Nauk AN SSSR. Fizika atmosfery i okeana 3:292–300 (in Russian)

Dubravin VF (1971) Geostrophic flows in the Guinea Current. Tr Atlant NIRO 37:81–96 (in Russian)

Dymnikov VP, Kurbatkin GP, Sarkisyan AS (1983) Some theoretical and experimental aspects of the "sections" programme. WCRP publications series 2, 1:201–214

Fiadeiro M, Veronis G (1977) On weighted-mean schemes for the finite-difference approximation of the advection-diffusion equations. Tellus 29:512–522

Friedrich HJ (1970) Preliminary results from a numerical multilayer model for the circulation in the North Atlantic. Dtsch Hydrogr Z 23, 4:145–164

Gamsakhurdia GR, Sarkisyan AS (1975) Diagnostic calculations of flow velocities at II horizons for the Black Sea. Okeanologiya 15, 2:239–244 (in Russian)

Gill AE (1975) Models of equatorial currents. In: Numerical models of ocean circulation, NAS Washington, DC, pp 181–204

Gurina AM, Demin YuL, Pavlova JV (1983) About hydrological conditions and the structure of the large-scale currents of the Sargasso Sea in summer period. In: Okeanologicheskiye issledovaniya, No 35, M, Radio i Svjaz, pp 43–54 (in Russian)

Haney RL (1980) A numerical case study of the development of large-scale thermal anomalies in the Central-North Pacific ocean. J Phys Oceanogr 4:541–556

Haney RL (1985) Midlatitude sea-surface temperature anomalies: a numerical hindcast. J Phys Oceanogr vol 15, No 6

Hellerman S (1967) An updated estimate of the wind stress on the world ocean. Month Weath Rev 95, 9:607–626

Holland WR, Hirschman AD (1972) A numerical calculation of the circulation in the North Atlantic ocean. J Phys Oceanogr 2, 4:336–354

Ilyin AM (1969) The difference scheme for the differential equation with the small parameter at the highest derivative. Matematicheskiye zametki 6, 2:237–248 (in Russian)

International oceanographical tables (1973) UNESCO, vol 2

Ivanov JA, Kamenkovitch NN (1959) Ocean floor topography as the main factor, which forms the non-zonality of the Atlantic circumpolar current. Dokl Akad Nauk SSSR 128, 6:1167–1170 (in Russian)

Joseph J, Sender H (1958) Über die horizontale Diffusion in Meer. Dtsch Hydrogr Z 11, 2:49–77

Kamenkovitch VM (1961) About the integration of equations of the sea current theory in multi-connected domains. Dokl Akad Nauk SSSR 138, 5:1076–1079 (in Russian)

Kochergin VP (1978) Theory and methods of calculation of oceanic current. M, Nauka, p 128 (in Russian)

Kolesnikov AG, Boguslavskii SG, Grigoryev GI, Ponomarenko GP, Sarkisyan AS, Felzenbaum AI, Khlystov NZ (1968) The discovery, experimental investigation and development of the Lomonosov current theory. Sebastopol, p 243 (in Russian)

Kozlov VF (1973) To the methods of the calculation of oceanic currents, according to the given density field. Meteorol Gidrol 1:79–84 (in Russian)

Kozlov VF (1975) To the problem of mutual adaptation of the field of masses and currents to the baroclinic ocean floor topography. Izv Akad Nauk SSSR. Fizika atmosfery i okeana 2, 1:43–52 (in Russian)

Kozlov VF (1977) About the application of monotonous difference schemes in diagnostic calculations of sea currents. Izv Akad Nauk SSSR. Fizika atmosfery i okeana 13, 7:728–737 (in Russian)

Krasnoselsky MA, Krein SG (1952) Iterative process with the minimal discrepancy. Matem sb, vol 31 (in Russian)

Kurihara G (1965) Numerical integration of the primitive equations on a spherical grid. Month Weath Rev 93, 7:399–415

Landau LD, Neuman NN, Khalatnikov IM (1956) Numerical methods of the equations integration in partial derivatives by the finite-difference method. Tr III Vsesojuznogo matematicheskogo sjezda, vol II, M, p 16 (in Russian)

Latif M, Maier-Reimer E, Olbers DJ (1985) The onset and evolution of El-Nino in ocean circulation experiments. IAMAP/IAPSO joint assembly abstracts, JS1–13, August 5–16, 1985, Honolulu, Hawaii

Levitus S, Dort H (1977) Global analysis of oceanographic data. Bull Am Meteorol Soc 58, 12:1270–1284

Lineykin PS (1957) The main problems of the dynamic theory of the ocean baroclinic layer. L, Gidrometeoizdat, p 139 (in Russian)

Mamayev OI (1963) Oceanographical analysis in the α-S-T-P-System. M, Izd MGU, p 223 (in Russian)

Manabe S, Bryan K (1985) CO_2-induced change in coupled ocean-atmosphere model and its paleoclimatic implications. IAMAP/IAPSO joint assembly abstracts, August 5–16, 1985, Honolulu, Hawaii

Marchuk GI (1967a) Numerical methods in the weather forecast. L, Gidrometeoizdat, p 223 (in Russian)

Marchuk GI (1967b) About the equations of the baroclinic ocean dynamics. Dokl Akad Nauk SSSR, 173, 6:1317–1320 (in Russian)

Marchuk GI (1969) About numerical solution of the Poincare problem for oceanic circulation. Dokl Akad Nauk SSSR, 185, 5:1041–1044 (in Russian)

Marchuk GI (1970) On the setting up of the problem of the structure of baroclinic sea currents with the account of the macroturbulent mixing. Meteorol Gidrol 3:12–17 (in Russian)

Marchuk GI (1972) Numerical solution of atmosphere and ocean dynamics problems on the basis of the splitting method. Novosibirsk: Nauka, p 170. Bibliogr pp 160–168 (in Russian)

Marchuk GI (1973) Methods of computational mathematics. Novosibirsk, Nauka, p 352 (in Russian)

Marchuk GI (1974a) Basic and conjugate equations for the dynamics of atmosphere and ocean. In: Difference and spectral methods for atmosphere and ocean dynamics problems. Proc of Symp Novosibirsk, 17–22 Sept, 1973. Novosibirsk, 1974, part 1, pp 3–33, Bibliogr: 5 ref (in Russian)

Marchuk GI (1974b) Numerical solution of atmosphere and ocean dynamics problems. L, Gidrometeoizdat (in Russian)

Marchuk GI (1975a) Methods of numerical mathematics. Springer, Berlin Heidelberg New York, p 313

Marchuk GI (1975b) Physics of atmosphere and ocean and the problem of weather prediction. Novosibirsk. CE SB Acad Sci USSR, p 17, Pre-Print (in Russian)

Marchuk GI (1976) On the formation of a thermocline in the ocean. Am Math Soc 104:172–178 (translation), Bibliogr: 9 ref

Marchuk GI (1980a) A mathematical model of general atmosphere and ocean circulation. Dokl Akad Nauk SSSR, 254, 3:577–581 (in Russian)

Marchuk GI (1980b) Methods of computational mathematics. M, Nauka, p 635 (in Russian)

Marchuk GI, Kochergin VP (1968) About vertical structure of currents in a baroclinic ocean. Meteorol Gidrol 2:3–10 (in Russian)

Marchuk GI, Sarkisyan AS (eds) (1980) Mathematical models of the ocean circulation. Novosibirsk, Nauka, p 341 (in Russian)

Marchuk GI, Sarkisyan AS, Kochergin VP (1973) Calculations of flows in a baroclinic ocean: numerical methods and results. Geophys Fluid Dyn 5:89–100, Bibliogr: p 99

Marchuk GI, Kochergin NP, Klimock VI, Sukhorukov VA (1976a) Mathematical modelling of surface turbulence in the ocean. Izv Akad Nauk SSSR, Fizika atmosfery i okeana, 12, 8:841–849 (in Russian)

Marchuk GI, Sarkisyan AS, Kochergin VP (1976b) Numerical methods of calculation of flows in a baroclinic ocean. Mathematical models in geoph. IASH-AISH publ No 116

Marchuk GI, Kochergin VP, Klimock VI, Sukhorukov VA (1977) On the dynamics of the ocean surface mixed layer. J Phys Oceanogr 7:865–875, Bibliogr: pp 874–875

Marchuk GI, Zalesny VB, Kusin VI (1979) A numerical model of calculation of thermodynamic characteristics of the world ocean. Papers submitted to the joint IOC/WMO seminar on oceanographic products and the IGOSS data processing and services system. Moscow, 2–6 Apr, 1979. Paris, 1979, pp 289–298 (Workshop report No 17, Suppl)

Marchuk GI, Sarkisyan AS, Dymnikov VP, Kurbatkin GP (1983) The world ocean monitoring and the "sections" programme. The 1st international symposium on integrated global ocean monitoring (MONOC). (USSR, Tallin, Oct 2–10, 1983): Abstr M, pp 6–7

Marchuk GI, Dymnikov VP, Kurbatkin GP, Sarkisyan AS (1984) The "sections" programme and the world ocean monitoring. Meteorol Gidrol No 8 (in Russian)

Marchuk GI, Dymnikov VP, Lykosov VN, Zalesny VB, Galin VYa (1985) Mathematical modelling of the general ocean and atmosphere circulation. Leningrad, Gidrometeoizdat (in Russian)

Mellor GL, Mechoso CR, Keto E (1982) A diagnostic calculation of the general circulation of the Atlantic ocean. Deep-sea Res 29, 10A:1171–1192

Munk WH (1950) On the wind-driven ocean currents. J Meteorol 7, 2:79–93

Neuman G (1955) On the dynamics of the wind-driven ocean currents. Meteorol Pap 2, 4:1–33

Neuman G (1958) On the mass transport of wind-driven currents in a baroclinic ocean with application to the North Atlantic. Z Meteorol 12, 4–6:138–147

Neuman VG (1970) New maps of the Indian ocean currents. Dokl Akad Nauk SSSR, 195, 4:948–952

O'Brien JJ (1979) Equatorial oceanography. Rev Geophys Space Phys 17, 7:1569–1575

Obukhov AM (1946) Turbulence in the temperatural inhomogeneous atmosphere. In: Tr Inst Teoreticheskoi Geofiziki Akad Nauk SSSR, 24 (151):3–42 (in Russian)

Ovsyannikov LV (1966) The theorem of uniqueness for the linearized system of ocean dynamics equations. Dokl na simpoziume po dinamike oz Baikal. Preprint NGU (in Russian)

Philander SGH, Pacanowski RC (1980) The generation of equatorial currents. J Geophys Res 85, C2:1123–1136

Phylippov DJ (1968) Circulation and structure of the Black Sea waters. M, Nauka, p 136 (in Russian)

Roache PJ (1976) Computational fluid dynamics. Hermosa Albuquerque, New Mexico

Romankevitch EA (1977) Geochemistry of organic matter in the ocean. M, Nauka, pp 43–44 (in Russian)

Rozhdestvensky BL, Yanenko NN (1968) Systems of quasi-linear equations and their application to the gas dynamics. M, Nauka, p 592 (in Russian)

Samarsky AA, Andreyev VB (1976) Difference methods for elliptical equations. M, Nauka (in Russian)

Samarsky AA, Nikolaev ES (1978) Methods of solution of finite-difference equations. M, Nauka, pp 93–97 (in Russian)

Sandström JW, Helland-Hansen B (1903) Über die Berechnung der Meeresströmungen. Rep Norw Fish Marine Invest, vol 2, No 4

Sarkisyan AS (1969) About drawbacks of the barotropic models of oceanic circulation. Izv Akad Nauk SSSR. Fizika atmosfery i okeana 5, 8:818–836 (in Russian)

Sarkisyan AS (1977a) Numerical analysis and forecast of sea currents. L, Gidrometeoizdat p 181 (in Russian)

Sarkisyan AS (1977b) The diagnostic calculations of a large-scale oceanic circulation. In: The sea, marine modelling. J Wiley, New-York, vol 6, pp 363–458

Sarkisyan AS, Demin YuL (1983) A Semidiagnostic method of sea currents calculation. Large-scale oceanographic experiments in the WCRP, WCRP publications series, Tokyo, 2, 1:201–214

Sarkisyan AS, Ivanov VF (1971) Joint effect of baroclinicity and ocean floor topography as an important factor in the sea currents dynamics. Izv Akad Nauk SSSR. Fizika atmosfery i okeana 7, 2:173–188 (in Russian)

Sarkisyan AS, Kochergin VP, Klimock VI (1972) A theoretical model and calculations of the density field in the ocean with an arbitrary ocean floor topography. Izv Akad Nauk SSSR, Fizika atmosfery i okeana, vol 8, No 7 (in Russian)

Sarkisyan AS, Demin YuL, Gurina AM (1980) To the calculation of the sea surface topography. Meteorol Gidrol 9:71–80 (in Russian)

Sarkisyan AS, Demin YuL, Trukhchev DI (1983a) A hydrodynamic model of currents and the field of density of the sea coastal zone. Dokl Bolg Akad Nauk 36, 3:341–344

Sarkisyan AS, Demin YuL, Trukhchev DI (1983b) To the calculation of vertical velocity in sea currents models. Izv Akad Nauk SSSR. Fizika atmosfery i okeana, 19, 7:730–740 (in Russian)

Sarkisyan AS, Demin YuL, Ibraev RA, Usychenko IG (1984a) The effect of bottom topography on the equatorial Indian ocean circulation. Trop Ocean-Atmosph Newsl 23:11–13

Sarkisyan AS, Demin YuL, Ibraev RA, Usychenko IG (1984b) On the circulation of waters of the equatorial zone of ocean. Dokl Akad Nauk SSSR, 276, 3:724–728 (in Russian)

Sarkisyan AS, Demyshev SG, Korotaev GK, Moiseenko VA (1985) Numerical experiments on a four-dimensional analysis of POLIMODE and the "sections" programme oceanographic data. In: Nihoul JCJ (ed) Coupled ocean-atmosphere models. Elsevier oceanographic series, pp 659–673

Sarkisyan AS, Demin YuL, Brekhovskikh AL, Shakhanova TV (1986) Methods and results of calculations of the world ocean circulation. L, Gidrometeoizdat (in Russian)

Sarmiento JL, Bryan K (1982) An ocean transport model for the North Atlantic. J Geophys Res 87, C1:394–408

Semtner A, Holland WR (1980) Numerical simulation of equatorial ocean circulation. J Phys Oceanogr 10:667–693

Seidov DG (1977) A numerical model of inhomogeneous ocean circulation. Izv Akad Nauk SSSR. Fizika atmosfery i okeana 18, 8:867–875 (in Russian)

Shtokman VB (1946) The equation of the total mass transport stream function, caused by the wind in an inhomogeneous sea. Dokl Akad Nauk SSSR 54, 5:407–410 (in Russian)

Stepanov VN (1974) The world ocean. The dynamics and properties of waters. M, Znaniye, p 256 (in Russian)

Stommel H (1948) The westward intensification of the wind-driven ocean currents. Trans. Am Geophys Union 29, 2:202–206

Stommel H (1965) The Gulf Stream. University of California Press, p 243

Stommel H, Schott F (1977) The beta spiral and the determination of the absolute velocity field from hydrographic station data. Deep-Sea Res 24:325–329

Sverdrup HV (1947) Wind-driven currents in a baroclinic ocean with application to the equatorial currents of eastern Pacific. Proc Natl Acad Sci USA 33, 11:318–326

The atlas of oceans. VI. The Pacific ocean, L, GUNiO MO SSSR, 1974 (in Russian)

The atlas of oceans. VII. The Atlantic and the Pacific oceans. L, GUNiO MO SSSR, 1977, pp 196–202 (in Russian)

Trukhchev DI, Stanev EV (1983) A numerical model of currents and water density of the Black Sea western part. Okeanologiya 23, 1:17–22 (in Russian)

Welander P (1968) Wind-driven circulation in one- and two-layer oceans of variable depth. Tellus, 20, 1:1–16

Wyrtki K (1975) Fluctuations of the dynamic topography in the Pacific ocean. J Phys Oceanogr 5:450–459

Yanenko NN (1967) The splitting method of solution of mathematical physics multi-dimensional problems. Novosibirsk, Nauka, p 195 (in Russian)

Zalesny VB (1984) Modelling of large-scale motions in the world ocean. M, VINITI, p 158 (in Russian)

Zeng QC, Yuan CG, Zhang X-h (1985) A numerical coupled atmosphere-ocean model for the simulation of long-term variations. IAMAP/IAPSO joint assembly abstracts JS4–33, August 5–16, 1985, Honolulu, Hawaii

Zenkovitch VP (1974) The Black and the Azov Seas. In: Okeanograficheskaya entsiklopediya. L, Gidrometeoizdat, p 586 (in Russian)

Subject Index